AUSTRALIA AND NEW ZEALAND

Economy, Society and Environment

Guy M. Robinson

Professor of Geography, Kingston University, Surrey

Robert J. Loughran

*Associate Professor in Geography and Environmental Science,
University of Newcastle, Australia*

Paul J. Tranter

Senior Lecturer in Geography, University of New South Wales, Australia

A member of the Hodder Headline Group
LONDON
Co-published in the United States of America by
Oxford University Press Inc., New York

First published in Great Britain in 2000 by
Arnold, a member of the Hodder Headline Group
338 Euston Road, London NW1 3BH

Co-published in the United States of America by
Oxford University Press Inc.,
198 Madison Avenue, New York, NY10016

British Library Cataloguing in Publication Data
A catalogue record for this book is available from the British Library

Library of Congress Cataloging-in-Publication Data
A catalog record for this book is available from the Library of Congress

ISBN 0 340 72032 8 (hb)
ISBN 0 340 72033 6 (pb)

1 2 3 4 5 6 7 8 9 10

Production Editor: Liz Gooster
Production Controller: Sarah Kett
Cover designer: T. Griffiths

Typeset in 10/12 pt Palatino by Scribe Design, Gillingham, Kent
Printed and bound in Great Britain by The Bath Press, Bath, UK.

What do you think about this book? Or any other Arnold title?
Please send your comments to feedback.arnold@hodder.co.uk

CONTENTS

ACKNOWLEDGEMENTS

Collaboration between the three authors first began when Guy Robinson was in receipt of a William Evans Visiting Fellowship at the University of Otago in 1997. Subsequently, both Australian-based authors took sabbaticals in the UK, affording further opportunities to plan the book and the division of tasks. Acknowledgements are due to Coventry University (Loughran), Liverpool John Moores University (Tranter) and the University of Otago (Robinson) for the hospitality offered during these sabbaticals.

The authors' three home institutions have provided much assistance and encouragement, especially the University of Newcastle, New South Wales, where Paul Tranter was formerly an undergraduate, Robert Loughran a member of staff (and former head of the Department of Geography), and where Guy Robinson was based on sabbatical in 1993.

Special thanks go to Claire Ivison at Kingston University for drawing all the maps and diagrams. Claire's translations of our garbled instructions and hasty scribbles were truly miraculous. Linda Parry, also at Kingston, produced black-and-white prints from the colour slides and photographs provided by the authors.

The authors and the publisher would like to thank the following for their permission to reproduce figures: Sydney University Press for Figures 2.1, 2.2, 2.3, 2.5, 3.6 and 5.3; CSIRO Land and Water for Figure 2.9; Department of Land and Water Consevation NSW, for Figure 2.10; AusInfo for Figures 3.1, 4.2, 4.5 and 16.2; New Zealand Geographical Society for Figure 3.2; Harcourt Australia for Figure 3.4; Cambridge University Press Australia for Figure 3.5; Victoria University Press, New Zealand for Figure 3.8; the Ministry for the Environment, New Zealand for Figure 4.1; Landcare Research for Figure 4.4; Ecos and CSIRO for Figure 4.6; Addison Wesley Longman Australia Pty Ltd. for Figure 4.9; Boolgrong Publications and the Royal Geographic Society of Australia Queensland Inc. for Figure 5.2; the Waikato Branch of the New Zealand Geographical Society for Figure 7.3; Oxford University Press for Figures 8.3, 14.5 and 14.6; Blackwell Publishers for Figure 15.2; Australian Academy of Science for Figure 16.1; Oxford University Press Australia for Figure 16.7; the Department of Conservation, New Zealand for Figure 16.8. Every effort has been made to trace copyright holders of material reproduced in this book. Any rights not acknowledged here will be acknowledged in subsequent printings if notice is given to the publisher.

Various colleagues have provided advice and encouragement, notably Kay Anderson, Rochelle Ball, Lisa Charters, Jack Doyle, Ken Johnson, Ged Martin, Eric Pawson, Ray Towse, Peter van Diermen and Hilary Winchester, and special acknowledgement is due to Susan Robinson who helped compile the reference section.

The most heartfelt thanks are due to our wives and families for their unstinting support. Finally, acknowledgement must be made for the significant diversions provided by the Australian cricket team, consistently humbling their English counterparts (not welcomed by all the authors!), and to infuriatingly unsuccessful West Bromwich Albion and Worcestershire CC, beloved of Loughran and Robinson.

Guy Robinson,
Robert Loughran
and Paul Tranter

1

INTRODUCTION

On the cover of the report by the Independent Commission on International Development Issues (1980), *North–South: A programme for survival*, the line on the globe separating 'north' from 'south' plunges dramatically across the Pacific Ocean to include a substantial portion of the Southern Hemisphere in the Commission's definition of 'north' or 'first world' (Figure 1.1).

This region, comprising Australia and New Zealand, is therefore symbolically removed from its south-east Asian and Pacific Island neighbours. It is a removal or transplantation that corresponds with the view of the world held before 1939 by many Australians and New Zealanders of European origin. Despite being over 12 000 km from the 'mother country', the

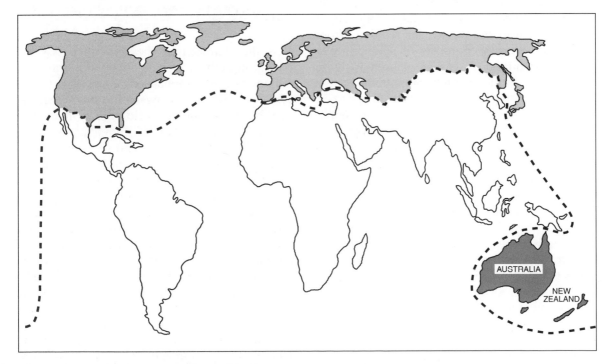

FIGURE 1.1 Map of the World from *North–South: A programme for survival* (1980)

United Kingdom (UK), they were tied irrevocably by cultural and economic links that rendered them islands of the Northern Hemisphere cast adrift in the vast seas of the southern ocean.

During the nineteenth and early twentieth centuries the view of the world as seen from 'Down Under' corresponded closely to that in *North–South*. It was Eurocentric, often Anglocentric, with a shared heritage that was reinforced on the sporting field and supported by economic and military alliances with the UK which had emerged tried, tested and victorious in the 1914–18 war. The presence of a small and declining number of indigenous Aborigines and Torres Straits Islanders in Australia, and a larger but still relatively insubstantial number of Maori in New Zealand, did little to disturb the Anglocentrism. However, fierce sporting rivalries between the two 'Down Unders' reflected differences between them conferred by their separate experiences of European colonisation, and also their distinctive physical environments.

This book focuses on the economic and social development of Australia and New Zealand in the twentieth century as they have both sought a new 'place in the world', far removed from nineteenth-century 'outposts of Empire' to a position identified increasingly closely with the broad Asia–Pacific region through trade, investment and immigration, and in which, in certain respects, the *North–South* line may be highly misleading. Although recent developments occupy most of the book's content, scope is given to the evolution of the economy, society and environment that has taken place since initial human occupation, and especially since the first European settlements. Hence, there is a mixture of various systematic aspects of the geography of Australia and New Zealand, or 'Australasia' as the two countries are sometimes collectively termed herein.

Economic, social, cultural, political and environmental dimensions all feature in the ensuing analysis which deliberately seeks to avoid dominant use of any one particular theoretical framework. Instead eclecticism is consciously employed to help convey the range of themes and approaches that have been used to study the changing geography of Australasia. Full use is made of reference to geographical research on key economic, social and environmental topics, so that issues are dealt with in some depth, but with direction to more detailed studies that can be consulted. The book is introductory in the sense that it provides a guide to the main components of the two countries' changing geographies, but without encompassing the depth of specialist studies focusing upon one aspect of geographical change or dealing with just one of the two countries. Hence, it can be used for introductory purposes in undergraduate study, and provides a detailed context from which to pursue further enquiry. Chapters 2, 3, 4 and 16 were written by Robert Loughran; Chapters 9–12 were written by Paul Tranter; and the remaining Chapters were written by Guy Robinson. This opening Chapter gives further guidance regarding the book's content and the main topics considered.

AUSTRALASIAN ENVIRONMENTS

It is the character of the distinctive physical background that forms the first part of this book (Chapters 2–4), emphasising aspects that have often shaped the image of the two countries held by the rest of the world. It is an image that has associated Australia indelibly with its assemblage of distinctive and endemic marsupials and the emptiness of its 'dead heart', or 'red centre', typified by the widely appropriated image of Ayers Rock in Uluru National Park. The great age of the rocks forming Ayers Rock and its much-photographed near neighbours, the Olgas, is referred to in Chapter 2, which contrasts the age and stability of the Australian continent with the comparative youth and instability of New Zealand, which lies over 2000 km to the east, separated by the Tasman Sea (Figure 1.2). The spectacular scenery of the Southern Alps and the geothermal activity on the North Island are typical images of the 'youthful' New Zealand. However, as mentioned in Chapter 3,

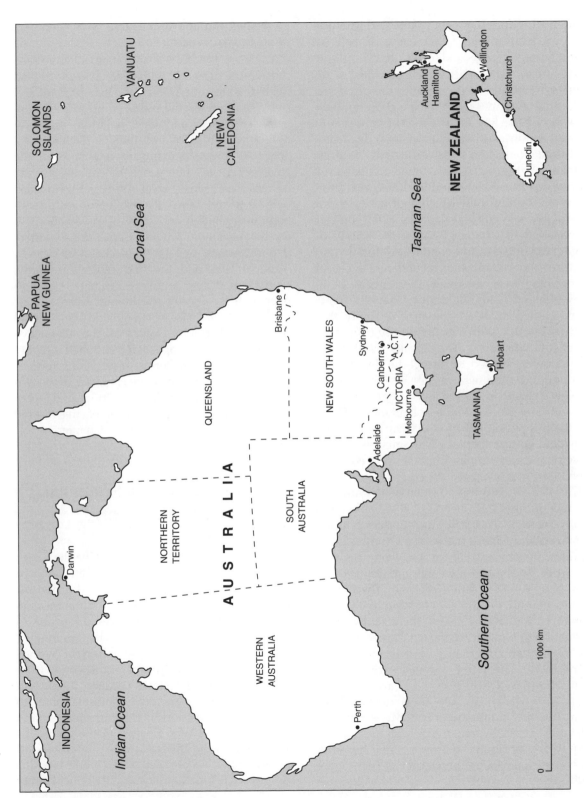

Figure 1.2 Australia and New Zealand

there is also a unique flora and fauna for which New Zealand is well renowned, especially its flightless birds, podocarp forests and tree ferns.

The relationships between landforms, soils, climate, hydrology and vegetation are discussed in Chapters 2 and 3 as an introduction not only to some of the key environmental concerns in the two countries, but also as points of departure for the economic and social issues that comprise the core of the book. For example, Chapter 2 emphasises the shallow and leached character of many Australian soils, which limit the extent of productive agricultural land, a point further emphasised in Chapter 3 when the aridity of large parts of the island continent is outlined. In contrast, the juxtaposition of different types of topography over relatively short distances in New Zealand is highlighted in Chapter 2. This reinforces pronounced climatic variability that contrasts with the greater uniformity over large areas of Australia. The greater aridity of Australia has helped to restrict the diversity of economic development in the interior, whereas in New Zealand elevation and topography have tended to be more prominent limitations. These are themes pursued further in Chapter 4, which considers environmental constraints on human activity and key human impacts upon environment.

Chapter 4 raises the question of the impacts of the indigenous peoples: the Australian Aborigines and Torres Straits Islanders, and the New Zealand Maori. Just as stark contrasts are evident in physical terms between the two countries, so the histories and impacts of these two peoples also differ sharply. There is a strong analogy to be drawn between the youthfulness of the landscape and the recent human habitation of New Zealand, and the great age of the Australian continent and of human settlement there. The latter covers at least 50 000 years compared with, at most, around 1000 years for Maori in New Zealand. In both countries the indigenous populations reduced the forest cover by fire, which was used deliberately by Aborigines to improve their hunting. However, the Maoris' much shorter occupancy led to several animal extinctions and to agricul-

tural modifications whereas the Aborigines did not cultivate the soil.

The impacts of European settlement have been much more dramatic, with various forms of land degradation, soil erosion, salinisation, plant and animal extinctions, the effects of introducing nonnative species, air and water pollution, modification of watercourses and damage to the coastal zone. As a result, there are few parts of the environment left untouched by human activity. Nevertheless, the great size of Australia (7 682 300 km^2) has left vast areas with seminatural vegetation and close approximations to climax vegetation, proving tremendously attractive to tourists, e.g. the 'wilderness' of Kakadu National Park and the 'red centre'. Similarly, tourist advertising for New Zealand is able to refer to the country's 'clean and green' image, with substantial areas designated as national parks. Furthermore, some of the distinctive human-made landscapes have obtained a highly positive response in terms of attracting tourists and adding to the distinctive character of the two countries, e.g. Sydney Harbour Bridge and Opera House. The economic impacts of this tourist industry are discussed further in Chapter 14.

THE LEGACIES OF EUROPEAN COLONISATION

The second part of the book (Chapters 5–8) focuses upon the development of the two countries following the arrival of Europeans. Again, there is emphasis on key differences in the nature of this 'arrival'. As discussed in Chapter 6, Australia had a convict 'beginning', with the creation of a penal colony at Sydney Cove in 1788 for prisoners transported from 'hulk ships' in the Thames Estuary. It also had a strong Irish component in its population, both from the composition of the transportees and from the miners who flocked to the goldfields of Victoria in the 1850s. At this time the island continent had several separate colonies, which did not come together in a single political entity as the Commonwealth of Australia until 1901. No treaty or military conflict had ever conferred legitimation on the European usurping of the Aboriginal occupation that had been uncontested by outsiders for over 50 000 years (Chapter 5).

In contrast, New Zealand was 'born' in 1840 from the Treaty of Waitangi, signed between representatives of the Crown and Maori leaders. Its principles were studiously ignored by the settlers once the colony had self-government in 1852. However, the ensuing conflict in the 1860s and 1870s between Maori and Europeans (called 'Pakeha' by Maori) did gain the Maori greater respect as fierce and resourceful fighters, and earned some concessions regarding the right to vote and representation in parliament. The colonists were largely English but with a strong Scottish contingent drawn to the 'New Edinburgh' of Dunedin. Like Australia, the young economy depended upon wool exports until the discovery of gold in Otago paralleled the Victoria goldrush across the Tasman.

Chapter 5 highlights the different experiences of the indigenous populations, emphasising how the presence of the Treaty of Waitangi has eventually helped to secure Maori more rights and recognition, even though many indicators testify to a strong degree of economic marginalisation and social deprivation within contemporary Maori communities. In contrast, mistreated and deprived of access to the environments in which they had been hunters and gatherers for thousands of years, the indigenous Australians had to survive on reserves in central Australia in the first half of the twentieth century. Post-1945 their numbers grew through rising birth rates and small-scale improvements in health care and disease prevention. Their move to the towns was less extensive than that for the Maori, but nevertheless there was a pronounced shift by working-age Aborigines to country towns and state capitals. Discrimination remained rife, but a referendum in 1967 granted Aborigines voting rights, and there were more concerted efforts to provide economic assistance to Aboriginal communities. There have followed three decades of growing politicisation of Aborigines, symbolised in the tented Aboriginal 'embassy' outside the federal parliament in Canberra and in the struggles for land rights. The latter have yielded important legally ratified advances in the 1990s, but as part of an ongoing argument involving vested economic interests and compromise at both state and federal level. Recently the emergence in the political arena of views attacking 'progressive' attitudes towards the indigenous peoples, and against Asian immigration, has cast doubts about the views of many Anglo-Australians towards the non-white population in general. Federal government multicultural policies, discussed in Chapter 10, have been questioned, as has the growth of strong economic links to south-east Asia.

Chapter 5 also considers significant developments in the relationship between Pakeha and Maori in New Zealand. Whilst some changes have been contested by sections of both communities, developments have generally been more progressive and consensual than their Australian equivalents. The Treaty of Waitangi has re-emerged as a 'living document', with its principles being incorporated in contemporary legislation. Moreover, the agreement between Maori and Pakeha, as enshrined in the treaty, has been addressed via the Waitangi Tribunal, established in 1975, to consider Maori grievances at non-adherence to the treaty's principles. Extended in 1985 to address grievances dating back to 1840, the tribunal's rulings have ensured that Maori tribes have received some financial compensation and reinstatement of rights over traditional sources of livelihood. However, there has been no wholesale redistribution of land, despite the initial submissions of substantial land claims. Although the various settlements have drawn upon the Exchequer and have by no means met with universal approval, there has been the establishment of an ongoing basis of mutual reconciliation between Maori and Pakeha. This has also been accompanied by a Maori cultural renaissance, so that the Maori language, educational initiatives and artistic development have been prominent within a more assertive and confident Maoridom. Yet, Maori (and recently established communities of Polynesian Islanders) are economically disadvantaged compared with Pakeha, and this differential has been exacerbated by the economic developments of recent years.

Chapter 6 examines the first steps towards nationhood in Australasia and discusses the characteristics of the emergent economies. Initial ties to the UK were reinforced by fresh supplies of emigrants and through trade. Britain supplied manufactured goods and received agricultural produce in return. From the 1880s the staple of wool was accompanied by meat and dairy produce transported in refrigerated ships. For New Zealand this three-fold export trade to the UK sustained the economy until the 1960s. It was also of great importance to Australia, and the problems associated with extensive pastoralism became written into Australian folklore: water shortage, rabbits, dingoes, low unit area productivity, mistreatment of Aborigines, and the rough and tough life of the stockmen and shearers. This has given a particular image to the Australian character out of all proportion to the number of Australians who actually had (or have) any direct connection to this way of life. In fact, the larger Australian population and greater diversity of production, including a significant minerals sector, provided a broader economic base than in New Zealand, though with wool and meat production playing a dominant role throughout the nineteenth century.

Both countries were bastions of the Empire, with public attention frequently focused on events in London or some other part of the UK. This dual focus on home and the 'mother country' continued longer in New Zealand, so that at times it seemed that mentally the Pakeha were located off the southern coast of England rather than in the South Pacific. Even as recently as 1984 the national miners' strike in the UK coalfields supplanted domestic news from the front-page headlines in the New Zealand daily papers, despite the dramatic domestic political events of the time.

Chapters 7 and 8 focus on key components of the two emergent economies that have remained vital components of economic development in the twentieth century: the agro-commodity sector in New Zealand (Chapter 7) and the minerals sector in Australia (Chapter 8). These have been singled out both because of their historical importance and their contemporary economic role in which their fortunes continue to play a highly significant part in overall economic well-being. In both cases, major changes during recent decades have affected the structure of production, with various geographical consequences. For New Zealand, Chapter 7 charts the transition from its role as 'Britain's farm in the South Pacific' to the development of new markets and new products since the 1960s, once the loss of the assured UK market was apparent through the UK's entry to the European Economic Community (EEC) (which was ratified in 1973). Although the three staples of wool, meat and dairy produce have continued to occupy a prominent role within primary-sector production and exports, new value-added products, exploiting 'niche' overseas markets, have been part of a general diversification process. However, since 1984 the farming sector has had to survive with greatly reduced state subsidy, as part of a far-reaching set of measures that has deregulated the economy and hastened the impact of world market forces. These forces are examined in more detail in Chapters 13 and 14, but also feature in Chapter 8 on the Australian minerals and energy industries, as Australia too pursued economic liberalisation measures in the mid-1980s. These helped introduce more foreign investment (especially from the Asia–Pacific region) and foreign control of these and other industries.

The first large-scale Australian trading ties to Asia were closely associated with the post-war Japanese 'economic miracle'. The coincident discovery of vast reserves of iron ore and coal in Australia and rapid expansion of the Japanese economy in the 1960s led to large-scale exports of these minerals. This has sustained an export-led development of the Australian minerals sector, though the diversity of resources within the country has prevented over-reliance upon a single mineral source. Australia has become one of the few countries to enjoy a positive trade balance with Japan. However, many of the major mineral developments are controlled by foreign-based multinationals. This high level of overseas control can also be seen in other sectors. The UK has retained a high level of investment, but is now matched by American and Japanese money. The latter has been especially prominent in the tourist, real-estate and financial sectors (Chapter 14). The level of foreign control has prompted concerns that the Australian economy is too reliant upon decision-makers belonging to companies located overseas, and therefore not subject to Australian government regulation. Structural characteristics of the economy, such as the relatively undeveloped minerals processing sector, have caused the greatest concern.

One characterisation of the growing reliance on unprocessed and semi-processed minerals was that Australia had been transformed 'from a farm to a quarry'! This highlights the relatively weak development of a home-based manufacturing sector, analysed in more detail in Chapter 13, whilst Chapter 8 focuses on other aspects of the minerals and energy industries: the nature of the minerals resource base, the development of mining towns, and some of the environmental conflicts associated with resource exploitation.

DEMOGRAPHY AND SOCIETY

The third part of the book (Chapters 9–12) deals with demographic and social characteristics, again emphasising the many similarities between the two countries whilst illuminating the nature of significant differences. Both Australia and New Zealand have many demographic indicators (e.g. growth rates, life expectancy, fertility rates, family characteristics and age composition) comparable with those elsewhere in the developed world. However, this disguises differences for the indigenous populations, reflective of their more marginal position in society. For example, Chapter 9 notes that whilst society as a whole is becoming more elderly (or 'greying'), this is not the case amongst indigenous Australians, for whom decreased infant mortality and high birth rates are helping to create a very youthful population. This Chapter also observes the changing distribution of population, in which a growing 'drift north' to the 'sun-belt' of Queensland and the northern New South Wales coast can be observed in Australia. In New Zealand there has been a similar pull exerted by the northern half of the North Island (to Auckland, the Bay of Plenty and Northland), but with a noticeable counter-flow to the northern South Island.

Chapters 9 and 10 emphasise that one vital aspect in which the two countries have differed has been with respect to the nature of their post-war migrant streams. Changing outlooks on the rest of the world were more pronounced in Australia from the 1930s onwards, by virtue of changes in the composition of the immigrants (Chapter 9). Pre-1939, some Italians were added to the dominant British and Irish stock to supply additional rural labour, especially in Queensland. In the 1950s their numbers grew and were swelled by other Mediterranean immigrants, as Australia followed the dictum of 'populate or perish'. There were simply insufficient numbers of British migrants to meet Australia's demand for people to work in the expanding economy, and hence the increasingly cosmopolitan character of the immigrants. The majority landed by boat or air in Melbourne and Sydney, and this was where the major concentrations of immigrant groups remained, with inner suburbs like New Brunswick (Melbourne) and Leichhardt (Sydney) being transformed by the addition of ethnic shops, food outlets and churches. As elaborated in Chapter 10, the government's stated aim was for the new immigrants to be assimilated into the majority population. But with clear cultural, and especially language, differences between immigrant groups and the host population, this was not immediate. Moreover, the majority of the new immigrants were unskilled and so worked on assembly lines, in manual labour occupations and low-skill ancillary jobs. This reinforced clear distinctions between Anglo-Celtic communities and the various ethnic groups, so that a more heterogeneous society was being created. However, it was not until the 1970s that an official policy of multiculturalism was instated, and diversity rather then assimilation could be considered more acceptable.

In contrast, New Zealand made no such conscious attempt to maintain high levels of immigration. Hence the UK remained the main supplier of new citizens, supplemented by smaller numbers from other north-west European countries. In the 1960s temporary migrants (guest-workers) from the small Pacific Island nations were permitted entry to meet labour demands, usually in low-paid jobs. Despite much opposition, many of these Pacific Islanders were then allowed to become permanent residents, bringing over their families and so creating a small but distinctive group alongside the Europeans and Maori. They were concentrated primarily in the rapidly growing largest city, Auckland, and by the 1970s so were many Maori, having been part of a substantial rural–urban migration in the 1950s and 1960s.

Both countries have significantly increased their trading ties to south-east Asia in the past two decades, both in terms of direct trade and through foreign-direct investment (FDI). This represents an 'Asianisation' paralleling a change in the migrant stream whereby both countries have accepted migrants on the basis of income and educational qualifications. Hence, immigrants from Hong Kong, Vietnam, South Korea and India have become leading groups amongst the new immigrants. Especially in Australia this has meant that there will be a substantial Asian minority (up to 8 per cent of total population) by 2020 if current inflow rates are maintained. Diversification of the immigrant stream has been less pronounced in New Zealand, but nonetheless amongst the leading suppliers of immigrants in recent years have been Hong Kong, China, South Korea and India. For Australia, the growth in Asian immigration has replaced the migrants from southern Europe who came

during the 1950s and 1960s, at which time New Zealand's migrants still came largely from the UK or the Pacific Islands. Continued immigration from the latter has contributed to the growth of sizeable Islander communities in Auckland and Wellington.

Chapter 10 examines the debates and policies associated with the concepts of monoculture, biculturalism and multiculturalism. In the early 1970s Australia took its first faltering steps towards establishing a new set of relationships with its Asian near neighbours. It did so by revoking the 'White Australia' policy, which it had pursued in both *de facto* and statutory form since Chinese miners were expelled from the Victoria goldfields in the 1850s. Attitudes towards different ethnic groups, and significantly those from the 'teeming masses' in Asia, had led to an exclusivity only partially relaxed through the influx of immigrants from the Mediterranean, including Lebanese and Turks, in the 1950s and 1960s. Now, in theory, there was no official racist bar to Asians. This was tested very quickly with the arrival of refugees fleeing from the Vietnam War. Australia became the largest recipient of these refugees, giving rise to a distinctive Vietnamese quarter in Cabramatta, inner Sydney. Accompanying moves to eliminate ethnic bias from immigration policy were measures according more civil rights to the indigenous population whose role within the new multicultural policies has been, at best, uncertain.

Biculturalism, rather than multiculturalism, has been pursued in New Zealand from the mid-1980s, recognising the special status of Maori and the rediscovered desire by Pakeha to respect the Treaty of Waitangi. However, this raises questions regarding the position of Asian immigrants within the bicultural society, a problem shown in sharp relief by anti-Asian sentiments voiced in some quarters and paralleling that occurring in Australia. One result has been recent measures taken by the New Zealand government to reduce Asian immigration.

Chapter 11 investigates the nature of the 'new societies' in Australasia. It considers the dynamics of social changes in response to the increasing influence of global economic forces and neo-liberal policies pursued by governments in both Australia and New Zealand. Key changes considered include employment patterns, growing social polarisation (with the emergence of disadvantaged minorities), housing reforms and the changing role of the welfare state. The latter reflects the 'rolling back' of the state from the early 1980s as part of a suite of reforms intended to develop both economic efficiencies and reduced government expenditure (and therefore less taxation). The economic consequences of these measures are discussed in Chapters 13 and 14, revealing both 'winners' and 'losers' in society as one of the policies' outcomes. In terms of welfare provision, it has been successive New Zealand governments in the 1990s that have reduced state support for health care and social security to a greater extent than in other OECD countries. This has contributed to greater inequalities in society, adding to other forces encouraging this, including huge income disparities in the dominant and growing service sector, changing family structures, spatially based variations in access to services, rising unemployment and falling incomes in farming. Particular attention is given to public housing reforms in both countries, in which state intervention is increasingly being replaced by subsidies and more reliance on the private sector. It is notable that housing disadvantage and other forms of social deprivation are most marked amongst the indigenous peoples and certain groups of immigrants, especially Pacific Islanders in New Zealand.

Whilst the ethnic composition of immigrants and the flow of trade have both shifted towards Asia, the character of Australasian cities remains unmistakeably Western (Chapter 12) and the skyscraper has become dominant in the central business districts (CBDs) of all the major cities. Very high levels of urbanisation were recorded at an early date compared with elsewhere in the world. This has contributed to an urban structure dominated by the suburbanisation process and the dominance of privately owned motor vehicles. In locations lacking major physical constraints on urban development this has produced substantial urban sprawl, often aggravated by post-war planning policies that emphasised centrifugal growth. More recently the trend towards out-of-town retailing and dispersion of both office- and factory-based employment has contributed to the maintenance of the dispersed form despite several initiatives emphasising revived CBDs and waterfront locations, and some improvements to selected inner areas (gentrification). In the case of restrictive physical features for some cities, e.g. Wellington and Hobart, more compact forms have emerged, whilst Canberra represents the best example of the garden city/city beautiful concept in Australasia.

Chapter 12 also discusses the character of those parts of Australian and New Zealand cities associated with particular ethnic groups. Concentrations of southern Europeans formed in both Melbourne and Sydney in the 1960s, but recently attention has focused on a similar development for Vietnamese in Cabramatta (Sydney). However, the levels of concentration have not approached those associated with certain ethnic groups in US cities, and dispersion over time to other parts of the city has also been a feature for Australia's European immigrants. It remains to be seen whether this will soon be the case for relatively recently arrived Asian immigrants.

RECENT ECONOMIC DEVELOPMENT

The final part of the book (Chapters 13–17) deals with various aspects of economic development, stressing the changes brought about through the combination of globalising economic tendencies and reforms pursued from the early 1980s in both Australia and New Zealand. These reforms, emphasising deregulation and encouraging restructuring, have affected all sectors of the two economies and have brought about substantial changes to economic life. Chapters 13 and 14 concentrate, respectively, on manufacturing industry and the service sector; Chapter 15 focuses on key spatial outcomes and regional planning; Chapter 16 refers to environmental dimensions of recent change; and the concluding Chapter (17) explores broader geopolitical issues and, especially, emerging Asia–Pacific regional economic ties.

Although small-scale industry developed in the nineteenth century to secure the growing domestic market, manufacturing establishments remained small, with two limited and widely dispersed home markets to serve. Protective tariffs and subsidies were used consistently as a means of fostering domestic manufacturing but often at the expense of rendering it uncompetitive in international markets. Overseas control in certain sectors was high and also benefited from protective measures. Despite post-1945 production increases to serve the rapidly growing domestic market, manufacturing industry remained relatively small when compared with the fast-growing service sector. Moreover, manufacturing was skewed towards the processing of primary produce. It is not surprising, therefore, that policies favouring economic liberalisation in the 1980s have impacted sharply upon this sector, generally pursuing a strategy to make domestic manufacturing 'leaner and fitter' by exposing it to greater overseas competition.

In the mid-1980s both Australia and New Zealand took measures to transform their economies by pursuing 'new right' policies of deregulation, privatisation of state-owned enterprises, reductions of protective tariffs and a general opening-up of all sectors to overseas investment and competition (Chapter 13). In New Zealand this process was termed 'Rogernomics', after its chief architect, Treasurer Roger Douglas. In Australia the process was begun under the Hawke government and its Treasurer, Paul Keating. In both countries, reforms have been continued in the 1990s under governments of different parties, with sweeping social reforms in New Zealand under the National Party, first led by James Bolger and then Jenny Shipley. The reforms have enabled the impacts of global economic processes to be substantial, transforming some sectors of the

economy by affecting production processes, ownership, labour management, input sources, markets and distribution. In short, many of the characteristics of Australasian-based production have been altered irrevocably as the impress of global forces has swept through the economies. This has not been a smooth or uncontested process, but has been achieved at the expense of traditionally low rates of unemployment, some loss of social welfare provision, and growing disparities between rich and poor despite increased wealth generation in particular sectors of the economy.

The biggest growth areas have been associated with various elements of the service sector, especially tourism, retailing, financial services, and personal, community and professional services (Chapter 14). This has generated significant spatial outcomes, including further concentrations of employment in the major cities, and especially Sydney and Auckland, but with new growth areas developing related to the advance of particular services, e.g. south-east Queensland for tourism, and Canberra for professional services (Chapter 15). Meanwhile, loss of jobs in manufacturing has had a disproportionate effect upon centres most reliant upon this sector, e.g. Melbourne, Adelaide and the 'heavy' industrial towns of Newcastle and Wollongong. There have also been important intra-urban developments, with further moves of economic activity to outer suburbs and greenfield sites. New locations for consumption activities have been developed and have become closely identified with the creation of new jobs. In contrast, old inner-city areas, especially those associated with manufacturing and poorer housing, have often become run-down and sources of out-migrants. City and regional planning policies have had to address the resultant urban imbalances, in part by attempting to revitalise city centres and channelling investment into inner suburbs. Three case studies of these planning policies are considered in Chapter 15: Melbourne, Brisbane and Auckland.

Chapter 16 continues the themes of planning and management, but focusing on the need to combat environmental hazards and also to conserve the two countries' rich environmental heritage. The hazards considered cover a broad spectrum from geophysical hazards, such as earthquakes and volcanoes, to extremes of weather associated with tropical cyclones, drought, bushfires and floods. Despite the fact that four-fifths of the population live in the 'safety' of the human-created environment of the city, they can still be threatened by devastating natural forces, as seen in such graphic examples as the Newcastle and Napier earthquakes and cyclone Tracy (which hit Darwin in 1974). These events have prompted disaster management planning and hazard assessments to help prediction of catastrophic events and to mitigate against their impacts. Alongside this move towards a pro-active framework for disaster management, there has also been an evolution in the attitudes towards management of land and water resources.

Environmental management in Australasia has moved from an exploitative phase associated with pioneering European settlement, to a more careful use of resources aimed at extending economic utility post-1945, and now to modern environmentalism. The latter represents a conflicting set of attitudes towards resource development in which various groups have adopted different positions, but with widely varying degrees of effectiveness. Thus management of the environment has become a highly contentious political issue, often putting government in conflict with conservation groups, e.g. over proposals for oil exploration on the Great Barrier Reef. However, various forms of accommodation have frequently been reached regarding less exploitative forms of resource use, as seen in Landcare programmes and agreements on the logging of native forests. Pressure grows to extend conservation principles into new areas of management, often supported by the desire to preserve attractive environments for the benefit of tourists. This utilitarian view of conservation has contributed to increased designation of conservation reserves and national parks, though even here conflicts between development and conserva-

tion interests can occur, as in the case of proposals for uranium mining in Kakadu National Park, Northern Territory. Also, over 90 per cent of Australia's protected reserves are less than 10 000 ha which may mean they are not ecologically viable systems.

Further population growth and economic expansion will certainly continue to threaten distinctive environmental systems, though conservation measures have established various controls on some destructive tendencies. The rhetoric of sustainable development, which is included in much recent legislation, especially in New Zealand, is yet to be translated into sustained regulation of rapid resource consumption. In Australia's case the vast mineral wealth perhaps masks the need for conservation. However, in both countries, the large measures of deregulation and pursuit of 'economic rationalism' have made the economies highly responsive to world economic trends rather than strongly controlled domestic agendas (which might be more supportive of conservation interests).

The final Chapter (17) extends the coverage of the geography of Australia and New Zealand into the geopolitical realm. Although other sections of the book examine the implications of globalisation for the economy and society of Australasia, this Chapter deals explicitly with the formal linkages developed with the rest of the world. In part this is another facet of the evolving story of how the two countries are creating a new 'place in the world', increasingly distancing themselves from a colonial image. Three components in particular are discussed here. The first is the post-1945 emergence of closer defence ties with the United States through the ANZUS Treaty. This has been the mainstay of military thinking for nearly half a century, though New Zealand's participation has been downgraded following its banning of nuclear-powered and -armed ships since 1984. Secondly, economic links are examined with respect to the establishment of multilateral agreements. Economic deregulation and restructuring have accompanied increased multifaceted linkages with Asia–Pacific countries through trade and investment. This has been underscored in official form by both countries' membership of the Asia–Pacific Economic Co-operation (APEC) group, established in 1989. One of APEC's stated aims is the advancement of free trade in the region, and hence the expansion of trade and other links between member countries. On a smaller scale this can be seen through trans-Tasman developments fostered in the Closer Economic Relations (CER) agreement, signed in 1983 and aimed at creating a single economic market between Australia and New Zealand. Finally, the impacts of the greater emphasis upon the Asia–Pacific region are discussed with respect to the changing patterns of international trade. The declining significance of trade with the UK since the 1950s is noted, with Japan, the USA and other Asia–Pacific trading partners coming to the fore in recent decades.

The concluding section of this Chapter, and of the book itself, discusses the ongoing debate about republicanism. For many Australians, the fact that Queen Elizabeth II is still the head of state is symbolic of bygone days, out of keeping with a modern, multicultural, Asia–Pacific country at the dawn of a new century. The desire to remove this vestige of a different era is strong, but there are others who wish to maintain a monarchial head of state as an essential part of continuity and maintenance of the country's heritage. In New Zealand there appears to be less pro-republican sentiment, but the debate is bound to continue there too if the drive towards furthering a distinctive Asia–Pacific identity continues.

FURTHER READING

There are a number of general texts on the geography of Australia, including the second editions of Heathcote (1994) and Walmsley and Sorensen (1993). The latter concentrates on economic and social issues. There is no compa-rable introduction for New Zealand, but there are two excellent detailed studies of the trans-formation undergone by the country since the mid-1980s, in Britton *et al.* (1992) and Le Heron and Pawson (1996). Both of these are edited collections comprising short studies of key issues and case studies by leading New Zealand geographers. There is no recent Australian equivalent, but see the second edition of the two-volume compendium edited by Jeans (1987). A more recent series of books produced by Oxford University Press provides geographical studies across a wide range of topics (e.g. Dargavel, 1995; Fagan and Webber, 1994; Forster, 1995; Kirkpatrick, 1994; A.R.M. Young, 1996). Other general work on Australia by geographers includes collections edited by Heathcote and Mabbutt (1988) and Heathcote (1988). An excel-lent collection of essays on New Zealand's physical environment is Soons and Selby (1992). A fine addition to the geographical literature on New Zealand is provided in the *New Zealand historical atlas* (McKinnon, 1997): superbly illus-trated, it has been compiled by leading scholars representing a number of disciplines.

Both countries are well served by journals catering for geographical research: *New Zealand Geographer, New Zealand Journal of Geography, Australian Geographical Studies* and *Australian Geographer. Asia Pacific Viewpoint* provides a broader focus whilst the *British Review of New Zealand Studies* includes articles from throughout the humanities and social sciences. Other journals covering various aspects of Pacific affairs include *The Contemporary Pacific, Pacific Affairs* (an interna-tional review of Asia and the South Pacific), *Pacific Economic Bulletin, Pacific Studies* and *Journal of the Polynesian Society.*

Those seeking more general and introduc-tory background reading should begin with Flannery's (1994) *The future eaters,* charting the impacts of human occupance upon the Australasian environment.

2

THE PHYSICAL SETTING: LANDFORMS AND SOILS

The Australian continent is a generally low-lying, ancient and stable landmass with low-energy environments, in contrast with more dynamic environments of high relief, and tectonic and volcanic activity in New Zealand. These differences can be largely explained by Australia's position within the Australian Plate compared to that of New Zealand on a compressional boundary between the Pacific and Australian Plates.

LANDFORMS OF AUSTRALIA

Lineaments (faults and major joints) traverse the continent, giving rise to uplifted blocks and ranges, and areas of subsidence (Twidale, 1968: 49–51). Three contiguous north–south basins form a lowland separating the plateau of Western Australia from the uplands of eastern Australia, dividing the continent into three physiographic divisions (Jennings and Mabbutt, 1986): the Western Plateau Division, the Interior Lowlands Division and the Eastern Uplands Division (Figure 2.1). This simple tripartite sub-division obscures the complexity and variety of Australian landscapes, for Jennings and Mabbutt (1986) identify and map 23 'Provinces' and 227 'Regions', each with distinctive assemblages of landforms and geo-

morphic history. For present purposes, however, major landform features are outlined within the framework of the three Divisions and the Australian Coast.

The Eastern Uplands Division

The western boundary of this division runs roughly parallel with the east coast of the continent from Cape York to Tasmania (Figure 2.1). It comprises the eastern section of the geological Tasman Fold Belt (Gale, 1992), an area which has experienced several phases of orogenesis during the Phanerozoic (Palaeozoic, Mesozoic and Cenozoic). Granitic intrusions into a variety of sedimentary and metamorphic rocks, volcanic rocks of Cenozoic age, and silcrete and ferricrete duricrusts form the broad lithological types (Ollier, 1986).

Topographically, the Division comprises a series of almost level surfaces approximately 1000 m in elevation that have been uplifted and dissected by fluvial and slope processes. This is best seen on the Great Escarpment which runs north-south along the length of eastern Australia (Ollier, 1982) (Figure 2.1). The Great Escarpment, formed by erosion after the New Zealand subcontinent rifted from the eastern Australian margin approximately 80 million years ago, lies to the east of the Great Divide

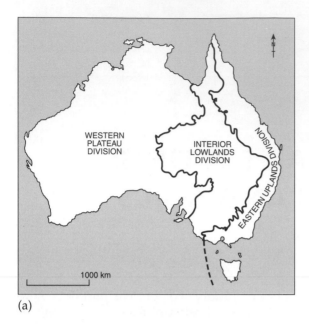

(a)

FIGURE 2.1 (a) The physiographic zones of Australia; (b) The Eastern Uplands Division (adapted from Jennings and Mabbutt, 1986)

which separates the easterly flowing rivers from those flowing inland. Rivers flowing to the Pacific coast, and associated slope processes, have eroded valleys, many of which are gorge-like in their upper reaches with steep waterfalls at the point where streams leave elevated tableland surfaces. The headwaters of these rivers are at relatively high elevations (>1000 m), but are in areas of low relative relief on the Great Divide inland of the Great Escarpment (Figure 2.1). The evolution of drainage networks is complex, with reversals appearing to have taken place along some systems (Haworth and Ollier, 1992). The geomorphology of streams in eastern Australia appears to be regulated by relatively infrequent flood events of high magnitude which can strip floodplains and mobilise bed sediments (Nanson and Erskine, 1988; Tooth and Nanson, 1995; Erskine, 1996), although the direct and indirect effects of post-European colonisation on vegetation and land use must also be considered.

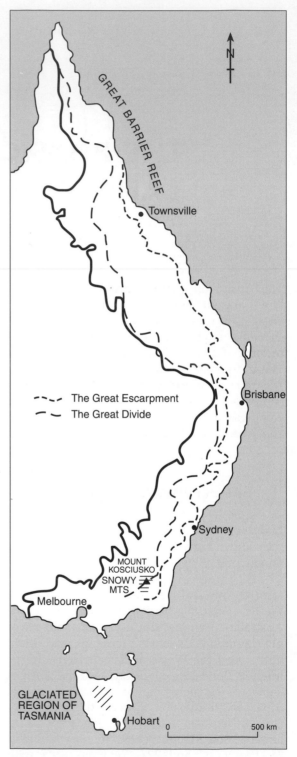

(b)

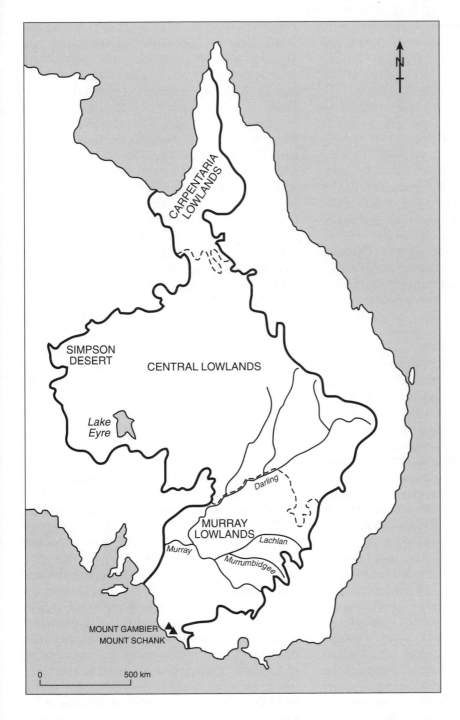

FIGURE 2.2 Australia: the Interior Lowlands Division (adapted from Jennings and Mabbutt, 1986)

The highest points on the Australian mainland, in the Mt Kosciusko (2228 m) area of the Snowy Mountains of south-eastern New South Wales (Figure 2.1), were affected by no more than 52 km² of glaciation during the late Cenozoic (Galloway, 1963). In Tasmania, where the glaciated high country was at higher latitudes, ice cover was much greater (Colhoun *et al.*, 1996), occupying approximately 2000 km², although the maximum extent is not known in detail (Figure 2.1). Both ice-cap glacial features and cirque and valley glacial features are seen in Tasmania.

Interior Lowlands Division

This lowland area is a structural depression that suffered shallow marine transgressions in the Mesozoic. Sedimentary rocks of low resistance to weathering and erosion are generally present within the Division, which has its eastern margin against the Eastern Uplands Division and its western against the Western Plateau (Figure 2.2). From north to south, the Division comprises three Provinces: the Carpentaria Lowlands, the Central Lowlands and the Murray Lowlands, respectively (Jennings and Mabbutt, 1986).

The Carpentaria Lowlands, generally below 200 m, contain a mix of clay floodplains, sloping sandy plains and laterite-capped plateaux.

To the south, the more extensive Central Lowlands broadly correspond with the internally drained Lake Eyre basin and the northern portion of the Darling River catchment (Figure 2.2) (Jennings and Mabbutt, 1986).

In the Lake Eyre basin, erosional (stony desert) zones lie on the margins, with depositional riverine and clay plains, sand dune systems and salt lakes developed to the west (Mabbutt, 1986) (Figure 2.2). The upper sectors of the rivers in this part of Australia consist of channels with sandbanks and gallery woodland. Channels are of generally low sinuosity with a high width-to-depth ratio. Distributaries and associated basins have finer deposits within them, while outer plains have aeolian sand dunes. Lower and terminal sectors of the riverine systems exhibit decreasing channel capacities because of transmission losses and the development of distributaries. At the terminus, distributary channels and floodouts (inland deltas) occur (Mabbutt, 1986). Lake Eyre salina, 16 m below sea level, is the focus of much of the drainage. Longitudinal sand dunes cover extensive areas of the lowland, particularly the Simpson Desert (Figure 2.2). The dunes are longitudinal and trend, with variations, SSE to NNW and there is evidence that the dune crests are currently active (Twidale and Campbell, 1993). Dune summits are between 10 and 38 m above the desert plain, and can be continuous for many kilometres. The Simpson Desert dunes form part of a large anticlockwise arc throughout arid Australia, from the Great Victoria Desert (west–east), to the Simpson Desert (south–north) and the Great Sandy Desert (east–west) (Figures 2.2 and 2.3) (Mabbutt, 1986).

In the south-east of the Interior Lowlands Division, drainage is to the upper Darling River system (Figure 2.2). Here, the landscape consists of silcrete-capped mesas and tablelands generally below 500 m elevation, stony plains, sandplains and floodplains. Alluvial flats and distributary channels are features of the fluvial system.

The Murray Lowlands Province forms the most southerly part of the Division (Figure 2.2) and has landscapes of silcrete, longitudinal aeolian sand dunes, gravelly and sandy plains, crescent-like lake-bordering dunes (lunettes), alluvial plains, sinuous rivers, parallel (former coastal) dune limestone ridges, karst and volcanic forms (Jennings and Mabbutt, 1986). The alluvial plains of the Darling, Lachlan, Murrumbidgee and Murray Rivers and their tributaries form the greater part of the Province, and contain lakes, lunettes, channels and deposits which reflect climatic and hydrologic changes of the Quaternary (Bowler, 1986). Between 60 000 and 32 000 years ago there was a phase of high water levels in lakes with large river discharges in channels with high capacities. A periglacial and glacial drier period followed in the highlands of south-eastern Australia, the catchment area for the southernmost streams of the region. This drier period was one of channel shrinkage and high aeolian activity which lasted until approximately 15 000 years ago. After the glacial phase, channels became the suspended-load type (Schumm, 1977: 156),

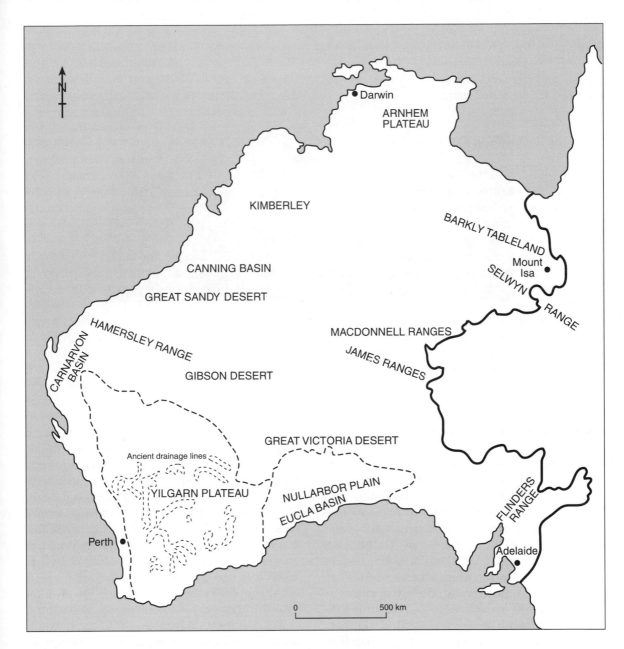

FIGURE 2.3   Australia: the Western Plateau Division (adapted from Jennings and Mabbutt, 1986)

with lower width–depth ratios and higher sinuosities (Bowler, 1986). These largely fluvial landscapes contrast with the coastal forms of south-east South Australia, which consist of abandoned foredunes providing a record of change in Quaternary sea level and slow tectonic uplift (Schwebel, 1983; Twidale and Campbell, 1993). This part of South Australia was also one of volcanic activity at Mt Schank and Mt Gambier between 9000 and 4000 years ago (Figure 2.2) (Sheard, 1983; Twidale and Campbell, 1993).

FIGURE 2.4   View southwards towards Wilpena Pound, Flinders Ranges, South Australia. Differential weathering and erosion have produced a series of structurally controlled ridges (resistant sandstones) and valleys (shales and siltstones) on folded strata. (photo: R.J. Loughran)

Western Plateau Division

Approximately the western two-thirds of Australia comprises the Western Plateau Division (Figures 2.1 and 2.3) (Jennings and Mabbutt, 1986). Much of the area has experienced tectonic stability, deep weathering and erosion during the Phanerozoic. The Yilgarn Plateau, a granitic shield block, the folded resistant rocks of the Macdonnell and Flinders Ranges, the uplifted and dissected plateaux with marginal escarpments of resistant sedimentary rocks (the Hamersley and Arnhem Plateaux), and the plains and low tablelands of horizontal sediments provide a varied landscape. Many of the surfaces are mantled with duricrusts, boulders, stones, alluvium and sand dunes.

The shield desert landscape of Western Australia (Yilgarn Block) is an undulating plain below 500 m altitude that has been laterised. The surface of the laterite is overlain by a restricted and thin layer of reworked sands. For the greater part, the laterite has been stripped and has an escarpment, or 'breakaway', at its margin, below which is a pediment (Mabbutt, 1986). Downslope, sediments moulded by wind and water are present, while at the lowest part of the system, calcreted valley fill and river-filled lakes occur. Another feature of the Yilgarn Block is a system of ancient drainage lines, now occupied by salt lakes. It is thought that the broad valleys were an integrated system of drainage towards the south, truncated by the break-up of Australia and Antarctica (Figure 2.3) (Ollier, 1986). In central Australia, the shield desert comprises granitic hills with pediments leading into wash plains traversed by ill-defined waterways (Mabbutt, 1986). Here, also, there is calcreted valley fill.

Landforms developed on folded resistant strata are prominent in the Flinders Ranges of South Australia, up to 1100 m high (Figure 2.4), and in the ranges of central Australia. In the Flinders, ridge and valley topography is formed on Proterozoic (2500 to 540 million years) and Cambrian (540 to 505 million years) rocks (Twidale and Campbell, 1993). The structures, etched out by the differential weathering and erosion of less resistant shale and siltstone compared with resistant sandstone and limestone, have a predominantly north–south grain in the central Flinders Ranges (Figure 2.4) (Corbett, 1987).

Within the Western Plateau Division, the Kimberley Plateau (<750 m), the Arnhem Plateau (<400 m) and the Hamersley Ranges (up to 1250 m) are uplifted blocks (Figure 2.3). Areas of subsidence, or negative tectonic regions, include the Carnarvon, Eucla and Canning Basins (Twidale, 1968). The basin plainlands are occupied by longitudinal dunes of the Great Victoria and Great Sandy Deserts and the stony desert of the Nullarbor (Figure 2.3). In the latter, the limestone surface of the Bunda Plateau rises inland from the cliffs on the Great Australian Bight to the Great Victoria Desert dunefield (Mabbutt, 1986).

The Australian coast

The length of the Australian coastline is approximately 70 000 km (Department of the Environment, Sport and Territories, Australia (DESTA), 1996) and may be divided into four process environments (Davies, 1986): warm temperate humid coasts, warm temperate arid coasts, tropical arid coasts and tropical humid coasts (Figure 2.5).

Warm temperate humid coasts extend from southern Queensland, through New South Wales, Victoria and Tasmania (Figure 2.5) (Davies, 1986). While there are rocky headlands and shore platforms, the coastline is dominated by siliceous sands of terrigenous origin. The sediments form barrier systems of Pleistocene and Holocene age created by wave and wind action on a relatively high-energy coast. There appears to be little supply of sediments by present-day rivers, the barrier sediments having been derived from the onshore movement of sands from the continental shelf.

The warm temperate arid coasts are developed westwards from Victoria to the North West Cape in Western Australia (Figure 2.5) (Davies, 1986). There is high wave energy but, because of the arid hinterland, very little supply of sediments from the continent. Barrier sediments are therefore of marine origin, now lithified into calcareous sandstones. Old barrier (Pleistocene) dune rocks have been eroded into cliffs, and estuaries and embayments are few. Coral reefs are present in the northern part of this section of coast.

From the North West Cape to Broome, Western Australia (Figure 2.5), there is a short reach of tropical arid coast (Davies, 1986). Low wave energy and a wide, shallow coastal shelf characterise this portion of the coastline. The lithified sediments of the barrier systems are less extensive than in the temperate arid zone.

The tropical humid coast is the longest section of Davies' (1986) classification, extending from Broome in Western Australia to Fraser Island in southern Queensland (Figure 2.5). Wave energies are generally moderate to low, but storm surges and tropical cyclones can raise levels. Sediment loads from rivers are the highest in Australia, yielding fine-grained materials to the coast. Coral reefs within Western Australia and Queensland (Great Barrier Reef) are the largest in the world (DESTA, 1996).

The Great Barrier Reef, which has an area of 230 000 km^2, extends over 2300 km from near Cape York to approximately Gladstone, Queensland (Figure 2.5). The reef becomes wider in a southerly direction, from 23 km in the north to 260 km in the south (Twidale and Campbell, 1993).

LANDFORMS OF NEW ZEALAND

Most of New Zealand's landforms have been created over the last two million years and over wide areas climate and relief produce high rates of erosion (Pillans et al., 1992). The distinctiveness and youthfulness of New

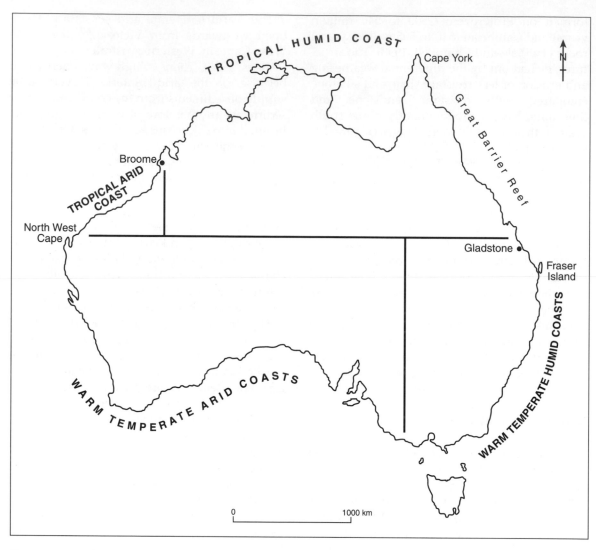

FIGURE 2.5   The coastal zone of Australia (after Davies, 1986)

Zealand's landforms largely reflect its position on a crustal plate boundary. Where the Pacific and Australian Plates converge there is a subduction zone under North Island, because the Pacific Plate is heavier than the continental crust. A subduction zone has not formed under South Island because both sides of the plate boundary are composed of buoyant continental crust, creating a transform boundary where the plates slide past each other (Kamp, 1992a; 1992b). Horizontal and vertical movements within the plate boundary zone over a 25-million-year period have given rise to the present shape of New Zealand (Kamp, 1992a). 'How and where this deformation has been distributed throughout New Zealand are contentious issues' (Kamp, 1992a: 19).

Therefore, from Hawke's Bay on the east coast of North Island to Fiordland in south-west South Island there is a zone of severe crustal deformation. This so-called Axial Tectonic Belt (Walcott, 1978) is between 70 and

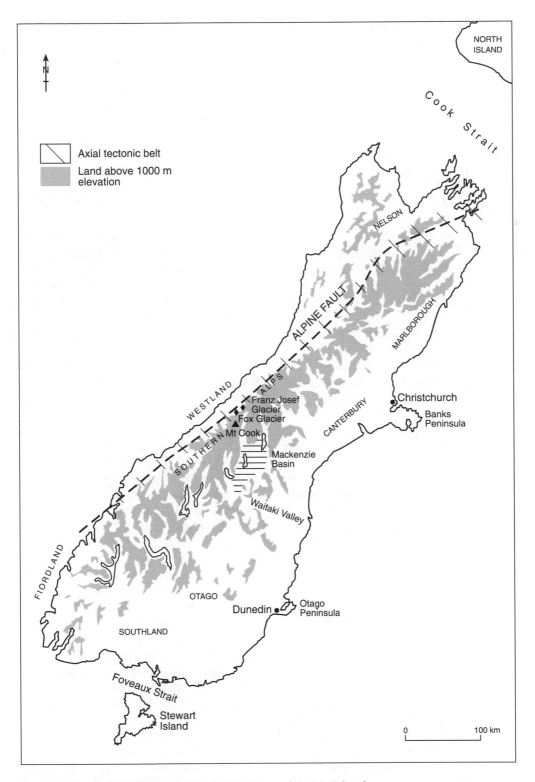

FIGURE 2.6   New Zealand: physical features of South Island

100 km wide and trends north-east to south-west. Rates of plate convergence increase from 30 mm y^{-1} south of South Island to 60 mm y^{-1} off East Cape, North Island (Selby, 1985: 81).

South Island

With altitudes in excess of 3000 m, the Southern Alps form the distinctive backbone of South Island, created by continuing uplift since the Miocene (Selby, 1985; Riddolls, 1987; Kamp, 1992a; Whitehouse and Pearce, 1992). Hard sandstones and mudstones (greywackes) of Late Palaeozoic and Cretaceous age are the dominant rock types, while the rocks of Fiordland are a Gondwanaland remnant after separation approximately 80 million years ago. The Alpine Fault (Figure 2.6) experienced transform motion in post-Miocene times, when the west coast moved northwards relative to the east coast. Rocks on the eastern side of the Fault in Otago and Southland are to be found on the western side of the Fault in the northernmost part of South Island in Nelson over 450 km away (Figure 2.6).

During the Pleistocene, South Island was subjected to several glaciations, producing glacially eroded valleys which now dominate the landscape in the mountains and south-westwards into coastal Fiordland (Pillans, *et al.*, 1992; Whitehouse and Pearce, 1992). Present-day glaciers are to be found in the Mackenzie Basin on the eastern side of the Alpine divide in the Mt Cook area (Figure 2.6). These valley glaciers are remnants of their more extensive Pleistocene valley and piedmont counterparts, and many are wasting away (Whitehouse and Pearce, 1992), with downstream surfaces covered with morainic debris. Other glaciers descend to the west coast from the same icefield, notably the Fox and Franz Josef glaciers (Figure 2.6) (Soons, 1992). There are also present-day glaciers further south in the Alps.

On the greywackes of the Southern Alps, mass movement has produced scree slopes, many of which were formed during late glacial or in immediately post-glacial times. Slope evolution has been restricted in Fiordland because the crystalline rocks there are less well jointed and possess greater shear strength (Whitehouse and Pearce, 1992).

Mountainous regions, peripheral to the Southern Alps, occur also in Otago, Canterbury, Nelson and Marlborough (Figure 2.6). Of these, the greywacke-argillite mountains of Marlborough were unglaciated in the late Pleistocene (O'Loughlin and Pearce, 1982). This region is tectonically very active and rapid mass movement is common. Central and eastern Otago, and north-west Nelson were affected by the late Pleistocene glaciation to a small degree; the Otago regions, composed of uplifted, faulted and tilted blocks, were affected by periglacial conditions, however. North-west Nelson has a similar geological structure, lithologies and tectonic history to Fiordland. Finally, glaciated, mountainous topography on jointed and fractured greywackes and argillites occurs in northern Otago, Canterbury and Marlborough (Figure 2.6). This region, because of its high susceptibility to weathering processes, experiences major landslides and alluvial fans spread into valleys (O'Loughlin and Pearce, 1982).

The two most extensive lowland areas within South Island are the Canterbury Plains and the Plains of Southland (Figure 2.6). The former extends along the east coast for 180 km and inland as much as 70 km. The Canterbury Plains consist of large alluvial fans laid down by the major rivers during the Pleistocene glaciations when great quantities of gravel were released into the streams. At the point where the rivers leave the foothills of the Alps, spectacular flights of terraces are visible on the alluvial deposits (Fitzharris *et al.*, 1992). On the Canterbury Plains, the rivers have developed wide, braided channels. The Miocene volcanoes of the Banks Peninsula are tied to the mainland by the sediments of the plain, which are modified by longshore drift and wave action on the coast north and south of the peninsula (Fitzharris *et al.*, 1992). The lowlands of Southland are, by contrast, geologically more complex than the Canterbury Plains.

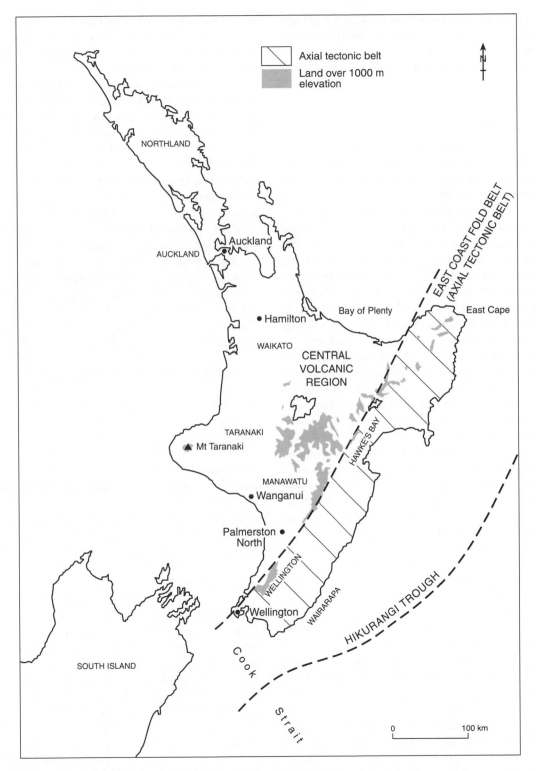

FIGURE 2.7   New Zealand: physical features of North Island

Hard rock uplands surround and segregate the lowlands, the higher parts of which are composed of early Pleistocene gravels, while deposits of loess occur across lowland surfaces (Fitzharris *et al.*, 1992).

North Island

Descriptions of the landforms of North Island are given in order from the east coast, to the Central Volcanic Plateau and to the west coast, the latter from north to south.

The eastern side of North Island from East Cape through Hawke's Bay to Wairarapa, has been affected by vigorous and continuing tectonic activity associated with the subduction zone (Kamp, 1992a; 1992b) (Figure 2.7). The oceanic crust moving from the east has delivered sea-floor sediments to the Hikurangi Trough where they were either subducted or sheared off and attached to the overlying continental plate (Figure 2.7). The accretion of sediments on the eastern edge of the Australian Plate has led to progressive upward thrusting and deformation of the rocks, creating parallel faults, ranges and lowlands trending north-east to south-west throughout the Hawke's Bay and Wairarapa regions (Figure 2.7) (Kamp, 1992b; 1992c). In this East Coast Fold Belt of North Island, large-scale landsliding has occurred, particularly on steep mudstone slopes during wet periods (Crozier *et al.*, 1992).

The Central Volcanic Region lies to the west of the East Coast Fold Belt, and has been produced by rifting of this part of North Island in response to the subduction of the Pacific sea-floor beneath the margin of the Australian Plate on which this part of New Zealand sits (Healy, 1992) (Figure 2.6). Both quiescent and active volcanoes are present in the region and the constructional landforms have only been slightly affected by erosional processes because the rocks are less than one million years old (Healy, 1992). The western part of the region is an ignimbrite plateau, ignimbrite being the name given to rhyolitic pyroclastic rocks. The central unit of the Volcanic Region is the Taupo

Volcanic Zone, which has experienced volcanism, rifting and subsidence. It is a region of complex topography, with craters, calderas, lakes and fault scarps, and includes all the volcanic activity which has occurred in the past 300 000 years (Healy, 1992). Ngauruhoe is the youngest of the volcanoes of the Tongariro Centre, beginning only 5000–2500 years ago and is still active (Nairn *et al.*, 1976; Healy, 1992) (Figure 2.8). Its neighbour, Ruapehu, erupted in 1995 and 1996, sending volcanic ash hundreds of kilometres downwind of the volcano in October 1996. Ensuing claims for damage exceeded NZ$98 million (SNZ, 1998: 6–7). The most easterly part of the Central Volcanic Region has an almost flat surface and is underlain by ignimbrite (Healy, 1992).

The north-westernmost peninsula of North Island, Northland, and Auckland to the south (Figure 2.7), have landforms which reflect earth movements and volcanicity from 15 to 25 million years ago. The present topography can be explained by block-faulting, inland erosion, deposition on the coast and Miocene and Holocene volcanic activity. There are approximately 60 volcanic centres in the Auckland area, comprising circular craters, lava fountains and lava flows. Accelerated erosion has occurred on the slopes within the region, with resultant valley-floor aggradation (Ballance and Williams, 1992).

To the south of Auckland lies the middle Waikato Basin centred on Hamilton (Selby and Lowe, 1992). The basin is surrounded by hills and ranges over 300 m in elevation, developed in response to regional tectonism with the addition of volcanic tephra beds. Farther to the south-west, in Taranaki and Wanganui, landforms were created by tectonic uplift, giving rise to elevated marine terraces, and outbreaks of andestic volcanic activity. Mt Taranaki (formerly known as Mt Egmont), which dominates Taranaki, is a cone of lava flows, while the surrounding landscapes have been mantled with deposits of volcanic ash. Debris flows and mudflows have spread onto the lowlands, and rivers have reworked volcanically derived sands and gravels (Neall, 1992). In contrast, the

FIGURE 2.8 Ngauruhoe volcano, North Island, New Zealand. An andesite stratovolcano, 2291 m elevation, in the Tongariro National Park (Central Volcanic Plateau). The cone commenced approximately 5000–2500 years BP. It had major recent episodes of eruption in 1948–49, 1954–5 and 1974 (Nairn *et al.*, 1976) (photo: R.J. Loughran)

area of Manawatu, centred on Palmerston North (Figure 2.7), is underlain by sedimentary rocks formed into fault-bounded ranges, coastal lowlands and plains containing river terraces, alluvial plains and coastal sand country (Heerdegen and Shepherd, 1992). Finally, the Wellington area is developed on greywacke rocks affected by faulting and earthquakes on the convergent plate boundary (Eyles and McConchie, 1992). The rigidity of the greywackes has meant that faulting rather than folding has occurred so that fault-guided valleys dominate the landscape.

Coasts of New Zealand

The evolution of New Zealand's great variety of coastal features depends on several environmental factors. During the period of the Pleistocene glaciations, weathering and erosion supplied sediments to the coastline, mainly outwash gravels on South Island and sands to Hauraki Gulf, the Bay of Plenty and the west coast of North Island. At this time, the sea level was lower and the sediments were delivered to the continental shelf. As the sea level rose after the glacial period, the sediments were swept onshore by wave action and were redistributed by longshore drift. On North Island, volcanic eruptions added further sediments to the system. Barriers, spits, sand and gravel beaches and coastal dunes are common forms created from these sediments (Healy and Kirk, 1992).

Tectonic activity has raised and tilted marine terraces, for example in Wellington harbour, while on the Marlborough coast, active depression has produced rias by drowning (Healy and Kirk, 1992). In Fiordland and southern Westland, fiords were created by glacial over-deepening of valleys during the Pleistocene.

There are a number of different cliff and platform types which reflect lithology. For example, high-tidal benches are developed on more indurated rocks, while broad intertidal platforms are on relatively unresistant sandstones and siltstones (Healy and Kirk, 1992).

SOILS

Soils of Australia

The scientific description, classification and mapping of Australia's soil resources began in the 1960s and continues today: 'by their very nature soils are difficult objects to classify' (Isbell, 1992: 826), not least because of their complex nature, the interplay of factors which have led to their genesis and the firmly held opinions of soil scientists. These difficulties and an increasing appreciation of the variety of Australia's soils have produced an evolving set of classifications.

According to Paton *et al.* (1995), Australian soils, developed on a plate centre of largely granitic (or granite-derived) rocks, have been affected by epimorphism and near-surface processes (Paton *et al.*, 1995: 127). Epimorphism may be defined as the reaction and adjustment of minerals, formed deep within the Earth, to

---

Box 2.1 Soil landscape mapping

Soil landscapes are 'natural areas of land of recognisable and specifiable topographies and soils that are capable of presentation on maps and of being described by concise statements' (Northcote, 1978). The soil landscape concept has developed from 'Land Systems' (Cooke and Doornkamp, 1990) of common terrain attributes, but giving greater emphasis to the soil component.

The land-systems approach has been used in Australia by the Commonwealth Scientific and Industrial Research Organisation (CSIRO) since the 1940s. 'A land-systems map defines those areas within which certain predictable combinations of surface forms and their associated soils and vegetation are likely to be found' (Cooke and Doornkamp,1990: 20–1).

A programme to map soil landscapes within New South Wales (NSW) was begun in the 1980s by the then Soil Conservation Service of NSW, now the Department of Land and Water Conservation (Atkinson, 1991). The purpose of the maps (1:100 000 scale) and reports is for the broad planning of land use.

For each soil landscape the following attributes can, for example, be described:

*   location;
*   landscape: geology, topography, vegetation, land use, existing erosion;
*   soils: dominant soil materials, occurrence and relationships;
*   limitations to development: landscape limitations (e.g. foundation hazard, steep slopes, mine subsidence), soil limitations (e.g. low fertility, hardsetting surface, strongly acid), fertility, erodibility, erosion hazard, foundation hazard, urban capability and rural capability.

Figure 2.10 illustrates a schematic cross-section of the Warners Bay (NSW) soil landscape (Murphy, 1992).

The soils are described according to great soil groups (yellow podzolic soils) and the principal profile forms (Dy3.11; duplex primary profile form with a yellow-grey clayey subsoil, an A horizon which is hardsetting and a B horizon which is mottled, without an A2 horizon and an acid reaction) (Northcote, 1979).

Dominant soil materials are:

*   wa1   friable brownish black loam
*   wa2   hardsetting bleached clay loam
*   wa3   mottled yellowish grey clay.

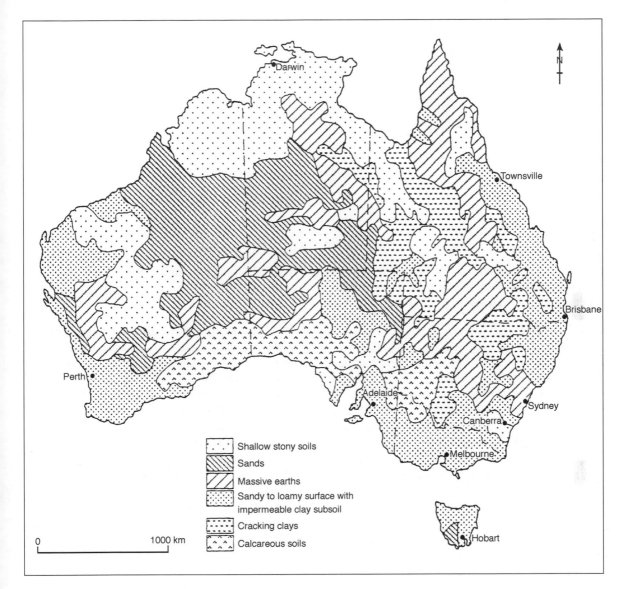

FIGURE 2.9 Australia: generalised soil types (adapted from CSIRO Land and Water)

near-surface conditions by weathering, leaching and new mineral formation. Also, some soil materials may have been unaffected by epimorphism, and their contribution needs to be assessed. Near-surface processes of bioturbation, rainwash, wind erosion and deposition and soil creep can all potentially contribute to the characteristics of the soil profile. On a continent of largely gentle gradient, the above processes have generated texture-contrast (duplex) soils, where a coarse topsoil of variable thickness overlies a finer textured subsoil. Consequently, the interplay between geology, long-term weathering, geomorphic and biologic processes will give rise to localised patterns of soil types which are described on the basis of the soil-toposequence for soil landscape mapping at scales of 1:250 000

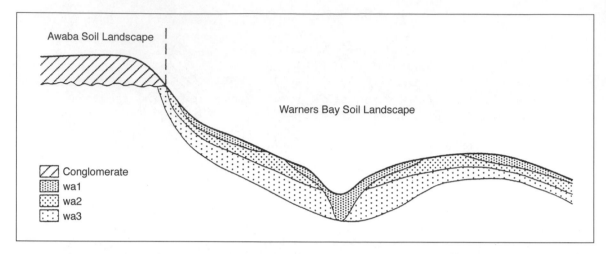

FIGURE 2.10 Schematic cross-section of the Warners Bay soil landscape showing the occurrence and relationship of dominant soil materials (Murphy, 1992)

and 1:100 000 (Atkinson, 1991) (Box 2.1). A description of Australia's soils on the scale approached here must, therefore, be extremely generalised.

The soil types depicted in Figure 2.9 are derived from the work of the Division of Soils, Commonwealth Scientific and Research Organisation (CSIRO). This simplified classification recognises six sub-divisions based on soil profile morphology, except for the calcareous soils mapping unit.

Duplex soils have a sharp texture contrast between a sandy to loamy topsoil and an impermeable clay-rich subsoil. This soil group occurs in a broad arc in eastern Australia, from South Australia to northern Queensland, and in Tasmania and Western Australia (Figure 2.9). The topsoil has a low to moderate organic content and is generally acid. Texture contrast soils are often associated with middle and lower slope positions, where bioturbated soil is sorted and moved downslope by rain splash and overland flow to provide the topsoil material (Paton *et al.*, 1995). The texture contrast can lead to waterlogging problems in these soils, and the topsoils can be hardsetting (McTainsh, 1993).

Soils broadly classified as massive earths generally lie inland of the duplex soils (Figure

2.9). They occur in a band running north to south from Cape York in northern Queensland through western Queensland and New South Wales. There are also occurrences in other states. These, like the duplex soils, have a texture contrast down the profile, but here the contrast is gradual and horizon differentiation is weak. Soil structure is generally absent, hence the name 'massive earths'.

The cracking clays cover large areas of inland Queensland and central western New South Wales, particularly the riverine plains. The soils are deep with a uniform clay texture (40–80 per cent clay) and colour profiles of black, grey, brown and red. The chief feature of these soils is that they crack deeply on drying and swell on wetting (Stace *et al.*, 1968). The black earths of this group have high fertility because they are derived from basalt and have a high humus content from the breakdown of grasses (McTainsh, 1993). Many of the soils of the inland plains are derived from the weathering and erosion of basaltic materials farther to the east.

A variety of shallow stony soils is developed in areas dominated by ancient folded and tilted strata and plateaux, on the Yilgarn Block and in the Hamersley and Kimberley Ranges of Western Australia, in Arnhem Land and the ranges running south-eastwards towards Mt Isa (the Selwyn Range), the ranges of central Australia (west of Alice Springs: Macdonnell and James Ranges), the Flinders Ranges of South Australia and the Barrier Range in the extreme west of New South Wales (Figures 2.3 and 2.9). This group includes soils with little or no profile development in places where the rate of soil erosion is high on mountainous and steep lands. Shallow sands, loams and clay loams contain fragments of rock (Stace et al., 1968).

The Great Victoria, Simpson and Great Sandy Deserts have soils with uniform sandy-texture profiles and a low water-holding capacity (sands: Figure 2.9; Figures 2.2 and 2.3). Two main sub-divisions may be recognised: the siliceous sands and the earthy sands. The former are characterised by their quartzose nature and lack of profile development, while the earthy sands have greater coherence due to the presence of iron oxides and clay minerals (Stace et al., 1968).

Finally, the calcareous soils are found on the Nullarbor Plain and on the lower part of the Murray–Darling Plain (Figure 2.9). The grey, red and brown calcareous soils which occur on the Nullarbor Plain are weakly structured loams to light clays with little profile differen-tiation, and are derived from the weathering of the underlying calcareous rock (Stace et al., 1968). On the lower Murray–Darling Plain, soils containing large amounts of calcareous material (calcium and magnesium carbonates) are present. Profiles show a gradual fining of texture with depth, and colours/textures vary from grey-brown, brown to red-brown sands, loamy sands, sand loams or loams (Stace et al., 1968).

Soils of New Zealand

New Zealand's position on a compressive plate margin produces circumstances which lead to the production of new materials by volcanic activity and high relief because of tectonic activity. The genesis of soils in such regions is dominated by epimorphism, bioturbation and lateral surface processes (Paton et al., 1995). The first scientific classification of New Zealand soils was by Taylor, who developed a 'grand soil-landscape model' (Hewitt, 1992a: 855), which recognised the three sets of processes listed above (Paton et al., 1995: 168). While this model was very useful at the recon-naissance stage of understanding the genesis of New Zealand's soils, it was unable to cope with detailed information that was being gathered in the 1970s and beyond, nor was it able to aid a finer-scale understanding of soils 'in the paddock' (Hewitt, 1992a). There was, therefore, a need for a new approach. In the 1970s and 1980s, the US soil taxonomy approach (e.g. Donahue et al., 1983) was examined, which emphasised soil morphology rather than soil processes as a basis for classi-fication. This was unsuccessful because it made 'inadequate provision for important classes of New Zealand soils' (Hewitt: 1992a: 848). A new classification was then developed specifically for New Zealand soils which was based on measurable soil properties rather than pre-sumed genesis, and emphasising the soil profile rather than its position in the landscape (Hewitt, 1992a).

The New Zealand soil classification has 15 orders (Table 2.1), four of which are associated with rocks of volcanic origin. Pumice soils cover large areas of the central volcanic plateau of North Island and are formed from volcanic deposits of 700 to 3500 years ago (McLaren and Cameron, 1996). They are coarse-textured and potentially erodible. The allophanic soils are also found on volcanic tephras on North Island, around Mt Taranaki (Egmont) and Waikato (Figure 2.7). They have a well-developed structure which aids the deep rooting of plants, and a generally moderate to high organic matter content. Strongly weath-ered tephras (>50 000 years old) are the parent materials for the granular soils which occur on the lowlands of Waikato and south Auckland on North Island (Figure 2.7). They have a high clay content and are difficult to work when wet (McLaren and Cameron, 1996). Oxidic soils are

TABLE 2.1 Orders of the New Zealand Soil Classification

Soil order	Features
Organic soils	Litter accumulation in forests and wetlands
Gley soils	Poorly drained, reducing conditions, greyish colours dominant
Ultic soils	Strongly weathered with clay rich subsoil, acid reaction
Podzols	Has eluvial horizon, below which is horizon containing sesquioxides and organics
Allophanic soils	Dominated by type of alumino-sillicate inorganic soil colloid; generally well drained
Pumice soils	Sandy or gravelly soils dominated by pumice
Melanic soils	Well structured with very dark topsoils
Semi-arid soils	Weakly weathered and leached; low rainfall regimes
Oxidic soils	Strongly weathered, with iron and aluminium oxides in subsoil; clay 50–90 per cent
Granular soils	Strongly weathered with well-developed structure; clay 40–80 per cent
Pallic soils	Pale-coloured subsoil, weakly weathered and low in sesquioxides
Brown soils	Yellow-brown subsoil; well drained
Anthropic soils	Substantially modified by human activity
Recent soils	Minimal profile development; distinct topsoil
Raw soils	Topsoil absent or fluid subsoil

Sources: Hewitt (1992a; 1992b); McLaren and Cameron (1996).

formed from clay products of andesites, dolerites and basalts, and occur only in the Northland and Auckland regions. They have a high clay content (Table 2.1), are low in potassium, magnesium, calcium and phosphorus, but are highly productive when fertilised (McLaren and Cameron, 1996).

The most extensively occurring soils are the brown soils, which are derived from the weathering of schist and greywacke or acid igneous rocks (rhyolites and granites). The melanic soils are developed in several areas of New Zealand, and are associated with calcareous rocks in Hawke's Bay, Wairarapa, north Canterbury and north Otago, and basalts on the Otago and Banks peninsulas and north Otago in South Island (Figures 2.6 and 2.7). Ultic soils also have both igneous and sedimentary rocks as parent materials. Siliceous sediments and acid igneous rocks in northern North Island and the Wellington, Marlborough and Nelson regions commonly give rise to soils of this order, which are prone to erosion when surface cover is removed (McLaren and Cameron, 1996). They are also poorly drained.

In the High Country of both islands, on the west coast of South Island and in Northland,

podzols are common. They are acid soils with an eluvial horizon. The downwards translocation of aluminium, iron and humus contributes to the formation of an illuvial horizon below (Table 2.1).

The semi-arid soils are confined to central Otago and the Waitaki Valley of South Island in areas of low annual rainfall (330–500 mm) (McLaren and Cameron, 1996). They are generally developed on colluvial, alluvial and aeolian deposits derived from schist and greywacke, and have a low clay content (<15 per cent).

Pallic, gley and organic soils are all affected by impeded drainage to some extent. The pallic soils of the downlands of Southland and Canterbury and the rolling and hilly lands of Manawatu, Hawke's Bay, Wairarapa, Marlborough and Nelson, are developed on loess and sediments derived from schist and greywacke (Figures 2.6 and 2.7). They have low organic matter (3–5 per cent) and pale-coloured subsoils which are dense, compact and subject to waterlogging, and have a high slaking potential. (Slaking is the partial breakdown of soil aggregates in water because of the swelling of clay minerals and the expulsion of air from voids.) The gley soils are even more poorly drained, with reduction producing the greyish colours that predominate. These soils occur in areas with high watertables, and are often drained for agricultural production (McLaren and Cameron, 1996). Finally, the organic soils of the wetlands and forests of Westland and Southland of South Island, and Waikato on North Island, are dominated by the accumulation of organic matter.

The remaining soil orders – recent soils, raw soils and anthropic soils (Table 2.1) all show minimal soil profile development, although the last-named are the more variable. Recent soils are found on alluvial flood plains, steep slopes and on young volcanic ash, whereas raw soils are developed on rocky areas, beach sands, active screes, and in lagoons and tidal estuaries (McLaren and Cameron, 1996).

SUMMARY AND CONCLUSION

Australia is a continent of low relief and very old landscapes. Its tectonic stability over long periods of geologic time has permitted weathering processes, chemical alteration, erosion and deposition to produce a landscape dominated by plains. In other places, mountain range landscapes exhibiting geologic structures, and uplifted plateaux with extensive erosional escarpments, occur. The stability of the Australian landmass has meant that broad coastal zones with similar characteristics can be delineated, as in the scheme of Davies (1986) utilised here.

The soils developed on Australian landforms tend to be shallow and leached of nutrients, although in restricted areas where recent deposition of alluvium has occurred, soils are deeper and more fertile. Several classifications of Australia's soils have been developed as knowledge has increased and needs changed.

In contrast, New Zealand is geologically young and subject to tectonic and volcanic activity because of its position on a compressional plate boundary. These factors and processes give rise to a high-energy geomorphic environment that has been severely affected by Pleistocene glacial activity on South Island and volcanic activity on North Island. Consequently both inland and coastal landforms show a juxtaposition of different types of topography over relatively short distances.

Weathering, erosion, associated sedimentation and the products of vulcanicity are important in the development of New Zealand's soils. As in Australia, soil classification has evolved with increasing knowledge and a realisation that specific approaches were required because of the very different environmental conditions that exist in New Zealand compared with other parts of the world.

Understanding of the detailed evolution and the processes currently shaping Australia's landforms is hampered by the size of the continent. There is, as yet, no region-by-region treatise on Australia's geomorphology, although Twidale and Campbell (1993) have produced a systematic text based on structure, process and time, using Australian examples. In New Zealand, by contrast, Soons and Selby (1982) edited what amounted to a regional geomorphology of that country, although acknowledging that the coverage was uneven. A second much expanded and updated edition appeared ten years later, but the editors noted that 'New Zealand is still wide open for further original studies of the development of its landforms' (Soons and Selby, 1992: xi).

FURTHER READING

The volume edited by Jeans (1986) has several Chapters on the Australian environment, including 'Physiographic outlines and regions' by Jennings and Mabbutt, 'Early landform evolution' by Ollier, 'Quaternary landform evolution' by Bowler, 'Fluvial landforms' by Pickup, 'Desert lands' by Mabbutt, 'The coast' by Davies and 'Soils' by Northcote. A comparable collection for New Zealand is Soons and

Selby (1992), which covers the evolution of New Zealand's landscape from the general view to a region-by-region description of the geomorphology by experts in the field. Although rather biased towards structural landforms (which abound in Australia), Twidale and Campbell (1993) is the only comprehensive text on Australia's landforms in one volume. McLaren and Cameron (1996) is a general text for soil science students that uses New Zealand examples throughout. Warner (1988) is a collection of essays on aspects of fluvial geomorphology in Australia, rather than a systematic text on the subject. Australia's fluvial systems differ from those of other regions, so this is a valuable and detailed survey.

3

CLIMATE, HYDROLOGY, THE OCEANS AND VEGETATION

The availability of water, temperature variations and seasonality, in addition to longer-term climatic fluctuations, has an important bearing on the natural ecosystems and economic activity in Australia and New Zealand. Often described as the driest permanently inhabited continent, Australia has approximately 70 per cent of its area classified as an arid zone, but 'no climate station records a mean annual rainfall below 100 mm, a value taken as marking the northern limit of the Sahara proper!' (Mabbutt, 1986: 181). The extent of this relatively moist desert is the major feature of Australia's climatic zones, rather than its absolute aridity. Better-watered areas lie on the continental fringe to the north, east and south of the arid core. New Zealand, because of its insularity and the presence of high mountain ranges, experiences a much moister climatic regime.

Situated in mid-latitudes between 10°S and 48°S, both countries are affected by the westerly wind circulation of the Southern Hemisphere, except the northern half of Australia, which comes under tropical influence. Excluding northern Australia, the region is dominated by the eastward passage of cyclonic depressions and fronts developed between air masses of the subtropical, high-pressure systems to the north and Southern Ocean low-pressure systems. In the southern winter, the high-pressure cells migrate northwards by 5°–10°, bringing generally dry conditions to the centre and north of Australia. At the same time, the southern part of the continent comes under the influence of rain-bearing frontal systems. In summer, the intertropical convergence zone (ITCZ), between easterly flowing tropical air in the equatorial region and the subtropical high-pressure zone to the south, migrates southwards into northern Australia due to continental heating. Over the continent, then, relatively low pressure tends to develop, interrupting the continuity of the high-pressure belt. Thus, in northern Australia, monsoonal conditions exist in the austral summer, with heavy rainfalls and tropical cyclones during 'the Wet'. The extensive desert region of Australia is a product of predominantly anticyclonic conditions, the lack of significant mountain barriers to promote orographic uplift, and distance from the ocean. As on other continents on both sides of the equator, an arid zone is centred on the tropic.

Because it lies farther south, New Zealand is influenced by westerly wind patterns for much of the year (Sturman and Tapper, 1996). These circulation patterns, with seasonal rainfall and temperature fluctuations, determine the broad climatic zones which exist, although local variations occur because of topography and other factors.

THE CLIMATIC ZONES OF AUSTRALIA AND NEW ZEALAND

Climate regions of Australia

Several classifications of Australia's climate have been produced (Sturman and Tapper, 1996). That by the Australian Bureau of Meteorology in 1989 is depicted in Figure 3.1. This six-region classification is based largely on amount and seasonality of precipitation (Colls and Whitaker, 1990: 30), but also on temperature zones.

The most northerly region, with summer rainfall and tropical temperatures, occurs across Western Australia, the Northern Territory and Queensland (Figure 3.1), and is characterised by the inflow of moist, unstable air during the summer months, particularly December–February. Median annual rainfalls range from over 1600 mm to 500 mm (Gentilli, 1986). Highest falls are experienced on the seaward eastern slopes of the ranges in north-eastern Queensland, with annual totals in excess of 3000 mm, and average rainfalls of over 20 mm per rainday (mean annual rainfall/mean number of rain days: Jennings, 1967). Average maximum and minimum monthly temperatures generally range from 35° to 12°C, respectively.

An arid zone, with mainly summer rainfall and subtropical temperatures (Figure 3.1), is present in central Western Australia, the southern part of the Northern Territory and south-western Queensland, with small areas within northern South Australia and north-western New South Wales. The rainfall, with median annual values between approximately 500 and 150 mm, is very variable and occurs mainly in the summer months. Average open-pan evaporation can be as high as 4000 mm per year,

with average maximum and minimum monthly temperatures between 39° and 6°C (Bureau of Meteorology, 1989).

To the south of this region occurs another arid zone, with winter or non-seasonal rainfall, and with warm to subtropical temperatures (Figure 3.1). It runs in a belt from Western Australia, through South Australia into western New South Wales. Rainfalls and evaporation rates are similar to the other arid region to the north, but average monthly maximum temperatures are approximately 6°C cooler.

The south-western part of Western Australia, the southern portion of South Australia, the western two-thirds of Victoria, extreme southern New South Wales and virtually the whole of Tasmania experiences a climate which has a marked winter rainfall distribution (Figure 3.1) due to the northward migration of the mid-latitude high-pressure belt over central Australia and the associated northward movement of the westerlies. Median annual precipitation is within the range 500 mm to more than 1600 mm, with some snowfalls in the highlands of Victoria and Tasmania. Average maximum and minimum monthly temperatures during the year can vary from 33°C to below freezing point, respectively.

In southern Queensland and northern New South Wales, the climate has been classified as a summer rainfall maximum, subtropical type (Figure 3.1). Annual rainfall amounts are generally between 500 mm and 1600 mm, with mean daily intensities as high as 15 mm per rainday. The average maximum monthly temperature is between 27° and 33°C, and the average monthly minimum varies from 9°C to below freezing point.

A zone of uniformly distributed rainfall with temperate conditions exists over much of New South Wales, eastern Victoria and a small area around Hobart in Tasmania (Figure 3.1). Onshore winds from the south-east and frontal rainfall approaching from the west ensure, on average, a year-round distribution of precipitation. Median precipitation is generally between 1600 mm and 800 mm annually, with relatively low intensities compared with regions in the

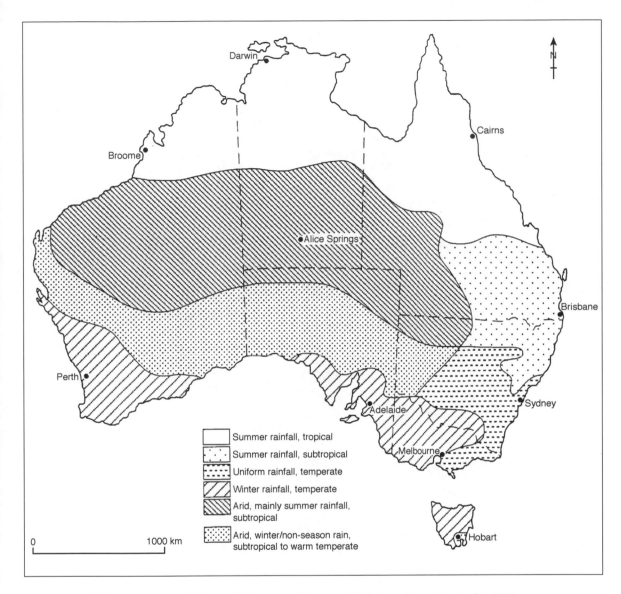

FIGURE 3.1 Climatic zones of Australia (source: Bureau of Meteorology, Australia 1989; Commonwealth of Australia copyright reproduced by permission)

north of Australia. Annual open-pan evaporation is at its lowest in Australia in this region: 1000 mm in south-eastern Tasmania to 2000 mm in central New South Wales (Bureau of Meteorology, 1989). Within this region, mean monthly maximum and minimum temperatures are between 33° and 3°C, respectively.

The chief features of Australia's climate are its unreliable rainfall, with marked fluctuations between periods of flood and drought, usually associated with the El Niño–Southern Oscillation, described below. The uneven geographical distribution of precipitation and high rates of evaporation also give rise to difficulties for agricultural production and urban water supply, topics dealt with in Chapter 4.

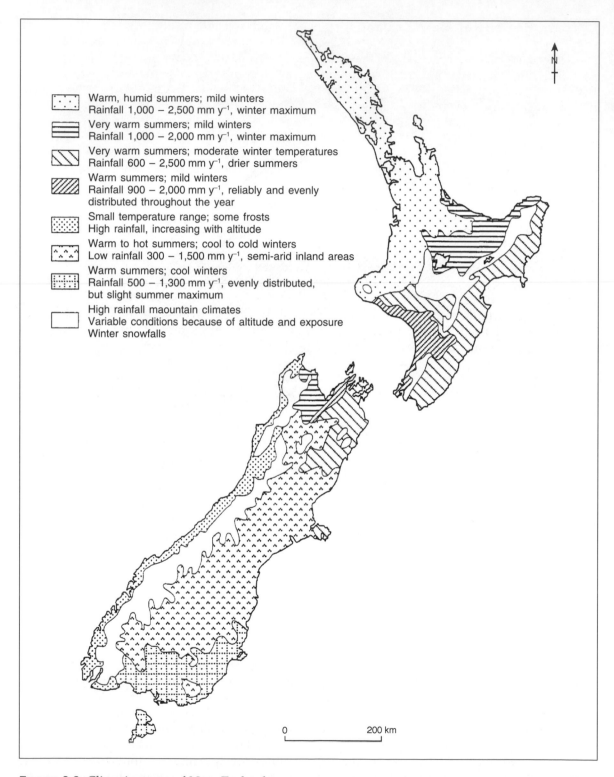

FIGURE 3.2 Climatic zones of New Zealand

Climate Regions of New Zealand

'New Zealand experiences much more rapid spatial change in environmental conditions than does Australia, where vast areas tend to have relatively uniform conditions'; also, New Zealand has a 'continuous variability in local weather on hourly, daily and weekly time scales' (Sturman and Tapper, 1996: 317). Thus, despite its smaller geographical area, New Zealand may be divided into many more major climatic zones than Australia. Figure 3.2 shows eight broad categories of climate classes for New Zealand, modified from the 17 regions mapped by the New Zealand Meteorological Service (1983) (Sturman and Tapper, 1996).

The western part of North Island is a region of warm summers and mild winters, affected by tropical maritime air masses from the Tasman Sea and Pacific Ocean (Sturman and Tapper, 1996). Mean annual precipitation is between 1000 and 1500 mm in the lower regions of Northland and Auckland, and 1500 and 2500 mm in the more elevated areas (Waikato, Taranaki and Coromandel).

Around the Bay of Plenty and the Central Plateau of North Island is a region of winter maximum precipitation (1000–2000 mm rainfall). Towards the coast, summers are very warm and winters are mild, whereas on the Central Plateau, winters are cooler with frequent ground frosts. Also included in this climatic region is part of Nelson, on northern South Island, noted for its long hours of sunshine, approaching 2500 hours per year, on average.

The east coast of North Island, but including inland Wanganui, Manawatu and South Island Marlborough, experiences warm summers and annual rainfalls between 1000 and 2500 mm. Along the coastal strip, summers are very warm (occasionally over 30°C) with dry foehn north-westerly winds. There is a tendency to a winter maximum of rainfall, and summer drought can be common in some areas. More elevated, inland locations generally have higher average annual rainfall (1500–2500 mm), with heavy falls from the south and south-east.

South-western North Island (Wellington, Manawatu and southern Taranaki, but including western Marlborough on South Island) experiences evenly distributed rainfall with west to north-west winds and relatively frequent gales through Cook Strait. Rainfalls are between 900 and 1300 mm annually, but in western Marlborough they range from 1300 to 2000 mm. Summers are warm and winters are mild.

The western coastal strip of South Island has high rainfall, rising rapidly with increasing elevation to over 5000 mm per year. The range of temperature is low due to the maritime situation; prevailing winds are from the south-west.

Central South Island, on the eastern side of the Southern Alps, has generally low rainfalls because of the foehn effect on the prevailing north-westerly winds on the leeward side of the mountains (Sturman and Tapper, 1996). Here, there is an increase in wind speed and temperature and a decrease in relative humidity. Consequently, temperatures in summer can exceed 30°C, but winters can be cold with frequent frosts. In coastal Canterbury rainfalls are between 500 and 800 mm annually, while in central Otago, semi-arid conditions prevail where annual precipitation is between 300 and 500 mm.

Southern Otago and Southland is a region of warm summers and cool winters, with rainfalls between 900 and 1300 mm annually. Generally, in Southland, conditions are windier with frequent showers in coastal locations.

Mountain climates prevail along the Southern Alps of South Island and central North Island. Much of the precipitation falls as snow, and there are permanent glaciers. Mean annual precipitation can exceed 10 000 mm in the south of the Southern Alps range, but on North Island precipitation is lower and less variable.

Climatic variability

Variations in climate on the global scale have occurred during geologic time, as in the

Quaternary, which was marked by major instability over the past two million years. During the period of instrumented record, approximately the past 150 years in Australia and New Zealand, observations have shown that there have been cool periods in the early 1860s and at the turn of the century, followed by a general stability of temperature from 1910 to 1950. Since 1950, there have been warmer temperatures (by approximately 0.5°C), consistent with the global trend, perhaps associated with the greenhouse effect (Sturman and Tapper, 1996). In New Zealand this trend is thought to be due to a change in wind patterns, with west-south-westerlies giving way to some extent to airflows from the north (Sturman and Tapper, 1996).

Perhaps the most well-recognised process influencing contemporary climate is the El Niño-Southern Oscillation (ENSO) of the Pacific (Allan *et al.*, 1996). The normal major circulation, called the Walker circulation, moves air between the eastern and western sides of the Pacific. There is rising air over Indonesia and descending air over the eastern South Pacific, which represent convection (cloud formation) and subsidence respectively, related to ocean circulation. When the pressure difference between Indonesia (low pressure) and the south-eastern Pacific (high pressure) is greatest, the Southern Oscillation Index (SOI) is high, and there is convective rainfall over Indonesia and dry conditions in the south-eastern Pacific. With a weakening of the circulation, the pressure difference decreases (a low SOI), leading to drier conditions over Indonesia. This alteration to the normal circulation, leading to a strongly negative SOI, is associated with the El Niño phenomenon – the occurrence of warm waters off the Peruvian coast. The oscillations vary in intensity, and occur every two to ten years.

During an El Niño phase (a negative SOI), Australia (especially the east) can experience extended dry conditions, as during the severe drought of 1982–3. In New Zealand, conditions tend to be cooler and windier, with drought on the eastern and northern coasts of both North and South Islands, but with more rain in the west and south (Sturman and Tapper, 1996; MfENZ, 1997). La Niña, or anti-ENSO events, occur in high phases of the SOI, and bring heavier rainfall and flood events to Australia.

Both long- and shorter-term variations in weather patterns have a very strong influence on the availability of surface and underground water, fluctuations in runoff and water quality. A discussion of the chief features of the hydrology of Australia and New Zealand follows.

HYDROLOGY OF AUSTRALIA AND NEW ZEALAND

The hydrological cycle of water transfer between the atmosphere, land and oceans is complex and therefore difficult to measure on a large scale. Nevertheless, a broad outline of its components is possible for Australia and New Zealand, especially the variability and character of the runoff component.

Precipitation distribution and variability are discussed in the description of climate zones above, while evaporation, groundwater and runoff are included in this section. Actual rather than potential losses due to evapotranspiration may be estimated on the continental scale by reference to open-pan evaporation rates from free water surfaces minus runoff. It can be argued that in unregulated river basins, the main outflow of water is due to evapotranspiration and runoff, because changes in near-surface water storages are minimal, and that over the long term changes in groundwater storages may be ignored (Ceplecha, 1971). In the arid zone of Australia (Figure 3.1), evapotranspiration is approximately equal to rainfall because runoff is almost negligible, although there are periods of flood when the potential rate of evapotranspiration may be achieved. In the near-coastal uplands of New South Wales, Victoria and Tasmania, and the extreme south-western corner of Western Australia, the actual evapotranspiration almost

FIGURE 3.3 Geologic-groundwater basins of Australia

equals the potential rate during all months and is controlled by the availability of heat energy (Ceplecha, 1971). In this zone, streams have almost perennial flow, although during El Niño drought periods, rivers in areas of lower rainfall do cease to flow. Throughout the remainder of the continent experiencing summer, uniform and winter rainfall regimes (Figure 3.1), actual evapotranspiration is variable. Only during some months of the year will the actual rate reach the potential rate, typically the summer months in the north and the winter months of the south (Ceplecha, 1971). Resultant runoff and groundwater recharge are therefore seasonal.

In New Zealand, estimates of mean annual tank evaporation of between 650 and 850 mm have been reported for North Island, and 700 to 950 mm for South Island (Finkelstein, 1973). Detailed studies of evaporation and transpiration under different ground covers have been carried out for well-watered pasture near Hamilton and mature *Pinus radiata* forest on North Island (Kelliher and Scotter, 1992). In the latter case, average rainfall exceeds total evaporation by 300 mm per annum, but in summer there is a water deficit (McAneney *et al.*, 1982). Tree transpiration was a large component of the water balance in the pine forest, estimated at 50 per cent of the incoming precipitation (Whitehead and Kelliher, 1991), whereas under tussock grassland, in an area of Otago with a mean annual precipitation of 1050 mm, annual transpirational losses were measured at 400 mm, or 38 per cent of rainfall (Campbell and Murray, 1990).

Groundwater is held beneath the surface within pores and fractures of rocks, and closer to the surface within unconsolidated sediments. The source of this water is precipitation, which has infiltrated surface soils and weathered rocks, and river waters (particularly during flood periods) entering recharge areas. Groundwater is important because supplies are large (approximately 25 per cent of all fresh water on the continents: Barry, 1969), it is largely free from evaporative loss, contributes approximately 30 per cent to total global runoff (Barry, 1969) and is relatively easy to exploit. However, it may be contaminated by pollutants entering the recharge areas, and may suffer over-exploitation, leading to the diminution of the resource. In a dry continent, therefore, groundwater is often a vital resource.

Much of the groundwater in Australia, particularly in the arid zone, is contained within aquifers of geological basins (Figure 3.3), the most important of which is the Great Artesian Basin covering nearly 23 per cent of the continent (Warner, 1986). In an artesian basin, water is held under sufficient pressure for it to flow to the surface when a well is sunk. The recharge area for the Great Artesian Basin is the western flank of the Eastern Highlands, with smaller zones in the north-east and north-west of the basin. The main aquifer, freshwater sandstones of Jurassic age, outcrops on the Eastern Highlands where mean annual precipitation is approximately 700 mm. It has been estimated by hydrologists at the Australian Nuclear Science and Technology Organisation that it takes waters one million years to reach the centre of the basin, indicating that wise management is required to maintain the resource.

Important surficial aquifers are present within the alluvial fill of the Burdekin River delta, in northern Queensland, and on the Swan Coastal Plain in Western Australia (Figure 3.3). Waters drawn from the former source have been used to irrigate sugarcane in Queensland, while the Swan Plain aquifer provides approximately 60 per cent of Perth's water requirements (DESTA, 1996).

Groundwater occurrences in New Zealand may be sub-divided into 'ambient' and geothermal types. The 'ambient', or normal aquifer type, is found in many parts of the country and constitutes 40 per cent of New Zealand's freshwater supplies (MfENZ, 1997). These supplies are found mostly in porous sediments overlying harder rock types in areas such as the Canterbury Plains. Where aquifers have been heated along faultlines, in deep fissures and in volcanic regions, geothermal groundwater will occur, classified as either low temperature (below 100°C) or high temperature (above 100°C). All but one of the 24 high-temperature fields occur in the Taupo Volcanic Zone on North Island. Surface features of these fields vary from warm ground, hot springs, boiling mud, geysers to fumaroles (MfENZ, 1997). The low-temperature geothermal fields are associated with deep groundwater circulation and faultlines on both islands.

Precipitation excess can lead to surface runoff over slopes and into stream systems, and constitutes the major source of water for agricultural, domestic and industrial use in Australia and New Zealand, as well as being an important provider of hydroelectric power (see Chapters 8 and 14).

In Australia, highly seasonal precipitation in the monsoon-affected north, the eastern

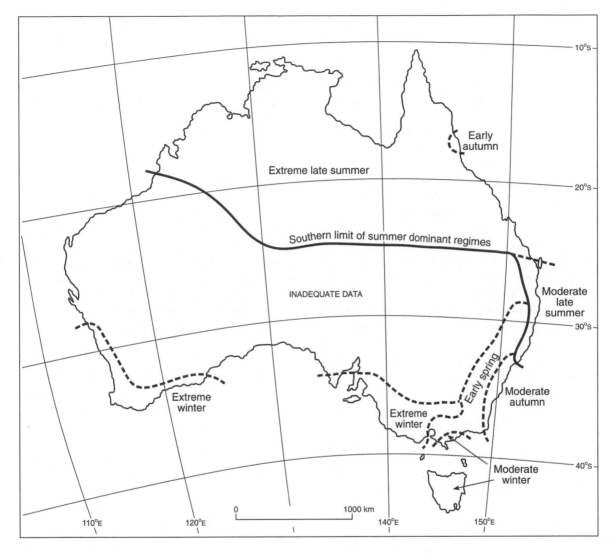

FIGURE 3.4 Runoff zones of Australia (after Finlayson and McMahon, 1988)

highland region of Queensland and the coastal fringe of northern New South Wales produces flood runoff with an extreme-to-moderate late summer regime (Finlayson and McMahon, 1988) (Figure 3.4). Highest mean annual rates of runoff in this zone are on the Queensland coast around Cairns, and on the Atherton Tableland (500 to >1250 mm: Warner, 1986). Along the central and southern coastal ranges and valleys of New South Wales, runoff has a moderate autumn regime, but, perhaps more importantly, there appear to be oscillations in the longer-term incidence of floods and droughts, particularly on the coastal rivers north of Sydney. These shifts are referred to as flood-dominated regimes, or periods (FDR), and drought-dominated regimes (DDR) (Erskine and Warner, 1988). In the period of record, an FDR was present between 1799 and 1820, a DDR occurred between 1821 and 1856, an FDR from 1857 to 1900, a DDR from 1901 to 1948, with an FDR from 1949 into the early 1980s at least (Nanson and Erskine, 1988; Erskine and Warner, 1988). Further research is required on the FDR/DDR phenomenon to examine its possible links to El Niño and La Niña events, and to see if the DDR/FDR pattern occurs elsewhere.

South-western Western Australia, southern South Australia, western Victoria and Tasmania experience an extreme to moderate winter runoff regime (Finlayson and McMahon, 1988) (Figure 3.4). In the western highlands of Tasmania, annual runoff is in excess of 1250 mm, but rates in other parts of this hydrological region are generally below 500 mm, reflecting lower precipitation amounts and higher insolation (Warner, 1986)

The Murray–Darling basin of inland eastern Australia has a drainage area of over 1 000 000 km², and its waters support a considerable proportion of Australia's irrigated agriculture (Pigram, 1986). Highly variable flood runoff, generated in the northern and eastern parts of the basin where annual rainfalls can be as high as 2400 mm, provides the bulk of stream flow. In the west of the basin, rainfall is approximately 200 mm. Riley (1988) reports 'clear evidence of a dry period between the mid 1930s and the late-1940s', with an increase in flow in the 'post-1940s' attributable to an increase in precipitation.

Surface runoff within the Australian arid zone is highly variable in space and time, and is usually produced by rainstorms over steep, stony ground with low infiltration capacities. On low-gradient alluvial surfaces, transmission losses occur due to infiltration and evaporation in flood periods. Typically, the runoff occurs rapidly and has short duration, as shown in the floods of the northern Lake Eyre basin in 1967 (Williams, 1970).

There are also seasonal variations in river flows in New Zealand (Duncan, 1992), although these are nowhere as marked as in Australia (or southern Africa), which have the most variable annual river flows of all the continents (Finlayson and McMahon, 1988). On the east coast of both North and South Islands summers are relatively dry, and there is associated low river flow at that season, except where snow-melt increases runoff in spring and summer.

For example, the Motu River, which drains to the eastern coast of the Bay of Plenty from the Raukumara Range on North Island, has a ratio of 2.2 between the month of greatest average flow (August), compared with the month with least flow (January). For the Manawatu River, on the southern part of North Island, the ratio is even more marked. Here the month of lowest average low flow is February and the month with greatest average flow is July, with a highest/lowest flow ratio of 3.4 (MfENZ, 1997). Generally, there is a trend to higher runoff in the winter months in the streams of North Island, in contrast with many rivers on South Island, which receive snow-melt, such as the Rakaia in southern Canterbury. The Rakaia has its highest average monthly flow in December and lowest in July, with a ratio of 2.2.

In terms of runoff per unit area (specific discharge), rivers of New Zealand have greater average flows than those of Australia, which reflects the greater humidity of the former. For six selected large river basins in New Zealand, specific discharges (million m³ km⁻²) range from 2.78 for the Grey River, on the north-west coast of South Island, to 0.87 for the Clutha River, in the south of South Island (Table 3.1) (MfENZ, 1997). For the major drainage divisions of Australia, runoff per unit area is very much lower, the greatest being in the zone of mixed seasonal rainfall in the basins of south-eastern Australia and Tasmania (Table 3.1) (Warner, 1986). In the Lake Eyre basin of central Australia, the specific discharge is only 0.0026 million m³ km⁻² (Table 3.1).

The sediment load of rivers depends on two sets of factors: hydrologic (rainfall and runoff characteristics) and catchment (relief, slope, geology, soils and land use). Studies of sediment transport have shown the importance of specific controls on river loads, and contrast the low-energy environments of Australia with the high-energy environments of New Zealand (Box 3.1).

There are large lakes in Australia and New Zealand that form an important part of the fresh water circulation, although some of them are artificial, or are natural lakes which have

Table 3.1 Specific discharges for selected rivers in New Zealand and drainage divisions of Australia

	Drainage area (km²)	Average annual runoff (m³.10⁶)	Specific discharge (m³.10⁶km⁻²)
New Zealand			
Clutha	20 580	17 976	0.87
Buller	6350	13 497	2.13
Waitaki	9760	11 547	1.19
Grey	3830	10 628	2.78
Waikato	11 400	10 312	0.90
Whanganui	6640	7064	1.06
Australia			
Summer rainfall region	1 636 215	303 000	0.19
Mixed seasonal rainfall	341 753	98 000	0.29
Winter rainfall region	369 390	8000	0.02
Murray–Darling	1 062 530	23 000	0.02
Lake Eyre basin	1 170 000	3000	0.003

Rainfall regions of Australia
- Summer rainfall: drainage areas to the Timor Sea and the Gulf of Carpentaria, and the east coast of Queensland.
- Mixed seasonal rainfall: coastal New South Wales, Victoria and eastern South Australia, and Tasmania.
- Winter rainfall: central-southern South Australia (draining to Spencer Gulf and Gulf St Vincent) and south-west Western Australia.

Sources: Warner (1986) and MfENZ (1997).

Box 3.1 Suspended-sediment loads of rivers in Australia and New Zealand

Suspended sediment is a major pollutant of water, creating turbidity and possessing absorbed chemicals and heavy metals. The yield of sediment from drainage basins has been measured mainly in the south-east of Australia (Wasson et al., 1996), but in New Zealand the coverage is much more uniform (Hicks et al., 1996). Despite the obvious effects of humans on land use in increasing soil erosion, there are marked contrasts in river sediment loads between the two countries. It is thought that the relative aridity and low relief of Australia are responsible for the low rates of sediment delivery to the oceans (Wasson et al., 1996). Inefficient delivery of sediment to the downstream part of large basins occurs because much of the sediment is deposited close to the source (Olive and Rieger, 1986). At Wagga Wagga on the Murrumbidgee River (26 400 km² basin area), the specific sediment yield is 20 t km⁻² y⁻¹, and at Balranald, farther downstrean (81 000 km²), the specific yield is reduced to 1 t km⁻² y⁻¹ (Wasson et al., 1996). In New Zealand there is great variability in sediment yield, which is related to annual rainfall and drainage basin geology (Hicks et al., 1996). The presence of erodible rocks and great levels of tectonic activity mean that the highest rates of sediment transport (1200–20 00 t km⁻² y⁻¹) are found near the plate boundary, in high mountains with orographic rainfall.

been raised artificially. There are many natural origins of lakes, including tectonic and volcanic activity, glacial, wind and fluvial action, and solution (Timms, 1992; Lowe and Green, 1992). One-quarter of the 40 largest lakes in New Zealand are of volcanic origin, the most extensive, Lake Taupo, being 623 km² in area (Lowe and Green, 1992). In the group of 40, glacial origins account for 20 lakes, all found on South Island. In Australia, episodically filled lakes are a feature of the arid zone, some occupy tectonic depressions, such as Lake Torrens, or are basins created by crustal upwarping, as in Lake Eyre, both of which are in South Australia. The latter occupied an area of nearly 10 000 km² after the floods of 1974 (Timms, 1992). Lakes of glacial origin are also found in Australia, in the Snowy Mountains of south-eastern New South Wales and in Tasmania.

Ocean currents off Australia consist of southerly flows off both the west and east coasts, which is unusual because most continents have a polar and a tropical current. In the Southern Ocean, a current flows from west to east. Because there is no upwelling of cold, rich water and because river flows to the oceans are small and low in nutrients, the oceanic fish harvest is depressed (DESTA, 1996). New Zealand has three ocean bioregions (MfENZ, 1997). The northern bioregion extends across North Island from Northland to East Cape and is influenced by the tropical East Auckland Current, which flows eastwards. The central bioregion is the most extensive, along the entire west coast and from East Cape to Otago on the east. There are ocean currents from the south along both these coasts, derived from the Tasman Current from eastern Australia. The southern bioregion is off

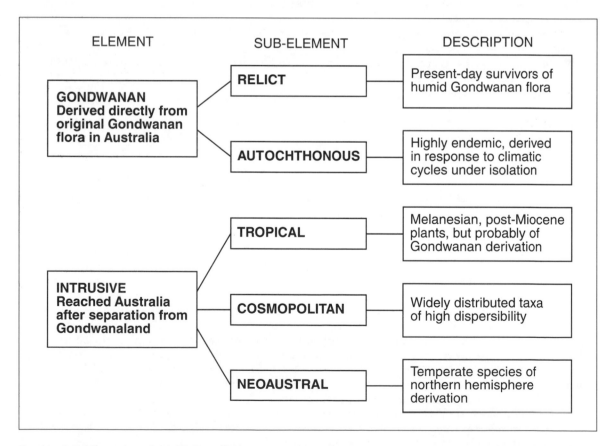

FIGURE 3.5 Elements of Australian flora

Southland and is affected by the easterly Tasman Current.

VEGETATION

The origins of the Australian flora can be divided into two elements, Gondwanan and intrusive (Figure 3.5) (Barlow, 1994). The Gondwanan element consists of a relict sub-element derived from a closed canopy rainforest which was thought to exist under moist and warm conditions at the beginning of the Tertiary. Retraction occurred as cycles of aridity became frequent from the mid-Tertiary onwards, with the eastern uplands of Australia perhaps being the only extensive area that has continuously had these moderately moist forests. A second, indigenous (autochthonous),

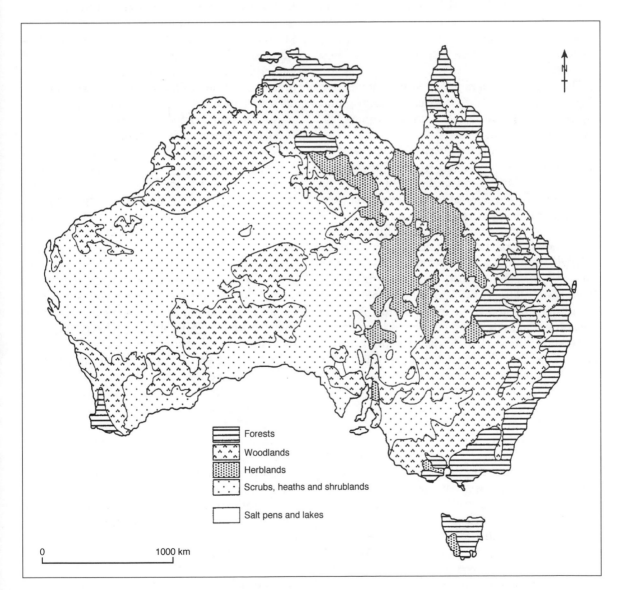

FIGURE 3.6 Vegetation regions of Australia (adapted from Carnahan, 1986)

Gondwanan sub-element of the vegetation has also been recognised, which became differentiated from the original flora. Increasing isolation after the break-up of Australia from Antarctica, until the Australian Plate collided with the Sunda Plate in the Miocene, increasing aridity and adaptations to nutrient-poor soils led to changes in the Australian taxa. A feature of these is scleromorphy (hard, rigid leaves), thought to be a response to nutrient deficiency rather than aridity (Barlow, 1994). In this sub-element, Eucalyptus and Acacia species are dominant in open forests and woodlands, but rarely penetrate the closed forests.

The intrusive element is sub-divided into three: the tropical, arid (Cosmopilitan) and alpine (Neoaustral). Melanesian contact in post-Miocene times has influenced the rainforests of north-eastern Australia, whereas the flora of the arid zones is a composite derived from other plant communities. There is some debate about the derivation of the alpine flora – immigrants by dispersal from cool, non-Australian sources versus a contribution from indigenous flora (Williams and Costin, 1994).

Vegetation regions of Australia, as they were thought to exist before European invasion, have been mapped by Carnahan (1986), and a simplified version appears in Figure 3.6. The basis for this map is the structural form of the vegetation, defined in terms of growth forms and projective foliage cover (Specht, 1970; Carnahan, 1986).

Forests are confined to the moist margins of the continent (Figure 3.6) and are sub-divided into closed and open forests, depending on projective foliage cover. Closed-forests are equivalent to 'rainforests', and most are evergreens with a canopy cover of over 70 per cent. The open forests have a canopy cover of between 70 and 30 per cent, and are dominated by Eucalyptus, Acacia and, to a lesser extent, Callitris (Carnahan, 1986).

Woodlands (Figure 3.6) have a projective foliage cover of between 30 and 10 per cent, and are floristically similar to the open forests, being extensions of the latter into areas of

lower rainfall. A sub-group, open woodlands, has a cover of less than 10 per cent.

Shrublands have a widespread distribution in the southern arid and semi-arid regions of Australia (Figure 3.6). Dominated by species of Eucalyptus and Acacia, these areas generally have less than 300 mm of rain per year. Understoreys can consist of a variety of species of grasses, herbs and forbs (Figure 3.7).

Masses of branching stems forming mound-like features between 0.3 and 1 m in height constitute the hummock grasslands (popularly known as spinifex). These usually occur as an understorey with trees and shrubs.

Herblands of grasses and forbs occur in areas of annual rainfall between 600 mm and 200 mm through western Queensland and parts of the Northern Territory and South Australia (Figure 3.6).

Kirkpatrick (1994) has identified a number of factors which influence the distribution of Australia's plant species: 'Australian plants have developed an awe-inspiring range of adaptations that mitigate the water balance problem' (Kirkpatrick, 1994: 12), including the ability to store water, close stomata, draw water from deep in the regolith and minimise water loss due to special anatomical features. The impact of fire varies with species, but most eucalypts have many mechanisms that ensure their survival, while other species require heat or disturbance for successful germination.

In New Zealand, four floristic assemblages are recognised (Wardle, 1991). Firstly, forest trees of the lowlands have their origins in the Miocene or earlier. Secondly, there are cold-tolerant plants of wet environments which also occur in other parts of the former Gondwanaland (e.g. Tasmania), while thirdly, trans-oceanic migrants, derived from Australian heath and sclerophyll vegetation, occur in heathlands and grasslands of low altitude. Finally, shrubby and herbaceous flora have evolved in response to environmental changes during the Quaternary.

New Zealand was covered mostly by rainforest 7000 years ago: tall conifer–broadleaved rimu (*Dacrydium cupressinum*) in

FIGURE 3.7 Acacia shrubland on the rocky slopes of the Barrier Range, near Broken Hill, New South Wales, Australia (photo: R.J. Loughran)

the north and west, with matai (*Prumnopitys taxifolia*) in the east (McKinnon, 1997). Beech-forest was rare. Due to a cooling of the climate by 1–2°C, cold-tolerant species survived well in the highlands, while species that were prone to frost damage lost ground in the lowlands. There was a spread of beechforest, although not as far south as Stewart Island, with broad-leafs and conifers becoming more common on North Island.

Vegetation zones of New Zealand prior to European settlement are shown in Figure 3.8 (Dawson, 1988). Broadleaf–conifer forest dominated North Island, while the beechforest (*Nothofagus* species) occurred extensively in the northern part of South Island and on the lower slopes of the North Island mountains. Lowland shrub and fernland, and short tussock grass-land, were present in the lowlands of North and South Island, respectively. Alpine vegeta-tion was dominant in the Alpine region of South Island, with the lower slopes covered with shrubs and herbs. Much of Stewart Island, the southernmost of the three largest islands of New Zealand and covering 1720 km², is covered in forest (Figure 3.9), but beech

(*Nothofagus*) has never been present, despite dominating some areas of both North and South Island (Wilson, 1994). During the glacial periods of the Pleistocene, conditions were too severe for tree growth and, afterwards, the slow advance of beech and the rise in sea level prevented its spread across Foveaux Strait.

FAUNA

Because of its long isolation, variety of climatic zones and size, Australia is considered one of the most biologically diverse countries of the world, with many endemic species (Table 3.2): 'Of the 12 nations in the world that contain major repositories of (total) species diversity, Australia is the only developed country' (DESTA, 1996). It has nearly 150 species of marsupials, plus two species of mammals that lay eggs (monotremes). New Zealand, in contrast, has no ground-dwelling native mammals, but over 130 'original' land bird species, 93 of which are endemic. Probably the best known of the birds is the flightless, forest-

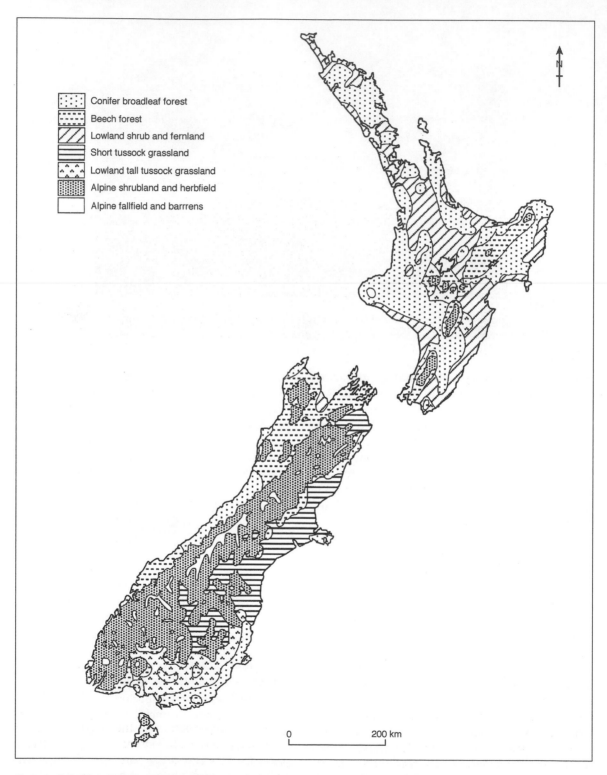

Conifer broadleaf forest
Beech forest
Lowland shrub and fernland
Short tussock grassland
Lowland tall tussock grassland
Alpine shrubland and herbfield
Alpine fallfield and barrrens

0   200 km

FIGURE 3.8 Vegetation regions of New Zealand (adapted from Dawson, 1988)

FIGURE 3.9 Forested landscape, Paterson Inlet, Stewart Island, New Zealand (photo: R.J. Loughran)

TABLE 3.2 Endemism in Australian mammals

	Total number of species	Endemic species	Percent- age endemic
Marsupials	141	131	93
Bats	69	40	58
Monotremes	2	1	50
Rodents	65	57	88
Seals	4	1	25

Source: DESTA (1996, Table 2.2).

dwelling, kiwi. It belongs to one of the Earth's oldest group of living birds (*Apteryx spp.*) and developed shortly after New Zealand broke from Gondwanaland about 80 million years ago.

CONCLUSION

The character and distribution of the climates of Australia and New Zealand differ markedly because of the greater area, latitudinal spread and generally low relief of the former. Tropical influences and lack of major mountain ranges to produce orographic rainfalls, except in the eastern fringe of the continent, contrast with New Zealand's cool temperate maritime climates on mountainous islands lying athwart the path of westerly rain-bearing air masses. Fluctuations between drought and flood in response to El Niño and La Niña´ events are seen both in Australia and New Zealand, however.

Climate, of course, has the major influence on the hydrological systems, but in Australia groundwater resources derived from long term recharge and storage are also a feature.

The transport of sediments in runoff waters are chiefly controlled by relief and energy, with New Zealand having the greater, particularly in the tectonically active parts of the country.

Vegetation history in both countries reflects Gondwanan influences, with later adaptations, migrations and introductions also playing a part.

The inherent geology, landforms, soils, climate and hydrologic characteristics have together influenced and constrained human activity, and human activity has left indelible marks on the landscape. These are the topics of the next Chapter.

FURTHER READING

A comprehensive review of the state of Australia's environment is provided by DESTA (1996). This includes content on biodiversity, the atmosphere, land resources, inland waters, estuaries and the sea. A comparable volume for New Zealand is MfENZ (1997), which is just as detailed and valuable. Groves (1994) is an edited collection on Australia's vegetation, type by type. Jeans (1986) has Chapters on Australia's climate (Gentilli), hydrology (Warner) and vegetation (Carnahan). Mosley (1992) is a detailed book on all aspects of hydrology in New Zealand, from precipitation to floods and from water quality to water use. Sturman and Tapper (1996) is a text on weather and climate in Australia and New Zealand, including processes, climates and climate change.

4

ENVIRONMENTAL ISSUES: CONSTRAINTS AND IMPACTS

Human activity is constrained by environmental conditions. The degree to which we can adapt to, or overcome, these constraints depends on our ingenuity, knowledge, education and resources – factors which have changed through history. In adapting and overcoming, we have had an effect on the environment, usually detrimental. This chapter, and Chapter 16, examine some of the major constraints placed by the environment on the human inhabitants of Australia and New Zealand, as they have endeavoured to survive and develop their way of life, and the ways in which the environment has been changed as a result of these activities.

The Aborigines of Australia adapted to the environment during their (at least) 50 000 years of occupance. The European invaders/settlers in both countries and, to a lesser extent, the Maoris of New Zealand were confronted by alien environments which, in the first instance at least, restricted their occupance and development of new territory. The vagaries of climate, flood and drought, infertile and thin soils, great distances and difficult terrain initially provided some problems. The ways in which these problems were overcome were, in many instances, environmentally unsustainable and led to great changes. But it is wrong to suggest that the damage was deliberate; very largely it was done in ignorance.

INDIGENOUS PEOPLES AND THE ENVIRONMENT

Before the European invasion of Australia, the Aborigines of Australia had developed a relationship with their environment that enabled them to prosper. This relationship included a spiritual attachment to the land and a knowledge which allowed them to survive; their hunting–gathering–gardening land use was sustainable, but it did affect the environment as both the Aboriginal economy and the environmental resources evolved (Aplin, 1998). It is debatable if Aboriginal activities created the extinction of marsupial species, but approximately 40 species disappeared after the advent of human occupation. Certainly, Aboriginal hunting did affect the number of (edible) wildlife, such as kangaroos, wallabies, possums and koalas. The vegetation increased its adaptation to fire as a result of Aborigines' use of fire to improve grazing conditions for game and to make hunting easier. Such fires may have triggered accelerated soil erosion by water by reducing protective ground cover in certain parts of eastern Australia (Hughes and Sullivan, 1981); the building of stone fish-traps may have slightly modified stream channels; but these impacts are part of a management strategy, rather than exploitative. Exploitation

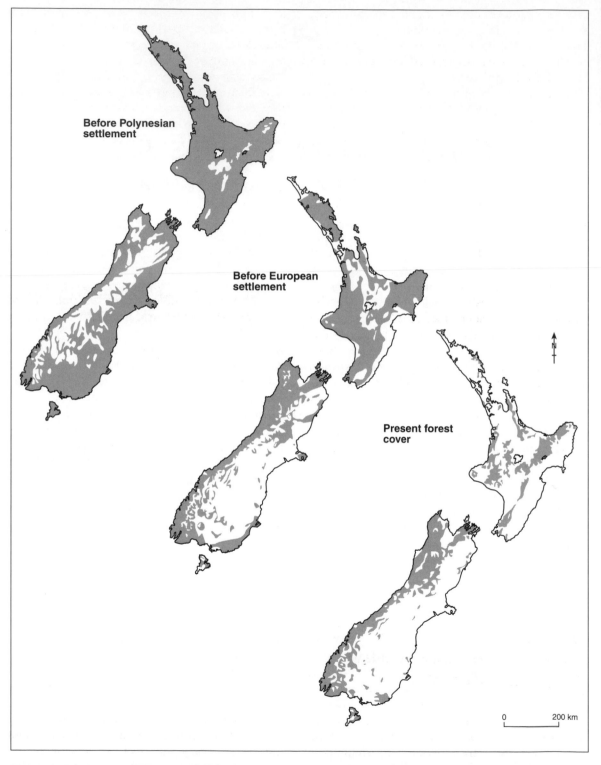

FIGURE 4.1 The distribution of natural forest and tall shrubland in New Zealand before Polynesian settlement, before European settlement and at present (from MfENZ, 1997)

creates long term environmental deterioration, whereas the effects of traditional Aboriginal land management are shorter-term and restricted in area because of low population densities (Kohen, 1995).

By contrast, the occupation of New Zealand by its first (Polynesian) colonists is much more recent: approximately 700 to 800 years ago. Hunting and gathering, particularly of seals and moas, exploited available food resources. The native plants were poor in carbohydrates, and the imported American sweet potato (kumara) was able to be grown only in restricted areas because of unsuitable climatic conditions. Maori population expansion created pressure on the land, with the extinction of many species of animals through hunting by humans, rats and dogs. Fires destroyed upland and lowland forests, which were replaced by bracken and tussock grasses (Anderson and McGlone, 1992). During the time of Maori occupation until the time of European settlement, native forests were reduced from about 85 per cent of the land area to 53 per cent, while tussock grasslands expanded from approximately 5 per cent to almost 30 per cent (MfENZ, 1997) (Figure 4.1). This reduction in forest cover created serious soil erosion, particularly in areas with erodible rocks.

Impacts on the environment of Australia by its Aborigines would seem far less than those by the Maoris in New Zealand, which together are far less than those imparted by the second major wave of settlers from Europe in the nineteenth century.

THE IMPACTS OF EUROPEAN SETTLERS ON THE ENVIRONMENTS OF AUSTRALIA AND NEW ZEALAND

The imposition of European styles of land use and farming methods began in Australia at the end of the eighteenth Century. Many settlers had little knowledge of farming techniques, and European methods were hampered by infertile soils and periods of flood and drought. The clearance of native forest and woodland for agriculture exposed soils which were susceptible to erosion by water and wind, the first documented signs of which appeared at the end of the nineteenth century during a period of expansion in farming in Victoria (Powell, 1976). The goldrushes of the 1800s in Australia and New Zealand also created environmental degradation, both directly and indirectly. Localised diggings and clearances of native vegetation laid bare the land surface and created the pollution of streams. From the mid-1880s to the beginning of the Second World War (1939) there was increasing demand for agricultural produce, which was in part met by the opening up of cropping land. Land under wheat increased three-fold in Australia in the 30 years after 1890, from 2.23 to 6.1 million hectares (McTainsh and Boughton, 1993). Between 1788 and 1995 in Australia, approximately 43 per cent of forests had been cleared and 99 per cent of the temperate lowland grasslands of the south-east had been lost (DESTA, 1996) (Figure 4.2). The clearance of native forests in New Zealand also accompanied the expansion of agriculture and was used for timber supplies, with the area of forest reduced to 23 per cent of the country's area within 100 years of European settlement. Irrigated farming also expanded, leading to problems of waterlogging and widespread soil and water salinity. The introduction of exotic species of plants and animals has also had effects in both Australia and New Zealand.

To summarise, European settlers in Australia and New Zealand have introduced agriculture, commercial hunting and fishing, exotic species of plants and animals, urbanisation, industrialisation and large-scale mining. Effects of these actions on the environment have been land degradation, a loss of biodiversity, a depletion and/or modification of water resources, air and water pollution, soil contamination, and changes to the coastal and marine environment (DESTA, 1996). Some of these effects are described below.

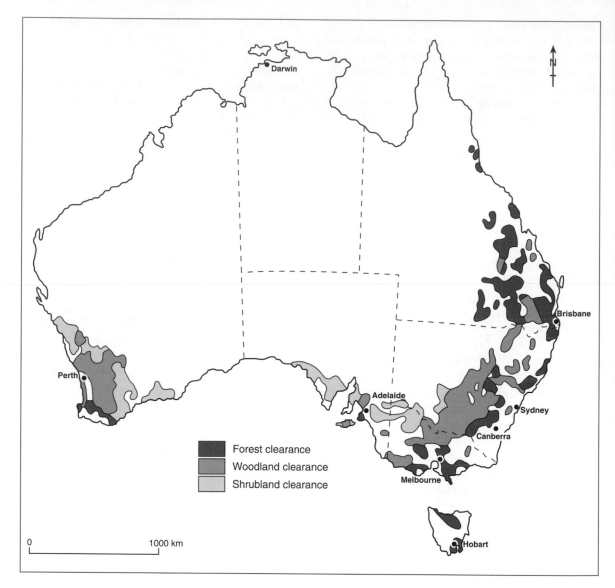

Figure 4.2 Vegetation changes in Australia since European settlement (source: ABS, 1992b Commonwealth of Australia copyright reproduced by permission)

Land degradation

Within Australia and New Zealand, several forms of land degradation have been recognised, including soil erosion, soil acidification, soil compaction, salinisation, waterlogging and vegetation decline. Of these, soil erosion is regarded as the most important because it is the most widespread, and because it affects the site at which it occurs (by reducing conditions for plant growth) and creates water and air pollution. While no recent figures for the extent of different types of erosion are available for Australia, it is known that rates of erosion in agricultural areas are probably much higher than rates of soil formation. The rates of soil formation are probably no greater than 0.5 t ha^{-1} y^{-1} under favourable conditions, while

Box 4.1 Soil erosion in Australia in relation to land use

A national reconnaissance of soil erosion is being conducted using the isotope caesium-137 (^{137}Cs). Caesium-137 is a product of nuclear weapons testing, and atmospheric fallout has labelled soils, enabling their movement to be traced (Morgan, 1995). Uneroded reference sites have levels of ^{137}Cs that can be compared with sites which have lost soil and ^{137}Cs. Because fallout commenced in the mid-1950s, soil loss estimates give time-averaged erosion rates for approximately 50 years. Furthermore, the depth distribution and greater-than-reference amounts of ^{137}Cs at depositional sites can be used to quantify sedimentation rates.

Empirical equations relating soil loss to ^{137}Cs loss from long term Australian runoff–erosion plots are being used to estimate soil erosion on selected hillslope transects throughout the country. The sites are representative of typical management practices, rather that worst-case scenarios. As may be expected, land uses which disturb the soil create the most erosion (Table 4.1), such as those under rotational potato cropping, vineyards and grain crops. Land used exclusively for sheep and cattle grazing appears to be the least affected by erosion, although some of the rangelands of semi-arid Australia have been severely eroded by wind and water. Because rates of soil formation are thought to be slow, all land uses with soil loss rates over $1\ t\ ha^{-1}\ y^{-1}$ may be regarded as unsustainable.

soil losses in important agricultural areas are often higher than $5\ t\ ha^{-1}\ y^{-1}$ (Box 4.1). Severe erosion events, while infrequent, can cause the bulk of this annual soil loss. For example, at the Cowra Research Centre of the NSW Department of Land and Water Conservation, soil erosion by rill and sheet runoff during two separate intense rain storms caused combined

Figure 4.3 Black earth soil deposited on a road crossing after storm runoff from cotton fields, near North Star, northern New South Wales (photo: R.J. Loughran)

TABLE 4.1 Soil erosion rates for selected land uses measured by caesium-137 (number of sites for each land use)

Net soil loss (t ha^{-1} y^{-1})	Pasture–cropping rotation	Potato–vegetable–pasture rotation	Pasture	Rangelands
30–49.9		1		
20–29.9		1		4
10–19.9	3	3		5
5–9.9	14	4	3	14
1–4.9	34	5	10	7
>1	31	1	46	15

Source: Loughran *et al.* (in preparation).

losses of 342 t ha^{-1} under conditions of traditional tillage (Hairsine *et al.*, 1993). The mean annual long term soil loss under the same management, predicted by a soil loss model, is 46 t ha^{-1} y-1. If no further soil erosion was to occur at this site, it would take approximately 900 years for the soil to recover (Hairsine *et al.*, 1993). The off-site effects of such events can also be severe, and lead to reservoir sedimentation, the burial of soil conservation ('contour') banks and roads (Figue 4.3), water pollution and eutrophication.

Direct observations, such as those described above at Cowra, provide a direct link between the factors responsible for the land degradation and the soil erosion which occurred. There are, however, greater difficulties when explanations of land degradation which took place in Australia 100 years ago are required. Because of lack of scientific observations at the time, interpretations involving records of production and investment, rainfall data and (Royal Commission) reports have been used to examine apparent causes of land degradation. Popular belief has it that much land degradation in the latter part of the nineteenth century was caused by overstocking by absentee landlords who were instructed to maximise profits for companies, many of which were based in London. Pickard (1993), in his research of conditions at the Momba Station in the Western Division of New South Wales

north of Wilcannia, is of the opinion that the overstocking (of sheep) 'was largely caused by the pioneer graziers misinterpreting two key components of the landscape (feed and rainfall)' (Pickard, 1993: 5). He says the sight of abundant biomass was mistakenly assumed to represent high productivity, and that water availability was the only limiting factor. 'Once the biomass was razed, the inherently low productivity was insufficient to maintain the flocks, even in the good years' (Pickard, 1993: 5). Also, at the time of European settlement, there was good rainfall, but at the turn of the twentieth century drought conditions prevailed: the 'biomass was destroyed and the land was degraded ... They did overgraze, and they did degrade the land, but primarily out of ignorance, not greed' (Pickard, 1993: 5).

Accelerated soil erosion because of human activity has occurred in New Zealand for a longer period than in Australia, and national statistics on the percentage of land affected by different types of soil erosion are available for the period 1975–79 from aerial photograph interpretation and field mapping (Table 4.2) (Eyles, 1983). The classification used in the survey for the New Zealand Land Resource Inventory, carried out between 1973 and 1979, was based on the form of erosion rather than process. Sheet erosion, where water had stripped a thin layer of soil leaving a bare surface, was prevalent on tephra slopes of the

TABLE 4.2 Soil erosion in New Zealand according to type: area affected 1975–9

Erosion type	North Island %	South Island %	New Zealand %
Surface erosion	23	74	52
Sheet erosion	19	55	39
Wind erosion	5	19	12
Scree	4	22	13
Mass movement	44	31	36
Soil slip	30	24	26
Earth slip	3	<1	1
Debris avalanche	11	11	11
Earthflow	9	<1	4
Slump	<1	<1	<1
Fluvial erosion	14	10	12
Rill	<1	<1	<1
Gully	10	5	7
Tunnel gully	3	<1	2
Streambank erosion	2	3	3

Note: Areas are those of the mapping units in which the erosion type occurs, not of the actual erosion features. The total area affected by erosion cannot, therefore, be obtained from the table because more than one erosion type may occur within the same mapping unit.
Source: Eyles (1983).

FIGURE 4.4 The distribution of soil slip erosion in New Zealand (source: MfENZ, 1997)

volcanic plateau of North Island and on the hill and mountain country of South Island. Generally, surface erosion was the most extensive of the erosion types, whereas mass movement (gravity induced movement) by flows, slips and avalanches was rather more frequent on North Island (Table 4.2) (Eyles, 1983). It was thought that the survey underestimated the extent of the fluvial erosion types, except for gully erosion.

Rates of soil formation in New Zealand are not known, but it is thought that average soil losses exceed gains. Clearance of forest cover is largely responsible for mass movement slope failure (slip erosion), particularly in areas of tectonic instability where erodible rocks are exposed, for example in the eastern parts of North Island and the north-western parts of South Island (MfENZ, 1997) (Figure 4.4). Significant surface erosion by north-west

winds occurs on South Island, particularly from springtime seed-beds in north Otago and south Canterbury, and single episodes of erosion of up to 100 t ha^{-1} have been reported. The caesium-137 technique (Box 4.1) has been used to measure net soil losses in south Canterbury downlands (Basher *et al.*, 1995). Results suggested no net soil loss from a cropped site, but there was soil redistribution. Generally, points of greatest soil loss were on interfluves, which suggest that wind erosion was responsible, rather than water erosion.

Soil salinity, or salinisation, is a major problem confronting farmers in some regions of Australia, but not in New Zealand. Primary soil salinity occurs as part of the natural

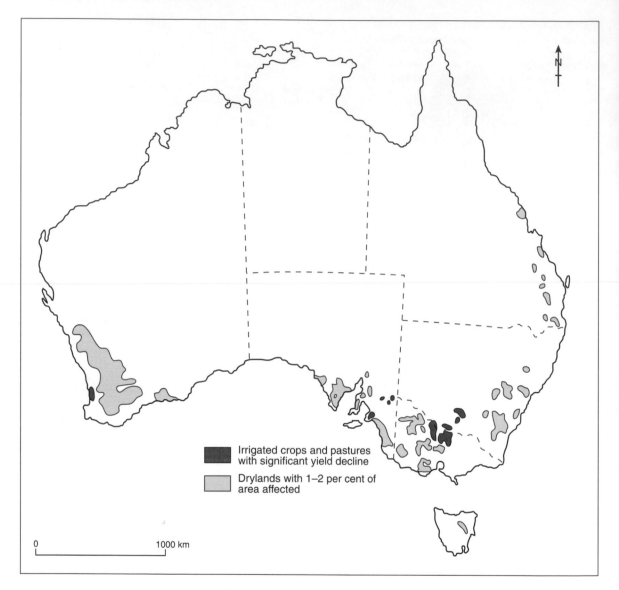

FIGURE 4.5 The distribution of soil salinity in Australia (source: ABS, 1992b Commonwealth of Australia copyright reproduced by permission)

landscape (salt marshes, salt flats and salt lakes), but secondary salinity has been induced largely by agricultural practices in irrigated areas or on non-irrigated farm land (dryland salinity) (Peck, 1993) (Figure 4.5). The addition of irrigation water at a greater rate than it is used will add water to the groundwater store, causing the watertable to rise. Waterlogging has occurred initially on a large scale in the irrigated areas of the lower Murray–Darling basin where the watertable has reached the ground surface. Maximum rates of watertable rise have been recently measured at 100 mm to 500 mm per year in the south-east of the basin. Eventually, evaporation has led to the accumulation of salt in the soils by the precipitation of salt derived from groundwater sources and the irrigation water applied from the rivers. In the

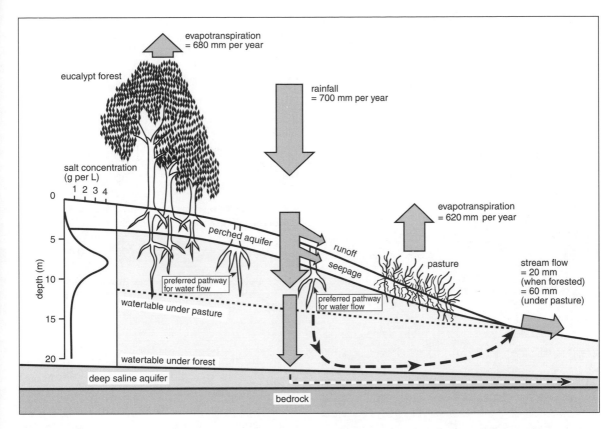

FIGURE 4.6 The effects of clearing native Australian forest on hydrology and salinity (adapted from Bell, 1988/89)

state of Victoria, it is estimated that nearly one-third of all irrigated land is affected by salinisation (DESTA, 1996).

Some southern regions of Australia which depend on precipitation for agricultural production have been severely affected by the accumulation of salts in the soil, resulting in dryland salinity and loss of productive land (Smith, 1998). Even districts with plentiful moisture, such as south-western Western Australia with over 1100 mm of annual rainfall and winter waterlogging, are affected by dryland salinity. Before the large-scale clearance of native forest and woodland for grain growing in the late 1800s, the watertable was at great depth (Figure 4.6). Although the soils and groundwater contained salts which had accumulated over geological time from precipitation

inputs, deep-rooted forest trees had high rates of evapotranspiration which kept the watertable below the root zone. Replacement of forest by shallow-rooted pastures and crops has meant a reduction in evapotranspiration and greater infiltration of water into the soil (Figure 4.6). This has led to a rise in the water table and the remobilisation of salts in the upper part of the soil profile which was previously unsaturated (Figure 4.7). Surface, throughflow and baseflow runoff has consequently produced waterlogging at the slope base, and the accumulation of salts by evaporation (Figure 4.6). It was estimated that approximately 2.5 million hectares of Australia were affected by dryland salinity in 1996, increasing at a rate of 3–5 per cent per year (Bennett, 1998). Of this total, 1.8 million hectares are in Western Australia (Table 4.3).

FIGURE 4.7 Soil salinity on a farm near Perth, Western Australia. Bare ground devoid of vegetation is the result of salt accumulation after the rise of saline groundwater to the surface. The white tube is a piezometer, which measures the height of groundwater. In this instance, the groundwater is under pressure, and the height of the watertable is approximately 1.7 m above ground level. Water in the piezometer tube is therefore 1.7 m high, approximately three-quarters of the way up the tube (photo: R.J. Loughran)

TABLE 4.3 Area of Australia affected by dryland salinity in 1996

State	Area affected by salt (ha)	Percentage of total
Western Australia	1 804 000	72.9
South Australia	402 000	16.3
Victoria	120 000	4.8
New South Wales	120 000	4.8
Tasmania	20 000	0.8
Queensland	10 000	0.4
Northern Territory	minor	
Total	2 476 000	100.0

Source: Land and Water Resources Research and Development Corporation, Canberra (1997), in Bennett (1998).

In both Australia and New Zealand there are serious threats to biological systems, largely as a result of direct and indirect human intervention. Increasing demand for natural resources and the adverse effects of technologies continue to contribute to a deteriorating situation, and few ecosystems remain in a natural condition. Loss of habitat, the decline in populations and decrease in genetic diversity all contribute to a reduced ability of plants and animals to compete, adapt and survive. There is also still an inadequate knowledge of the biodiversity that exists, which hampers the development of conservation strategies.

In Australia, it is estimated that there are 20 000 species of flowering plants, 85 per cent of which are endemic. Since 1788, 76 species are presumed extinct (not located in a natural habitat during the preceding 50 years); 301 species are considered endangered (i.e., in danger of extinction if threats to survival continue), and 708 are vulnerable (that is, will very likely be 'endangered' in the next 25 years if threats continue) (DESTA, 1996). In addition, there are introduced species of plants, now making up approximately 15 per cent of the total Australian flora. 'About half of them invade native vegetation and about one-quarter are regarded as serious environmental weeds' (DESTA, 1996: 4.18). These species destroy water-bird habitats by invading open water, and smother and displace native trees and shrubs.

The number of species of vascular plants in New Zealand is low (approximately 2500), but

FIGURE 4.8 Native forest and bushland (middleground and centre), coniferous forest plantations (background) and pasture land in Wairarapa, North Island, New Zealand (photo: R.J. Loughran)

about 80 per cent are endemic (MfENZ, 1997). Most of New Zealand's 200 threatened species are shrubs, grasses and herbs, but none of the unique timber-producing trees are threatened with extinction, despite the original forest cover being reduced in area from 85 per cent to 23 per cent, although planting of non-native species for conservation and timber supplies has been extensive (Figure 4.8). There are, however, nine plant species 'presumed extinct'. In contrast to Australia, introduced species of plants may now outnumber native plants in New Zealand's natural habitats: approximately 2500 native species compared with 25 000 introduced.

Australia has approximately 6000 vertebrate species of fish (both freshwater and marine), amphibians, reptiles, birds and terrestrial mammals, and nearly 160 000 species of arthropods (including beetles, moths, butterflies, ants, wasps, bees, flies, mosquitoes, spiders, mites, crabs and prawns). Within these groupings, 19 terrestrial mammals (including 10 marsupials and eight rodents) and 20 bird species have become extinct since European occupance. Conversely, there have been 18

introduced mammals which have established feral populations, including goats, dogs, cats, foxes, water buffalo, camels and horses, many of them having detrimental effects on populations of native animals and creating environmental damage, such as mud wallows and soil erosion. Rabbits, introduced by European settlers, reached plague proportions over much of Australia, competing with native animals for limited food resources and degrading soils and vegetation. It has been estimated that rabbits cost between A$60 million and A$90 million a year in lost production and reduced land values (DESTA, 1996). Since European settlement in New Zealand, 16 land birds and one native bat have become extinct, but 34 species of birds had gone prior to European arrival.

As in Australia, acclimatisation societies introduced birds and mammals to make New Zealand more like Europe; 54 mammalian species were introduced and, while not all survived, 14 have become widespread (including humans, rats, sheep, cattle and possums!) (MfENZ, 1997). Rabbits have been a problem in the High Country of South Island, causing the degradation of grasslands and contributing to soil loss. Other introduced mammals have created environmental degradation, including various species of deer and goats, but possums are 'public enemy number one'. They were introduced from Australia between 1837 and 1922 for their fur, but after being released into the wild they were classified as pests in 1947 because they cause severe damage to native trees.

Impacts on the atmosphere

Urban and industrial development has had some effect on the reduction in air quality in cities of both countries. Emissions of carbon dioxide (CO_2), carbon monoxide (CO), lead (Pb), methane (CH_4), oxides of nitrogen (NO_x), sulphur dioxide (SO_2) and particulates have created health problems and environmental deterioration. During the 1970s and 1980s, the scientific community and politicians began to understand that there were serious air pollution problems in some of the larger cities, particularly Sydney. A combination of increasing vehicular traffic, air circulation and geomorphology produces significant air pollution there. Sea breezes and air drainage within basin-like topography (high elevations to the north, west and east of the city, and the Pacific Ocean to the east) combine to transport, trap and circulate pollutants, to the extent that, during the summer, Sydney is called the 'Los Angeles of the southern hemisphere' (Bridgman *et al.*, 1995). High levels of tropospheric ozone are a particular problem in Sydney. Because of the above-mentioned topography and air circulation, high levels of tropospheric ozone occur in the outer western and southwestern regions of the city. Nitrous oxides and hydrocarbons, produced in central city areas, react with sunlight to produce ozone, and the air settles in the outer suburbs. In New Zealand, domestic fires of wood and coal, with some vehicular pollution in major cities and temperature inversions in winter, produce serious air pollution, notably in Christchurch (Sturman and Tapper, 1996).

While levels of SO_2 in Australian and New Zealand cities are within international guidelines, total suspended particulates in the air of Sydney, Brisbane and Newcastle are closer to World Health Organisation (WHO) standards, partly because of industrial fallout, but also because of significant additional contributions from dust derived from rural areas during drought (Bridgman *et al.*, 1995). Respiratory problems can often be linked to atmospheric pollution of this kind, although there is debate about direct linkages. Industries, such as smelters, have been identified as serious polluters at Port Pirie (South Australia) and Boolaroo (south of Newcastle in New South Wales), where long term lead emissions have created concern for children's health, because it is generally thought that ingestion of even small quantities of lead can impair intellectual development. Power production in both countries relies largely on coal- and oil-fired power stations, hydro power and geothermal sources (New Zealand), but there are no nuclear power stations. Atmospheric pollution from power generation is therefore confined to the burning of fossil fuels, which has been the cause of some concern over health problems on the eastern Australian coastal belt between Sydney and Newcastle. While acid rain impacts in Australia and New Zealand are rare compared with problems encountered in Europe and north America, there are warnings that thermal power stations in the Latrobe valley of eastern Victoria may create acid rain problems in the highlands of Victoria and NSW (Cocks, 1992).

Another consequence of urbanisation on climate is the production of heat islands, where the built-up areas are warmer than the surrounding countryside. While this is a universal phenomenon, it has been studied in several cities in Australia and New Zealand. In the 1970s and 1980s, research in Australia showed that temperatures were elevated in Newcastle by 1–2°C, Adelaide by 2°C, Hobart by 4–5°C and Sydney by 5°C (Bridgman *et al.*, 1995). Sturman and Tapper (1996) reported a maximum elevation of 5.5°C for the city of Christchurch, New Zealand.

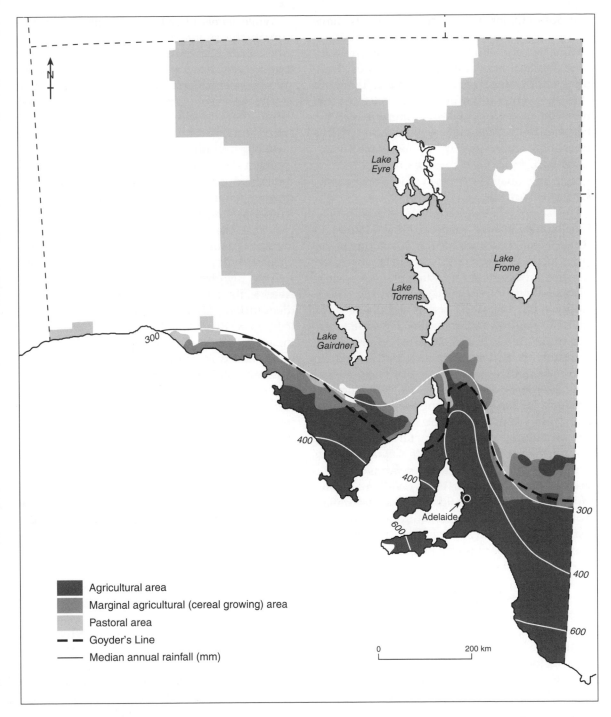

FIGURE 4.9 Goyder's Line in relation to mean annual precipitation and land use in South Australia (adapted from Harris, 1986)

Impacts on inland waters

During the post-European settlement phase in New Zealand and Australia there have been great modifications to the surface water and, to a lesser extent, the groundwater phases of the hydrological cycle. The harvesting of water for irrigation, hydroelectric production (HEP), domestic and industrial supplies, and flood mitigation, have been most influential on the timing and quantity of river flows and water quality.

The lack of reliable rainfall and surface water in Australia has been emphasised in Chapter 3. These factors constrained the development of agriculture in Australia, whereas in New Zealand, a wetter country, such problems were relatively small. In South Australia, the driest state on the driest inhabited continent, less than 250 mm rainfall is received over 82 per cent of its area (Jensen, 1986). Aborigines and Europeans settled near water supplies, and in 1865 the Surveyor-General of South Australia, G.W. Goyder, drew a line to mark the northern limit of 'safe' agriculture, based on the distribution of natural vegetation as an indicator of rainfall. Optimistic farmers settled beyond 'Goyder's Line' (Figure 4.9), but many were forced to abandon their farms after a series of droughts in the 1880s (Williams, 1974; Smith, 1998). In the arid zone of Australia, groundwater is a valuable resource, but its quality is generally too poor for human consumption because of dissolved salts derived from the rocks it has moved through.

Early European settlers throughout Australia collected rainwater from roofs, and built shallow earth tanks and embankments to pond water for stock and domestic use (Figure 4.10), but these methods were prone to failure during times of low precipitation and were too small to allow for the expansion of commercial enterprises. Bigger structures to impound water were obviously required to provide water more reliably for an increasing population and agricultural expansion. Pigram (1986) reported that the first 'large dam' was built in 1857 for Melbourne's water supply, followed by the Cataract Dam for Sydney. The unreliability of rainfall and runoff has meant that large stores are required to provide reliable supplies, so that Sydney has the largest per capita water storage in the world (932 kilolitres per inhabitant). New York has 250, London 182 and Birmingham 86 kilolitres for every inhabitant (DESTA, 1996). In major reservoirs in Australia total water storage is 81 000 gigalitres, 'which is equivalent to three Olympic swimming pools for every one of the country's 17.8 million people' (DESTA, 1996: 7.8). Of this, the ten largest reservoirs hold approximately 50 per cent of the national capacity, most of which are in the south-east of the continent along the Great Dividing Range and in Tasmania. 'This reflects availability of runoff, topography suitable for dam construction, and population distribution' (DESTA, 1996: 7.8).

In New Zealand, Dunedin was the first of the major cities to install a water supply system (1863) and was followed by Wellington (1874), Auckland (1877) and Christchurch (1909). HEP generation became important in New Zealand, and large structures were built during the period from the 1950s to the 1970s. In Australia, the Snowy Mountains hydroelectricity scheme construction-phase began in 1949 (see Chapter 8). These were the start of large-scale river regulations affecting many of the river catchments of Australia and New Zealand by drowning often productive land, reducing and redirecting river flows, trapping sediments, altering habitats and releasing low-temperature waters.

One of the consequences of water storage and the diversion of flows to another river basin, as in the Snowy Mountains scheme of southern New South Wales, is the reduction of flows in the beheaded stream (Smith, 1998). The Snowy River currently has only 1 per cent of its average 'natural' volume. This decrease in flow has reduced the quality of habitats along the river, increased water temperatures and allowed the encroachment of riparian vegetation into the channel. Furthermore, for the first time since the early 1980s, the mouth of the Murray River at Marlo, Victoria, is likely

FIGURE 4.10 Water harvesting at Mundi Mundi (Barrier Range, near Broken Hill) in the arid zone, New South Wales, in the late nineteenth century and early twentieth century employed a masonry dam (centre) across a small ephemeral stream (creek). Water was diverted into a masonry tank lined with mortar (immediately in front of the vehicle). The tank was originally covered with a sheet-iron roof, supported on wooden beams, to reduce evaporation, and keep native, feral and stock animals from entering it. A study is being made of the sediments which have accumulated behind the dam to elucidate the soil erosion history of the catchment (photo: R.J. Loughran)

to be closed by the accumulation of coastal sand. Environmental groups argue that 30 per cent of natural flows are required to rehabilitate the system. In New Zealand, where 98 per cent of all waters for human use are for HEP, impacts on flow regimes are much less than in Australia. Because of diurnal oscillations in electricity demand, short-term variations in river flow can be marked, however. For example, at the Roxburgh power station on the Clutha River, South Island, daily flows can vary between 200 and 600 m³ sec⁻¹ as a result (MfENZ, 1997). On a seasonal basis, flows remain largely unaltered because the power stations operate on 'run of flow', with storages for only a few days' production.

In the Murray–Darling River system, the extraction of water from the river for irrigation has had serious consequences for the riverine environment. It was reported that over 50 per cent of wetlands have been destroyed, six out of 36 species of native fish are threatened, and there was a major outbreak of toxic algal blooms along 1000 km of the streams in 1991 (Wahlquist, 1997). In addition, there are rising levels of salinity associated with vegetation clearance and rising watertables along the Murray River.

Natural wetlands have also suffered from human intervention, mainly through drainage for urban expansion (e.g. Christchurch), for increased agricultural production (mostly pasture), flood mitigation and the mining of peat. The wetlands of the west coast of South Island, New Zealand, one-third of those in the state of Victoria, 70 per cent of those on the Swan coastal Plain of Western Australia and in New South Wales, 89 per cent of those in the

TABLE 4.4 Perceived impacts on water quality in New Zealand

Source of impacts	Average rank
Agriculture	4.9
Human sewage	4.8
Urban stormwater	3.9
Industry	3.8
Agricultural processing	3.7
Mining	2.6
Forestry	2.6

Type of agricultural impact	Average rank
Sedimentation	6.4
Nutrient contamination	6.2
Alteration of physical characteristics	5.6
Faecal contamination – surface water	5.4
Nitrate contamination – groundwater	4.6
Pesticide contamination – surface water	2.8
Faecal contamination – groundwater	2.8
Pesticide contamination – groundwater	1.6

Ranked on a scale from 0 = no damage to 10 = severe damage, by regional officials of the Ministry of Agriculture and Fisheries, New Zealand: Sinner (1992), in MfENZ, (1997, Table 7.5).

TABLE 4.5 The condition of some coastal elements in Australia

Element of the coastal environment	Condition
Saltmarshes	Extensive loss near urban areas
Seagrasses	Area diminishing in the south
Beaches and dunes	Generally good, except near settlements. Some areas modified by mining
Estuaries	Most are degraded, particularly in the south-east
Mangroves	Extensively cleared near coastal settlements
Coral reefs	Signs of degradation
Fish and fisheries	Most stocks fully exploited

Source: DESTA (1996, Table 8.19)

south-east of South Australia and over 60 per cent of those in Tasmania have been destroyed, or severely affected by nutrient enrichment (Fahey and Rowe, 1992; DESTA, 1996).

Water pollution in rural and urban areas is a serious threat to water supplies. Agriculture is perceived as the greatest single threat to water quality in New Zealand, with animal wastes, fertilisers, pesticides and soil disturbance being major factors identified in a survey by the New Zealand Ministry of Agriculture and Fisheries (Table 4.4). In Australia, levels of contaminants in freshwater systems have not been studied for sufficiently long to reveal if increases are occurring, but there are local examples of nutrient enrichment in unsewered areas, from sewer overflows and storm runoff which have caused concern for public health. Sewer outfalls off the coast of Sydney and the occurrence of pathogens in the city's water

supply in 1998 have been the cause of much concern and debate amongst health workers, environmentalists and politicians (Bridgman *et al.*, 1995; Doherty, 1998).

Threats to the long term future of parts of the coastal environment have come from a number of direct and indirect human activities. The clearance of native vegetation and weed infestation on coastal sand dune systems, sand mining for heavy minerals and construction materials, the construction of urban areas and associated harbours, industrial sites, resorts, inlets and marinas have all had direct effects on water quality, the biota and sediment circulation. Indirect effects include the delivery of water-borne sediments and contaminants from largely agricultural inland catchments. The condition of selected elements of the Australian coastal environment is given in Table 4.5.

Because most Australians are near-coast dwellers and use the coastal zone for recreational fishing, boating and diving, surfing and relaxation, there are concomitant pressures, particularly in the south-east of the country. The use of four-wheel drive vehicles to gain access to beaches has created severe degradation of dunes systems, although regulations now restrict vehicle use. The building of resorts of different types has frequently been at the expense of Pleistocene and Holocene sand dunes, and buildings have been damaged and threatened by storm waves, particularly at high tide (Mahony, 1978). Groynes – walls usually built of boulders perpendicular to the coastline – interrupt the transport of sand along beaches. On the coast of New South Wales and southern Queensland, groynes have prevented the long term movement of sediments in a northerly direction. This has led to sediment starvation on the down-drift beaches, making them more vulnerable to net erosion. It has been estimated that the groynes (or breakwaters) of the Tweed River estuary, on the border between New South Wales and Queensland, deprived the Gold Coast beaches (southern Queensland) of up to 15 million m^3 of sand over a 12-to-14 year period (Mahony, 1978).

Runoff containing sediments and other pollutants has been deposited in coastal waterways, smothering sedentary plants, creating turbidity and generally reducing the habitats of marine plants and animals. Changes in land use in northern Queensland river catchments towards more intensive agricultural production, have increased the content of nutrients and sediments delivered to the coastline at the Great Barrier Reef. Increased water turbidity, reducing the amount of light that can reach the corals, coral mortality and the erosive effects of storms all create damage (DESTA, 1996).

The major problems confronting the coastal zone in New Zealand are associated with human activity, including the delivery of sediments and pollutants from urban areas and the accumulation of waste from seagoing vessels. Estuaries have been the most severely affected elements, as most have settlements near them, but there are no scientific assessments available (MfENZ, 1997). As in Australia, buildings and roads have been placed too close to the coastal fringe and, as a consequence, are under threat from erosion. Coastal stability has also been affected by aggregate mining and the destabilisation of coastal sand dunes.

CONCLUSION

There are few parts of the environment of Australia and New Zealand that are untouched by human activity. In the period before European settlement in Australia the environmental impacts of Aborigines were relatively minor, except for the increasing adaptation of vegetation to fire regimes. In New Zealand, by contrast, the Maori population created greater impacts, clearing and firing vegetation so that the area under forest diminished. Increasing populations in urban centres, associated rural and city-based industries and agriculture have played a major part in transforming the landscape of both countries. Firstly, immigrants faced an unfamiliar world which they did not treat with care, but this was probably not deliberate, and due to ignorance. Also, there were natural hazards, such as floods and droughts, earthquakes and volcanic eruptions. During the post-Second World War period, awareness of environmental damage and natural hazards increased, and mechanisms were gradually put in place so that rehabilitation could be effected. These topics are addressed in Chapter 16.

FURTHER READING

Bridgman *et al.* (1995) deals with urban environments in Australia, including impacts on climate, air quality and health issues, land and water, biological systems and urban bushland, concluding with a chapter on management and the future. McTainsh and Boughton (1993) is a detailed collection of chapters by specialists on all aspects of land degradation in Australia. It includes material on present environmental conditions, processes and conservation techniques. Smith (1998) is a text on water resources and their management in Australia, including chapters on the resource base, water quality, the history of water resources development, droughts and floods, community–state relationships and 'the future'. See also DESTA (1996), Mosley (1992) and MfENZ (1997).

5

THIRD WORLD IN THE FIRST?

Just as Australia and New Zealand have major differences in their physical characteristics and geological histories, so there are wide disparities in the history of human occupation in the two countries (Pearson, 1994). The parallels between geological formation and the history of human settlement are striking. Australia has some of the oldest rocks and most stable landforms on the planet, as well as having an indigenous population with evidence of people on the continent for at least 50 000 years. In contrast, New Zealand is youthful both in terms of its geology and human occupance, with barely 1000 years elapsing since forebears of the contemporary Maori arrived on the shores of the land they called Aotearoa. This Chapter focuses on the contrasts between the histories of the indigenous peoples in the countries, recognising the significance of their differential treatment by the European colonists which has had such significant ramifications for the nature of contemporary society.

The Chapter's title, 'Third world in the first?', is taken from a book of the same name by Elspeth Young (1995), which emphasises how European colonisation marginalised the 'first peoples' and how subsequent development disregarded the socio-economic structures of their communities. Although there are clear contrasts between Aboriginal and Maori experiences of these processes, both peoples today experience significantly greater poverty and social dislocation than the European majorities.

CATASTROPHIC CONTACT: ABORIGINAL AUSTRALIANS AND THE EUROPEANS

When the British first established settlements in eastern Australia at the end of the eighteenth century, it is estimated that there were only some 300 000 Aborigines scattered across the whole of the continent. There has been surprisingly little interest in their long history of settlement and occupance until relatively recently. In part this reflects general European attitudes to the Aborigines, but also the lack of evidence of their activities over a prolonged period of time. Until the 1980s it was thought that this period covered about 30 000 years, with the Aborigines having arrived from south-east Asia via New Guinea, either on foot at a time of lower sea level or as the first people in the world to develop seafaring vessels. However, recent discoveries have extended this to at least 50 000 years (Flannery, 1994). Even so, records of long term occupation in the form of artefacts are scarce, as the Aborigines were a nomadic people, hunting, gathering and fishing, who did not build substantial

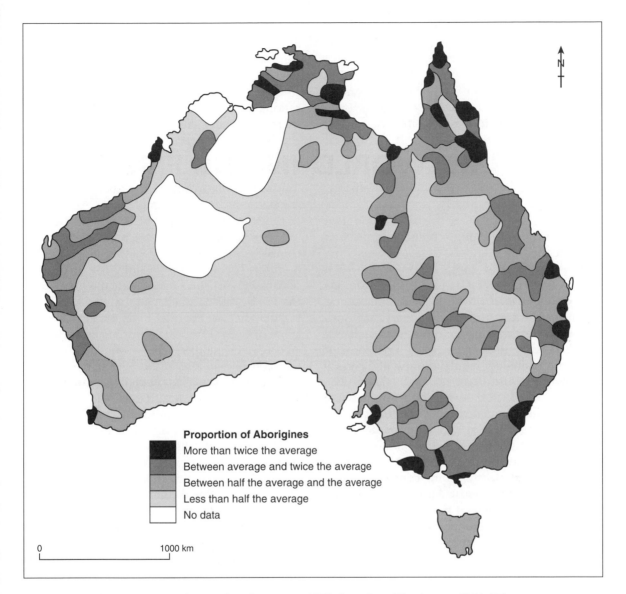

FIGURE 5.1 Aboriginal population distribution, *c.* 1770 (based on Heathcote, 1994: 57)

shelters or utilise more than hand-tool technology to construct weapons or domestic utensils. They did not cultivate or herd livestock, but they did use fire, and it is believed that this had a major impact on the environment by encouraging certain types of species that were fire-resistant or benefited from repeated burning (or 'firestick farming' as it is sometimes termed) (Head, 1989; Pyne, 1991). This favoured the spread of grasslands through the destruction of rainforest and also encouraged an increase in the sclerophyllous forest (Dodson, 1992; Flannery, 1994, Ch. 18).

Nomadism helped to ensure the maintenance of sufficient food supplies whilst helping to keep population numbers low. The main areas exploited were on the periphery of the continent, with the interior largely empty. In effect, the nomads occupied the richest ecosystems and the most heavily wooded areas, living in small extended family groups and speaking a wide variety of languages that reflected the disparate and widely separated nature of these groups. For example, for New South Wales, estimates of the Aboriginal population in 1788 are 40 000 in the coastal valleys; 10 000 on the western slopes of the tablelands; and just 5000 along the Murray–Darling river frontages (Heathcote, 1994: 56–9) (see Figure 5.1).

Prior to European settlement Aboriginal communities were living within territories that were defined by complex rituals of ceremony, dance, symbol and song. Elders within each tribe were the custodians of these traditions and of the tribes' frontiers and boundaries. There may at one time have been as many as 700 tribal groups of between 400 and 500 people each within their own territory, but split into extended family groups of 15 to 20 people (Davis and Prescott, 1992) (Figure 5.2). The key difference between the tribes was the economy they practised, which reflected available resources: hunting tended to predominate in the drier interior areas whilst gathering was more prominent in the forests.

There is evidence for some contact between Aborigines in north-west Australia and Malay fishermen for as long as three centuries before the first European settlement. Here the contact may have reduced Aboriginal numbers by spreading disease to which Aborigines had little or no resistance. However, this effect was far more pronounced when they came into contact with larger numbers of Europeans from 1788 onwards. Disease, the introduction of alcohol and the deliberate killing of Aborigines by Europeans rapidly drove the indigenous people away from the areas of European settlement. The Aborigines' nomadic, hunter–gatherer lifestyle and their unfamiliar physical characteristics to European eyes meant that no consideration was given to their right to occupy land until the 1820s and 1830s when 'protectors' were appointed to help organise the settlement of Aborigines on

reserves separate from colonial society. Some of these were created in the coastal lowlands of south-east Australia, but, as new settlers demanded more land in this area, so the reserves were moved into the interior. Food rations and blankets were handed out, but there was very little effort to improve the condition of the people. Indeed, it was not until the 1930s that a number of remote reserves in central Australia were established where 'traditional' life could at least be preserved to some extent, with access of white Australians to the reserves strictly controlled.

The reserves may have helped prevent the complete extinction of the Aborigines, whose numbers had dropped calamitously as a result of contact with Europeans, many of whom, until the 1950s, 'still half-believed that the Aboriginal population was doomed to extinction' (Powell, 1988: 209). By 1950 official estimates suggested Aboriginal numbers of around 80 000 or between one-quarter and one-sixth of the 1788 total. The reserves also offered opportunities for better provision of health care and education, but often these were rudimentary.

Some reserves run by church organisations were also established in other parts of the country, usually attempting to train Aborigines in techniques of cultivation and pastoral activity. However, this did not enable the development of Aboriginal farming communities, but simply provided a very cheap source of labour for graziers. This use of Aboriginal labour represented some of the first migrations from the reserves, but post-1945 a larger exodus occurred, first to country towns and then the big cities. Government-run hostels (Figure 5.3) and camping reserves near the towns were opened up to help enable the migrants to adjust to urban life. Yet Aborigines were not accorded the same rights as the rest of society; they were banned from public houses and it was an offence to sell them liquor.

The reserves may have preserved certain traditional aspects of Aboriginal life, but the imposed segregation has also promoted a deep sense of grievance in Aboriginal communities. In part this has been fostered by the lack of

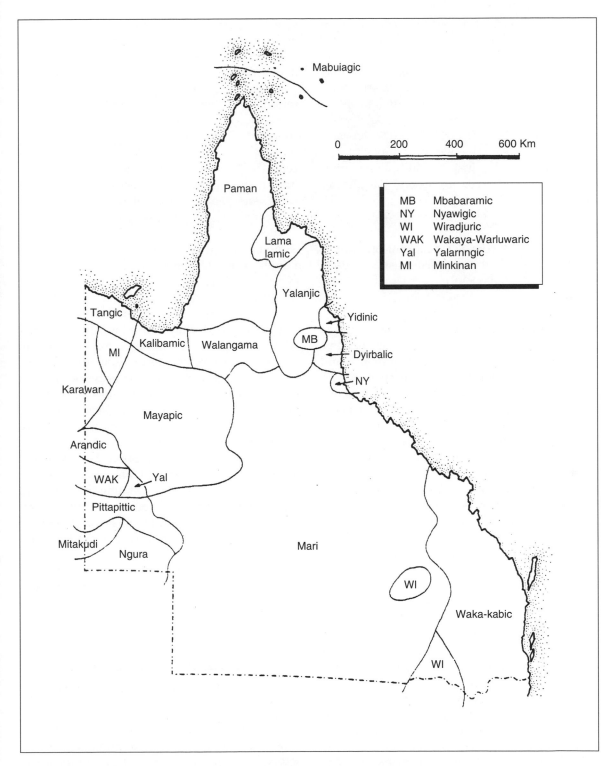

Figure 5.2 Aboriginal language areas in Queensland (Anderson 1985)

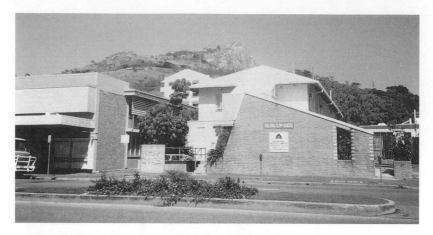

FIGURE 5.3 Government hostel for Aboriginal men at Townsville, north Queensland (with Castle Hill in the background) (photo G.M. Robinson)

decision-making accorded to Aborigines in the running of the reserves. The impact that this has had in squashing Aboriginal initiative and engendering dependence on and subservience to white authority has been long established. The conformity to white norms and beliefs at the expense of Aboriginal tradition and culture meant the undermining of the Aborigines' extended families and tribal groups, group 'ownership' and nomadism.

Gradually, during the 1960s, with Aboriginal numbers rising, there was an acceptance of some Aborigine-controlled management bodies and the recognition of a need for positive discrimination in the provision of health care, education and housing provision. However, positive measures often suffered from the confusion associated with the divisions of responsibility between federal, state and local governments. Federal schemes could be negated by actions at local level. Only in the last three decades has some change occurred through a growing awareness of the plight of Aborigines in the rest of Australian society. In part this reflects the greater visibility of Aborigines as more of their population has become urbanised, but it also reflects a shift in government policy.

In 1967 a federal referendum was held to determine the future status of the Aboriginal population and where responsibility for their affairs should lie. Of those voting, 91 per cent gave the Commonwealth Government the power to legislate on behalf of all Aboriginal people, overriding state authorities if necessary. It also recognised that Aborigines should be enumerated in the national population census. However, the reliance on people to self-identify as Aboriginal or Torres Strait Islanders has proved a continuing problem in interpreting census data. With the granting of full civil rights to Aborigines, the Commonwealth government allocated a budget of A\$10 million to Aboriginal affairs. This was increased twentyfold over the next two decades, though federal budget cuts in 1996 removed A\$400 million from funds allocated to the Aboriginal community.

URBANISATION OF THE ABORIGINAL POPULATION

In 1971 when Aborigines were first enumerated separately in the population census, official reserves covered 7 per cent of the continent, though 90 per cent were in the area

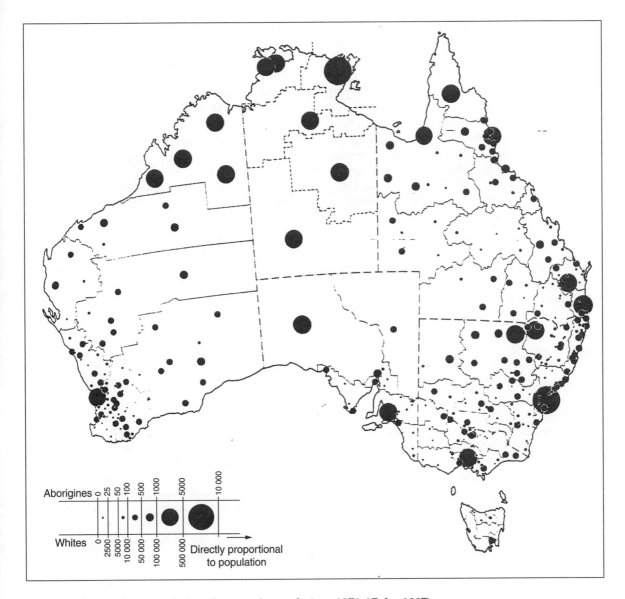

FIGURE 5.4 Distribution of the Aboriginal population, 1971 (Gale, 1987)

described by Heathcote (1994) as 'relict landscapes': in most cases very arid areas. The census recorded an Aboriginal population of 106 290, or 0.83 per cent of the population, of which 43.5 per cent were in urban areas. Most of the remainder were in harsh, remote areas on reserves (Drakakis-Smith and Hirst, 1981; Gale, 1990) (Figure 5.4). High fertility rates explain the rapid population growth post-1945,

though without accurate data in the early part of the twentieth century it is difficult to determine rates accurately (Gale, 1972).

Between 1971 and 1981 numbers rose by 50 000 (+3.2 per cent per annum which is double the national rate of increase). This brought numbers to 160 000 of whom 70 000 were classified as 'full-bloods'. This can be compared with 279 000 Asian-born enumerated in the 1981 census. Life expectancy for Aborigines at birth was only 50 years, and the population was a youthful one, with 41 per cent less than 15 years of age compared with 25 per cent for the country overall (Gale, 1987). Therefore the Aboriginal population pyramid is similar to that of populations in the developing world: there are high birth rates, high death rates, low life expectancy and high infant mortality rates.

The urban migration had first become apparent in the 1960s (Rowley, 1970) and its subsequent continuation has radically altered the distribution of the Aboriginal population (e.g. Gray, 1989; Taylor and Arthur, 1993). The majority of migrants left the reserves with no assets, little money and no facility for obtaining credit. Hence they were only able to obtain very cheap rental accommodation. In the larger cities this meant that they became concentrated in the poorerst inner city areas, the largest concentration being in Redfern in Sydney (Smith, 1980). However, after the initial urbanisation of Aboriginals, a substantial injection of Commonwealth finance into housing for urban Aborigines has promoted dispersion within the major cities. Another factor that has helped to restrict the development of concentrations has been the language differences between groups and strong differentiation of the Aboriginal population into separate groups.

In the remoter areas there has been some retention of traditional ways of life and the population numbers are still sufficient at present to enable communities to survive. However, in the urban areas many Aborigines are marginalised, with high levels of un-

---

Box 5.1 Aboriginal deaths in custody

In 1987 a Royal Commission was established to investigate Aboriginal deaths in prison. Its report, released in June 1991, represented a damning indictment of the treatment of Aborigines, especially by the police, and of the position of Aborigines within Australian society. The Commission surveyed every person taken into custody, Aboriginal and non-Aboriginal, across Australia during the month of August 1988. This revealed that Aborigines accounted for 29 per cent of the population taken into police custody. In Western Australia, where Aboriginal deaths in prison were highest, Aborigines were held in custody at a rate 43 times that of non-Aborigines. In the 1980s, 99 Aborigines died in police custody, the average age at death being 32 years. Of these, 43 had been separated from their natural families in childhood by state authorities or missions who were pursuing the long-established assimilationist policies. Only two had completed secondary schooling and 43 had been charged with an offence by the age of 15. It was revealed that in New South Wales, between 1928 and 1964, 15 000 Aboriginal babies were removed from their natural parents and fostered with white families as part of the policy of forced assimilation. In May 1998 a 'Sorry Day'

commemoration was held as a form of national apology for the programme of forcible removal of children from Aboriginal families.

The report stated that several Aborigines who were put in jail in August 1988 were suffering from ill health but had been jailed because police assumed they were drunk. No evidence of police foul play was found by the Commission, but various 'irregularities' were revealed, including lying under oath, fabricating evidence, incarcerating Aborigines without legal justification and neglecting Aboriginal prisoners. A series of criminal prosecutions of police and disciplinary charges were instigated. The 5000–page report made 339 recommendations and called upon Australia to honour its international obligations over the treatment of its Aboriginal minority. This was followed by calls from the federal government for a 'process of reconciliation' between black and white Australia, to be completed by 2001. A Council for Aboriginal Reconciliation was established to develop this process. However, there remains widespread scepticism amongst Aborigines that there can be a meaningful process of reconciliation whilst there remain such wide disparities in the standards of living between Aborigines and other Australians.

employment and a set of social characteristics and health problems commonly associated with those on the margins of Western society (Box 5.1). For example, the suicide rate amongst Aborigines is six times greater than that for other Australians, the act of suicide being seen by some Aborigines as a calculated act of freedom from daily oppression. Lack of participation in higher education has continued to be a characteristic of Aboriginal society, helping to restrict movement of Aborigines into professional and managerial positions. In 1990 5000 Aborigines were enrolled at Australian universities, a five-fold increase since 1980, but this represents a much smaller proportion of the Aboriginal population than the corresponding figure for Australians as a whole. One attempt to increase the numbers of Aboriginals attending tertiary education has been the creation of an Australian First Nations University, linking Aboriginal centres on 37 campuses in a similar fashion to the Saskatchewan Indian Federated College in Canada, which has operated for over two decades.

ABORIGINAL LAND RIGHTS

Academic writers on Aboriginal issues in the 1960s highlighted social injustice, especially relating to Aboriginal land rights and welfare, and there were several government enquiries on Aborginal issues, for example the Henderson Commission of Inquiry into Poverty, which reported in 1975. Land rights in particular have become part of a significant change in policy from assimilation to self-determination following the 1967 referendum. However, with the exception of the Northern Territory, land rights remained a state responsibility post-1967 as the federal government failed to override strong lobbies from the mining and pastoral industries. As Figure 5.5 shows, it has been the more remote areas of Northern Territory, Western Australia and South Australia where Aborigines have been

most successful in regaining control over large parts of their ancestral territory (Jacobs, 1988; Young, 1996: 59–60).

The battle over land rights was joined in the 1970s as Aborigines and their supporters sought Aboriginal control and ownership of reserves. Following the federal government's refusal in 1972 to grant land rights based on traditional association or to rescind assimilationist policies, a small group of Aborigines mounted a year-long protest outside parliament in Canberra. This raised political consciousness and helped encourage a more positive view towards Aborigines by the incoming Labor government. Battle-lines were drawn against large mining corporations and intransigent state governments keen to obtain revenue from exploiting the mineral wealth of many of the reserves (e.g. Dillon, 1991; Dixon and Dillon, 1990; Howitt, 1991a). The conflict was especially strong initially in the Northern Territory, which in proportional terms had the largest area of Aboriginal reserves and the largest Aboriginal population. An Aboriginal Land Rights Commission exposed the need for stronger legislation to protect land rights, and led to the 1976 Aboriginal Land Rights (Northern Territory) Act, through which 19 per cent of the Territory was converted to freehold status under Aboriginal ownership, and all remaining Crown lands were opened to 'native' claims to be made by the Aborigines themselves through their regional land councils (Dale, 1992; Young, 1992). The latter were created to take responsibility for conducting claims·procedures, with a government-funded Aboriginal Land Fund Commission to purchase leasehold land to meet Aboriginal needs (Baker, 1992; Young, 1996: 61–3). In effect, 35 per cent of the Territory passed into Aboriginal control under various tenures (Table 5.1). However, further Aboriginal acquisition has been restricted by legal constraints associated with negotiations involving the Territory's attempt to obtain full statehood.

An additional constraint has been the continuing conflict over Aboriginal rights to land rich in minerals, including land with uranium

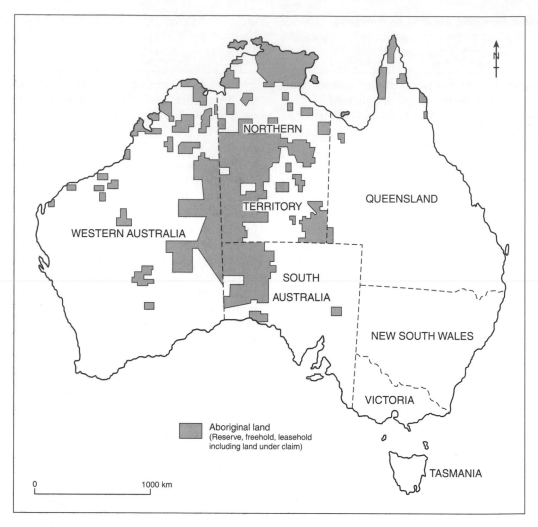

FIGURE 5.5 Aboriginal lands in the mid-1990s (based on Young, 1995: 60)

deposits. This has involved Territory and Commonwealth authorities, private mining interests, the Australian Atomic Energy Commission, Aborigines, pastoralists, the tourism industry, conservationists, peace activists and trade unions. A public enquiry in the mid-1970s, focusing on land rights and uranium extraction, yielded a two-volume report (the Fox Report, 1976/7) containing a wealth of information but not recommending formally against mining interests. It focused especially on the Ranger uranium project in the Alligator River region. Despite strong criticism, the

Commonwealth government retained a large stake (50 per cent equity) in the uranium industry, making it impossible for it to act as an independent arbiter between the mining companies and Aborigines. The report recommended that uranium mining should proceed in the South Alligator River basin, but under strict supervision and control (Young, 1988).

In 1978, in return for agreeing to allow uranium mining to proceed, the Northern Land Council (NLC) received compensation of A$200 000 per annum. A single payment of A$1.3 million was made to local communities

TABLE 5.1 Aboriginal land in Australia

	Aboriginal population		Aboriginal freehold		Aboriginal leasehold		Reserves		Total ATSI land	
	1991 census	%	km²	% of all land	km²	% of all land	km²	% of all land	km²	% of all land
NSW/ACT	70 709	1.18	492	0.06	842	0.10	–	–	1,349	0.17
Vic	16 570	0.39	31	0.01	–	–	–	–	32	0.01
Qld	67 012	2.24	5	0.00	31 990	1.85	95	0.10	33 955	1.97
SA	16 020	1.14	183 146	18.61	507	0.05	–	–	184 157	18.72
WA	40 002	2.52	35	0.00	103 227	4.09	202 223	8.01	305 485	12.10
NT	38 337	21.87	451 219	33.52	26 424	1.96	1	0.00	486 956	36.17
Tas	8683	1.92	2	0.00	–	–	–	–	2	0.00
Australia total	257 333	1.53	634 930	8.27	162 990	2.12	202 363	2.63	1 011 934	13.17

ATSI = Aborigines and Torres Straits Islanders
Source: Young (1995: 61) (based on data from the Heritage Division, Department of Aboriginal Affairs).

by the Ranger Co., and royalties of over 4 per cent on production were agreed with the Aboriginal Benefits Trust Account. The details of the agreement have since been challenged by the NLC, especially the environmental safeguards (Levitus, 1991). Subsequently the creation of Kakadu national park and designation of world heritage status has restricted mining, though some compromises have been reached. For example, in 1986 a 'conservation zone' was excluded from the national park to allow for possible mining developments, especially at Coronation Hill where there are gold, platinum and palladium deposits. Five years later, following a major environmental inquiry and strong support for local Aborigines from the Prime Minister, the conservation zone was listed as a world heritage area. Nevertheless, in late 1998 a team of advisors to the UN World Heritage Bureau was scheduled to visit the site of a proposed new uranium mine at Jabiluka, surrounded by the national park. The site is on Mirrar clan land, awarded under the Native Title Act, but is the subject of a proposal to extract 20 million tonnes of ore over a 28-year period.

In South Australia in 1981 the state government handed over its portion of the Great Central Reserve, covering 10 per cent of the state, to the Pitjatjanjara people. A similar accommodation was reached with some tribes in Western Australia, in contrast to the situation in Queensland where successive state governments have been reluctant to transfer ownership. Whilst the Pitjatjanjara have assumed substantial control over their lands (Toyne and Vachon, 1984), the Tjarutja, whose reserve was further south in South Australia, had a protracted struggle in seeking recompense for the effects of UK atomic bomb testing on their tribal lands at Maralinga between 1952 and 1957. Although people were moved out of the area during the tests, plutonium was left behind on the surface at a time when its contaminating role and long term effects were not adequately appreciated. A Royal Commission that investigated the circumstances of this contamination and grounds for compensation for the Tjarutja recommended that the UK government pay compensation and contribute to cleaning up the area. However, much greater contamination was revealed than originally thought, preventing the Tjarutja from returning to their tribal lands until April 1998 when, after a £39.3 million (A$86.5 million) clean-up, land was formally handed back.

During the 1980s Aboriginal land claims escalated (e.g. Larritt, 1995), with a number of substantial ones being made in New South Wales and Queensland. Perhaps the most significant claim was that launched by Eddie Mabo in 1983 on behalf of the Meriam people on Murray Island, 160 km off northern Queensland in the Torres Straits. The claim argued that, when the island was taken over by the state of Queensland in 1879, Mabo's people had never validly surrendered their rights to it. In 1992 the high court found that the Meriam people were entitled to ownership and occupation of the island. This ruling implied that there could be an extension of 'native title' to the rest of the country in specific cases where Aborigines could demonstrate a continuing connection with the land (Reynolds, 1996). This implies that acceptance of the concept of *terra nullius* is invalid, that is the continent was not unoccupied prior to European settlement and that Aborigines have rights to compensation for losing their occupance of the land. This raises a number of crucial questions regarding the extent of native title, recognition of Aboriginal occupance, and the determination of appropriate compensation. For example, can native title override freehold or Crown leasehold? Are Aborigines entitled to royalties from wealth drawn from the land and, if so, how far back in time can such claims be extended?

These are issues that are currently provoking great debate, bringing Aboriginal land rights to the forefront of public attention. Initially the Mabo decision was the catalyst for a number of land claims to be brought forward. For example, a claim for 1.1 million ha of the Simpson Desert national park in Queensland was just one of 15 claims in the state lodged under state land rights legislation passed by the state parliament in June 1991. This legislation allows for up to 2.9 per cent of Queensland to be made available for claim, including 2.3 million ha in national parks but excluding

Crown land in urban areas. Other claims include one made in 1992 for Brisbane city centre, which was not proceeded with, and in 1993 in New South Wales one by the Wiradjuru people laying claim to one-quarter of the state.

A federal government White Paper on the subject in 1993 proved inconclusive as it was deemed to be trying to appease all sides of the land rights debate. However, with 1993 designated by the United Nations as 'The Year of the World's Indigenous People', impetus was given to further initiatives from interested parties. For example, Aboriginal Legal Services, based in Sydney, proposed suing the UK government for trespass and crimes against humanity, which it has since pursued via the International Court of Justice. The legacy of harsh treatment by the colonial Europeans remains a key ingredient in continuing land rights claims via new legislation passed in 1993, in the form of the Native Title Act (Box 5.2), and more recent controversial legal rulings (Box 5.3).

The debate regarding Aboriginal land rights is progressing, with the courts continuing to grant some Aboriginal claims. For example, for the first time rights to sea were granted to Aborigines in July 1988 when five Aboriginal groups were awarded native title rights to the sea surrounding their traditional lands on Croker Island off the Arnhem Land coast of the Northern Territory. However, this does not give them power to exclude others, such as anglers, commercial fishing and mining operators. Moreover, just two days later the Senate passed the controversial land rights legislation proposed by Prime Minister Howard, thereby severely limiting the conditions under which Aborigines can make claims for native land rights over land on which pastoral leases apply. This now places the onus on state governments to mediate between claims to pastoral land made by Aborigines and graziers. Aboriginal groups have expressed the fear that in both Queensland and Western Australia, which have substantial areas under pastoral leases, past unsympathetic attitudes to Aborigines will be repeated, especially if

Box 5.2 The Native Title Act

Paul Keating, Prime Minister from 1992 to 1996, coupled the idea of an Australian republic to reaching 'an agreement' with the Aborigines as a central feature of his vision for the country. To this end he supported legislation, passed in December 1993, enshrining Aboriginal land rights in a Native Title Act. This was highly contentious, though, and was only passed by the upper chamber by two votes from the Green Party. It allowed Aboriginal claims referring to approximately one-tenth of the country's land mass. It also validated almost all the previously issued land titles whilst establishing a system of tribunals to give native title to people who can show customary use over land. However, Aborigines will be obliged to respect mining, ranching and other leases on any property to which they win claim (Dodson, 1996; French, 1996). They will be able to argue against commercial projects, especially mining, but will have no right of veto. Following passage of the Act, a proposed mining project at Coronation Hill in the Northern Territory was rejected. This was where Newcast Mining wanted to mine for gold, platinum and palladium, but on a sacred site of the Jaiwayn tribe.

Under the Act, state governments will retain the final word on land use. This is likely to produce the sharpest conflict in Western Australia where the state government had enacted legislation extinguishing all native title and replacing it with other rights to land use. The Native Title Act will have precedence over this. The Act was opposed by the Liberal/National opposition on the grounds that it was likely to damage the mining sector and foreign investment. The Liberal/National attitude, now that it is the government, will be vital to the ongoing development of Aboriginal land rights and to the

extent to which Aborignal grievances are addressed.

Effective from January 1, 1994, the Act attempted to turn the Mabo case into legislation. It has four key elements:

*   recognition and protection of native title;
*   validation of all land acts before 1 January 1994, if they had been invalidated by native title;
*   a 'just and practical' regime to govern acts that occur after 1 January 1994;
*   a 'rigorous, specialised and accessible' tribunal and court processes for determining claims to native title.

Key problems include lengthy timetables for negotiation and arbitration, difficulties of dealing with multiple and rival native title claims, and unclear guidelines on who can make an agreement on behalf of native title holders. State governments, notably Queensland, have tended to ignore it (Mercer, 1997).

By the end of 1996 Aborigines owned or controlled around 15 per cent of Australia's land mass and had native title claims over approximately 20 per cent. The debate about Aboriginal land rights has become even sharper, though, as ramifications of the Mabo ruling are revealed. More details regarding the mistreatment of Aborigines in the past have added to the sense of grievance felt by many Aborigines, and have been coupled with the ongoing debate about the need for greater recognition of Aboriginal rights in the new millennium and in a republican Australia. However, the stance taken by the federal government over land rights has been the prime focus for argument, emphasising divisions within Australian society.

'concessions' to Aborigines could be seen as conflicting with economic development. There are concerns that the legislation will put an end to negotiations between mining companies and Aborigines whereby, since 1994, in recognition

of the existence of native title, some companies have negotiated with the Aboriginal land councils, granting them royalties, jobs, a say in mine construction, and protection of sacred sites.

Box 5.3 The Wik Ruling

The issue of whether a pastoral lease, which is essentially rented Crown land, was claimable by Aborigines after the Mabo ruling was tested by the Wik people of Cape York, northern Queensland. The Wik people claimed that, in spite of the grant of pastoral leases over their traditional lands, they still held native title rights. They were defeated in the federal court, but appealed to the high court, where in December 1996 four of the seven judges held that a pastoral lease did not necessarily extinguish native title, though in conflicts over land rights between Aborigines and pastoralists, the latter's rights would always prevail. However, this cast doubts upon the rights of farmers occupying pastoral holdings (amounting to 42 per cent of the continent), especially regarding land uses beyond the scope of the pastoral lease. For governments there was a fear that they could be forced to pay compensation if native title rights on pastoral

leasehold land are affected. The federal government's response in 1997 was the issuing of a 10-point plan by Prime Minister Howard (Table 5.2), which left Aborigines protesting at its perceived threat to native title whilst antagonising pastoralists who still felt that pastoral leases were under threat. It is noteworthy that the pastoral lobby has featured strongly in support of the anti-Aboriginal rhetoric of the newly formed One Nation Party.

Prime Minister Howard's aim is to implement the 10-point plan as an amendment to the Native Title Act. It will retain native title but raises the threshold test for Aboriginal claimants and will provide a list of land titles which automatically extinguish native title. More power will be given to state governments to extinguish native title on all the different types of leases that were granted with the intention of giving their owners exclusive possession of the land.

THE MAORI SETTLEMENT OF AOTEAROA

The first human settlers in the islands later to be named New Zealand by the first European explorers arrived by sea from eastern Polynesia (Irwin, 1992). There is still dispute regarding the date of their arrival and over their relationship to later Polynesian arrivals, with the picture of New Zealand's pre-history being continuously modified. For some time the initial settlers were termed the Moriori (a name now given to Polynesians settling on the remote Chatham Islands) or archaic Maori or Moa-hunters (Davidson, 1984). They were long regarded as hunter–gatherers who arrived perhaps as early as the eighth century AD, though the weight of opinion currently favours a date in the tenth century (Anderson, 1991) perhaps even as late as the thirteenth century (Crawford, 1993). They settled throughout New Zealand hunting the moa (*Dinornis*), a giant flightless bird over 3 m tall, with all 11

species eventually hunted to extinction, unlike its much smaller relative, the kiwi (*Apteryx*). However, evidence now points to early settlement combining a diversity of subsistence economies with considerable local variation that included hunting moa, fishing, killing birds and seals, gathering shellfish and growing vegetables (Anderson, 1989). Popular tradition believes that the main wave of Maori settlers arrived in the land they termed Aotearoa ('land of the long white cloud') in a series of voyages from the tenth century onwards. Some legends refer to their main period of arrival as being the mid-fourteenth century, travelling from eastern Polynesia (the Society, Marquesas and Cook Islands) in a great fleet that landed in the Bay of Plenty. Today, a series of smaller arrivals is postulated, which may have become two distinct groups quite quickly: with more emphasis on hunter-gathering in the less hospitable South Island, and cultivators in North Island where classic Maori civilisation flourished (Starzecka, 1996). However, Maori society was tribal, with signif-

TABLE 5.2 Prime Minister Howard's 10-point plan

1. *Validation of acts/grants*
   Legislation will ensure that validity of acts or grants made on non-vacant Crown land between passage of the Native Title Act and the Wik decision is put beyond doubt.

2. *Extinction of native title on exclusive tenures*
   States and territories will be able to confirm that exclusive tenures such as freehold, residential, commercial and public works in existence on or before 1 January 1994, extinguish native title. Agricultural leases will also be covered if exclusive possession was intended when they were granted.

3. *Government services*
   Impediments to the provision of government services to land on which native title may exist will be removed.

4. *Native title and pastoral leases*
   As in the Wik decision, native title rights over pastoral leases and agricultural leases not covered under (2) above will be permanently extinguished where rights are inconsistent with those of the pastoralist. Activities relating to primary production will be allowed on pastoral leases including farmstay tourism, even if native title exists, provided the dominant purpose is primary production.

5. *Statutory access rights*
   Where registered claimants can demonstrate that they currently have access to pastoral lease land, continued access will be legislatively confirmed until the native title claim is determined.

6. *Future mining*
   For mining on vacant Crown land there would be a higher registration test for claimants, no negotiations on exploration, and only one right to negotiate per project. For mining on 'non-exclusive' tenures such as current or former pastoral leasehold land, right to negotiate would continue to apply until a state or territory provided a regime acceptable to the Commonwealth which included compensation taking account of co-existing native title rights.

7. *Future development*
   On vacant Crown land outside towns and cities there would be a higher registration test to access the right to negotiate, but this right would be removed in relation to the acquisition of native title rights for third parties for the purpose of government-type infrastructure. For compulsory acquisition of native title rights on other 'non-exclusive' tenures such as current or former pastoral leasehold land and national parks, the right to negotiate would continue to apply until a state or territory provided a statutory regime acceptable to the Commonwealth which included compensation taking account of co-existing native title rights. Future actions for the management of any existing national park or forest reserve would be allowed. A regime to authorise activities such as the taking of timber or gravel on pastoral leases would be provided.

8. *Water resources and airspace*
   The ability of governments to regulate and manage surface and subsurface water, offshore resources and airspace, and the rights of those with interests under any such regulatory or management regime would be put beyond doubt.

9. *Management of claims*
   In relation to new and existing native title claims, there would be a higher registration test to access the right to negotiate, amendments to speed up handling of claims, and measures to encourage the states to manage claims within their own systems. A sunset clause within which new claims would have to be made would be introduced.

10. *Agreements*
    Measures would be introduced to facilitate the negotiation of voluntary but binding agreements as an alternative to more formal native title machinery.

Based on Woodford (1997).

icant diversity between the tribes (*iwi*). As shown in Figure 5.6, 26 different tribal areas can be recognised today.

The lack of animals transported by the Maori from eastern Polynesia (apart from the Polynesian dog and rat) suggests that only a

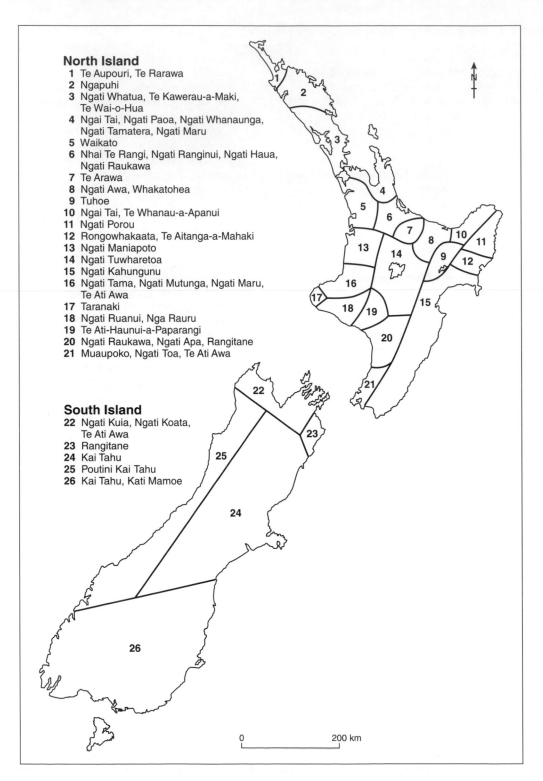

North Island
1 Te Aupouri, Te Rarawa
2 Ngapuhi
3 Ngati Whatua, Te Kawerau-a-Maki, Te Wai-o-Hua
4 Ngai Tai, Ngati Paoa, Ngati Whanaunga, Ngati Tamatera, Ngati Maru
5 Waikato
6 Nhai Te Rangi, Ngati Ranginui, Ngati Haua, Ngati Raukawa
7 Te Arawa
8 Ngati Awa, Whakatohea
9 Tuhoe
10 Ngai Tai, Te Whanau-a-Apanui
11 Ngati Porou
12 Rongowhakaata, Te Aitanga-a-Mahaki
13 Ngati Maniapoto
14 Ngati Tuwharetoa
15 Ngati Kahungunu
16 Ngati Tama, Ngati Mutunga, Ngati Maru, Te Ati Awa
17 Taranaki
18 Ngati Ruanui, Nga Rauru
19 Te Ati-Haunui-a-Paparangi
20 Ngati Raukawa, Ngati Apa, Rangitane
21 Muaupoko, Ngati Toa, Te Ati Awa

South Island
22 Ngati Kuia, Ngati Koata, Te Ati Awa
23 Rangitane
24 Kai Tahu
25 Poutini Kai Tahu
26 Kai Tahu, Kati Mamoe

FIGURE 5.6 Simplified tribal map of New Zealand (based on map by Neich and Taua in Starzecka, 1996)

FIGURE 5.7 A traditional carved *whare runanga* (meeting house) facing the *marae* at Rotorua (photo: G.M. Robinson)

small number of migrants made the journey. Furthermore, many of their traditional crops would not grow in the colder climate of Aotearoa, e.g. banana, breadfruit and sugar-cane. The introduced plants that survived to European times were the kumara, yam, taro, bottle gourd (*Lagenaria siceraria*), mulberry and a species of cabbage tree. In the unfamiliar cold climate the Maori adapted the growing of kumara to survive winter frosts, and they developed weaving and other craft skills to provide clothing and shelter, but did not have pottery nor any kind of metal (Belich, 1996: Chapters 2 and 3; Clark, 1949). There is dispute over how destructive the Maori may have been of the environment. For example, Cumberland (1949; 1962) argued that they destroyed large amounts of forest, especially in the east of the South Island, creating substantial areas of tussock grassland and scrub. One estimate is that by 1500 a few thousand Maori had destroyed 3.2 million ha of forest, reducing it to open scrub and grassland (Figure 4.1) (Crawford, 1993: 242). However, the extent of destruction in the South Island is disputed, with some regarding tussock as the climax

vegetation. Nevertheless, there is evidence of soil erosion and extinction of some plants and animals associated with Maori occupance. As with the Aborigines, although the Maori lived as an integral part of their environment and utilised plants and animals for every facet of their lives (incorporating 'nature' into their myths and beliefs), they were not merely passive components of that environment but were active agents of change and even destruction.

When Europeans first started to settle in New Zealand in the early nineteenth century, competition for land and reserves between Maori tribal groups (*iwi*) had led to frequent conflict, with huge fortifications (known as *pa*) erected on defensible positions, e.g. Auckland's One Tree Hill and the tops of the other small volcanoes that are a feature of the area. There is evidence that there was some cannibalism as part of the inter-tribal fighting. The Maori of the North Island were more prosperous, with subsistence production of kumara and taro and use of local rivers and seas to supply fish. Carving and weaving were prominent features of life (Figure 5.7). Estimates of the population

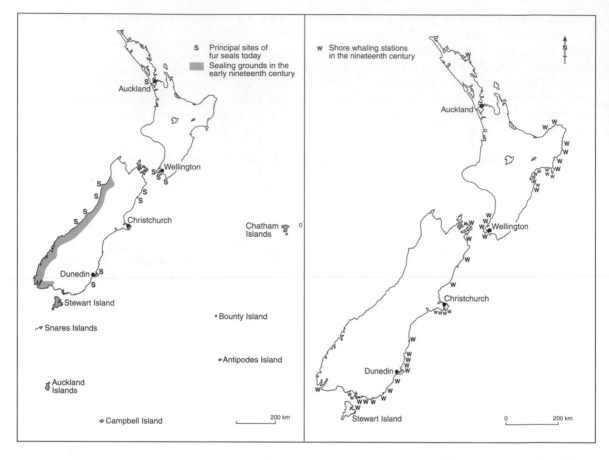

FIGURE 5.8 Sealing grounds and shore whaling stations, New Zealand, *c.* 1820s (based on Ell, 1995)

at the time of Captain Cook's voyage in 1769 vary from the oft-cited 100 000 to 250 000 or even higher, of whom at least 90 per cent were in the North Island (Pool, 1991). The most commonly accepted total is around 90 000 (Belich, 1996: 178).

The contact with Europeans proved disastrous to Maori development. The first Europeans who came to New Zealand were the whalers and sealers, using the country as an outpost for their main bases in Australia (Figure 5.8) (Ell, 1995). However, by the 1830s there were a series of whaling stations around the New Zealand coast. The country was also used by the Royal Navy as a source of supply of timber, especially using kauri for masts, and flax (*Phormium tenax*) for canvas, rope and

cordage. Gradually some of those involved in this 'economy of plunder' became permanent settlers, a principal focus being Kororareka (now called Russell) in the Bay of Islands, described in the 1820s by the country's first surveyor-general as having 'a greater number of rogues than any other spot of equal size in the universe' and being 'the very seat of Satan'. The permanent settlements were associated with the debasing of the Maori. Maori women were sold to the Europeans (termed 'Pakeha' by the Maori, probably referring to their white skin, like the flesh of turnips) in exchange for gifts; disease spread rapidly; land was seized or purchased for meagre amounts; and there was a rapid breakdown of traditional patterns of living. The use of European weapons in conflicts between Maori tribes further contributed to the destruction of Maori population and culture, the deeds of the tribal chief Te Rauparaha in the 1820s being especially bloodthirsty. The production and trading of European vegetables to supply the early Pakeha settlements disrupted the traditional subsistence economy and, more importantly, the Maori lost faith in the strength of their own gods so that Christianity was spread rapidly amongst the tribes.

THE TREATY OF WAITANGI (1840)

The jurisdiction of the New South Wales courts was formally extended to cover New Zealand in 1823, but this expression of authority was not accompanied by any means of enforcing law and order. In 1832 James Busby was sent from New South Wales to be the British resident in New Zealand and 'to protect Maori from British adventurers ... with the aim of curtailing lawbreakers, reassuring settlers and traders, and meeting the express wish of the Maori people that peace be preserved' (Martin, 1992). Again, though, no means of enforcing this protection was provided. In 1835, from his residence at Waitangi in the Bay of Islands, he persuaded 35 Maori chiefs to proclaim themselves the United Tribes of New Zealand, asking for protection from the British against claims of rival nations and perpetrating the fiction of the Maori as being a fully sovereign people rather than a collection of independent tribes. Maori had already expressed concern over possibilities of a scramble for land involving various countries and, in 1831, 13 northern chiefs had petitioned for greater protection by the British Crown.

By 1837 the British Colonial Office was receiving reports of inter-tribal conflict, crimes by British subjects and disagreements between Maori and Pakeha which threatened lives and trade. At the same time more British emigrants were planning to head for New Zealand. Given this situation, the British government decided

TABLE 5.3 A recent translation of the Maori version of the Treaty of Waitangi (1840)

The First
The Chiefs of the Confederation and all the chiefs who have not joined that Confederation give absolutely to the Queen of England for ever the complete government over their land.

The Second
The Queen of England agrees to protect the Chiefs, the sub-tribes and all the people of New Zealand in the unqualified exercise of their chieftainship over their lands, villages and all their treasures. But on the other hand the Chiefs of the Confederation and all the Chiefs will sell land to the Queen at a price agreed to by the person owning it and by the person buying it (the latter being) appointed by the Queen as her purchase agent.

The Third
For this agreed arrangement therefore concerning the government of the Queen, the Queen of England will protect all the ordinary people of New Zealand and will give them the same rights and duties of citizenship as the people of England.

Source: Orange (1990).

it was imperative to secure Maori acceptance and co-operation in establishing New Zealand as a British colony. In 1840, under orders from London to negotiate the transfer of sovereignty to the British Crown, Captain William Hobson was dispatched from New South Wales to join Busby. With the help of missionaries, a treaty was drafted in both English and Maori, and at Waitangi, on 6 February, a group of Maori chiefs signed the treaty. This was the Treaty of Waitangi (Table 5.3), under which the English version refers to Maori ceding sovereignty to the Crown in return for retaining peaceful possession of their lands and 'all the rights and privileges of British subjects'. The Crown took the sole right to buy from the Maori, though this Crown pre-emption was removed in 1862.

In the Maori text, chiefs ceded not sovereignty but *kawanatanga*, meaning governorship or governance; their *rangatiratanga*, or chieftain-ship, was guaranteed. As the latter had been used in the 1835 Declaration of Independence to refer to New Zealand's independence, as acknowledged by Britain, it is reasonable to assume that the Maori believed that their sovereign rights were being confirmed in return for a limited concession of power (Kawharu, 1989; 1992). The differing interpretation between *rangatiratanga* and the word 'possession' in the English text, has been a crucial element in more recent considerations of the meaning of the treaty (Orange, 1990; Sorrenson, 1992).

The signing of the treaty involved 40 Maori signatories, with a total of around 400 Maori in attendance. Subsequently the treaty document received 500 signatures as it was taken around the country to inform other Maori and to obtain their support. Significantly, 39 chiefs in the Waikato refused to sign and this subsequently became an area of Maori opposition to incursions by Pakeha settlers.

The treaty had common benefits for both sides. These included the proper control of the growth of trade, control of how and where new immigrants would settle, a common legal system and control of land sales. The Maori believed that in exchange for acknowledging the rights of Pakeha to settle in Aotearoa, their natural rights as the original occupants would be respected and upheld. On the British side, the treaty had secured sovereignty over the territory and land for new immigrants (Kawharu, 1989).

PAKEHA 'LOSS OF MEMORY' AND THE WAITANGI TRIBUNAL

The Crown's ability to control the purchase of land from the Maori was cast into doubt very quickly, as new settlers from Britain clashed with local Maori in a land dispute on the Wairau Plains near Nelson in 1843. There was further trouble in Northland in 1845, but disputes were only sporadic until 1860 when a

combination of factors on both sides led to more sustained fighting, subsequently termed the Maori Land Wars or, by more recent scholars, the New Zealand Wars (Belich, 1986; 1996: Ch. 10).

By the 1850s the contact with the Pakeha had contributed to a fall in the Maori population to around 60 000 according to the most reliable estimates at the time (Belich, 1996: 177). Undoubtedly this contributed to a reduced ability to resist any Pakeha incursions in defiance of the treaty. However, in the late-1850s the Maori King Movement (*Kingitanga*) emerged as a significant force dedicated to the retention of Maori-owned land. It was not avowedly anti-Pakeha, but was more concerned with redressing the government's neglect of the treaty and of Maori welfare. The movement's formation coincided with attempts by Pakeha settlers to expand pasture land for dairying in the Waikato, Manawatu and Taranaki, with the government being urged to acquire more Maori land.

In the fighting, although the Maori's mixture of guerrilla warfare and pitched battles achieved some successes (see Binney, 1995), the tribes were not unified, with some Maori fighting on the government's side against traditional tribal enemies. By the time that fighting had ceased in 1881, large areas of fertile land in the Waikato, Taranaki and the Bay of Plenty had been confiscated. It is estimated that the Maori lost 1.2 million ha as a result of the wars and the removal of Crown pre-emption in 1862. The removal of pre-emption also marked the point at which the European population first exceeded that of the Maori. A further 2.8 million ha had been sold by the Crown before 1862, and a similar amount was sold between 1862 and 1892, leaving just 4.45 million ha in Maori hands, one-quarter of which was leased to Pakeha settlers. In effect, much of the land remaining under Maori control was the least attractive to Pakeha because it offered the least opportunity for economic exploitation (Tutua-Nathan, 1992) (Figure 5.9).

Once the administration of the country passed to a settler government, under the

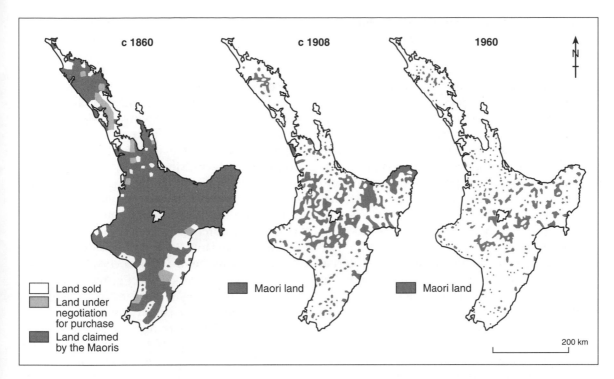

FIGURE 5.9 The loss of Maori lands, North Island, New Zealand (based on Orange, 1990)

Constitution Act of 1852, Pakeha respect for the treaty diminished rapidly and laws ignored its existence, e.g. the 1863 New Zealand Settlers Act. For a period of 123 years New Zealand courts continued to regard the treaty as having no legal status in domestic law. During this time the amount of land in Maori hands was reduced from 27 million ha to 1.3 million ha (3 per cent of the country) (Douglas, 1984). This long period has been referred to by the historian Claudia Orange (1990) as 'the Pakeha loss of memory'. Pakeha 'reawakening' began in 1975 with the passing of the Treaty of Waitangi Act (ToWA), establishing the Waitangi Tribunal which was to investigate Maori claims against the Crown dating from the tribunal's inception. In 1995 the tribunal's budget was $3.25 million, three-quarters of which was spent on claims processing and support services.

In 1985 the ToWA was amended to permit claims dating back to the signing of the Treaty of Waitangi in 1840. Two years later the court of

appeal was required to consider the principles and relevance of the treaty to contemporary New Zealand. This was encouraged by the tribunal's chairman in the early 1980s, Chief Justice Edward Durie, a Maori, under whom a series of influential reports and recommendations was produced (Pawson, 1991). The ruling by the five judges of the court confirmed that the treaty had established a partnership between Maori and Pakeha, and that it was the duty of both sides 'to act reasonably and in good faith' towards each other. So the Treaty of Waitangi has re-emerged as a living document and continues to act as a national symbol of unity and understanding between the two cultures.

Orange (1990) singles out four key factors that led to this 'rediscovery' of the treaty's significance:

(a) the emergence of 6 February as a national day to be celebrated. In 1932 the Governor-General bought the Treaty House at

Waitangi for the nation as a symbol of nationhood. In itself this may not have had much impact on the population as a whole. However, from the late 1960s National Day celebrations at Waitangi were televised, helping to raise consciousness of the significance of the events at Waitangi in 1840;

(b) the role of Maori within the Labour Party. In the 1930s the Labour Party established an alliance with the Maori Ratana Party. This encouraged Labour to be more sympathetic to Maori views and, during the postwar period, this developed into a stated commitment to legislative recognition of the treaty;

(c) the growth of Maori aspirations. External events such as the United Nations' work on human rights for indigenous peoples in the 1960s had an influence upon those Maori leaders not prepared to accept the status quo;

(d) urbanisation of the Maori population. The post-war 'urban drift' of the Maori brought their problems to the attention of a larger proportion of New Zealand society (Kawharu, 1968). Furthermore, Maori population numbers had never been reduced to such a small proportion of the overall population as had been the case with Aboriginal Australians. Numbers reached their nadir in 1896 (just under 40 000), but then grew rapidly from the early 1920s, with birth rates double that of the Pakeha population (though death rates were also twice as high) (Grey, 1994: 300–1). This disparity narrowed from the 1960s and has been all but eliminated. Nevertheless, the 1996 census recorded a total of 523 374 Maori (or 14.5 per cent of the total population: 273 438 Maori as the sole ethnic group and 249 933 Maori with other ethnic groups), over 80 per cent in urban areas. The economic and social disparities between Maori and Pakeha were clear and brought about a vocal Maori lobby for improvement and for some redress regarding long-held grievances over their treatment at the hands of Pakeha.

One outlet for this voice was a march on the capital in 1976 demanding that the government pay attention to Maori land claims. Another, building slowly from the early 1970s, was a Maori cultural rebirth in which the Maori language became a symbol for a rediscovery of Maori culture involving art, crafts and literature. This reversed a longstanding trend: between 1867 and 1927 the proportion of children who could speak the Maori language, living with Maori parents, fell from 90 per cent to 26 per cent, and for much of the first three-quarters of the twentieth century Maori children were punished for speaking Maori in schools (Walker, 1992).

In the 1980s Maori language nests (*kohanga reo*) were established in kindergartens, and Maori-language radio programmes were launched to help encourage tribal cohesion and renew pride in Maoridom. Maori was formally recognised as an official language of New Zealand in 1987 in the Maori Language Act when a Maori Language Commission was also established. This was part of a tribunal recommendation in an attempt to promote Maori as a living language. At the time of the Act, 80 per cent of Maori speakers were over 35 years of age, with 40 per cent over 55 years. There were 10 000 fewer fluent speakers of Maori in 1990 compared with 1980.

Some of the initial claims to the Waitangi Tribunal were substantial. For example, the Ngai Tahu sought a full honouring of the contracts by which early colonial governments obtained most of the South Island in return for guarantees of reserves adequate to sustain the various sub-tribes and clans. The tribe claimed that the guarantees were never honoured and that compensation paid by the government in 1946 was too small. In 1998 the 546–page Ngai Tahu Claims Settlement Bill recorded the Crown's apology for breaches of the Treaty of Waitangi, notably food-gathering rights (*mahinga kai*), which had ensured that Ngai Tahu would become marginalised. The Bill formalised payment of £56.6 million in compensation.

The tribunal ruled that it would not recommend the extinction of title that was valid under European law. Furthermore, the government declared that it would not be bound by the tribunal's recommendations if the costs of compensation were too high. In effect, the intention has been to pay some compensation, to return certain types of lands, such as those held by the Crown (which could include substantial parts of national parks in the South Island), to make special grants and to restore privileges, especially regarding rights of access and use of fishing grounds. Indeed, fishing rights have been one of the tribunal's major considerations, and were brought into sharp focus in 1986 when many Maori treaty rights to fish were lost when the government introduced a quota system of management for the inshore fishery, intending to give fishers property rights to defined quantities of catch.

Maori felt that these rights were primarily being conferred on Pakeha and represented a breach of the treaty. In arguing this point of view, Maori claimed half of the total one billion dollars per annum income from fisheries, raising the question of just what are 'Maori fisheries' in the treaty and which tribes should be compensated. In the protracted negotiations one problem was the lack of clear definition of the Maori 'fisheries' referred to in the treaty. Furthermore, different tribal groups have conflicting and overlapping claims.

From 1985, government has also been forced to write the principles of the treaty into its legislation. The most celebrated case was with respect to the 1987 State-Owned Enterprises (SOE) Act under which Crown assets were transformed into nine corporations set up to run what had previously been departmental trading operations. However, the appeal court ruled that where this contravened the principles of the treaty then the tribunal should have final jurisdiction. This decision was upheld over the sale of the vast forests owned by the Forestry Corporation (an SOE). Maori were guaranteed the proceeds from the sale of the forests or leases or cutting rights if the tribunal

upheld particular claims. One outcome is that the tribunal can no longer inquire into commercial fisheries. This restriction results from a deed of settlement signed between Maori and the Crown in September 1992, representing a permanent settlement of Maori fishing claims. The tribunal can still hear claims from Maori in respect of non-commercial fishing and traditional treaty rights in fishing. For example, one early notable case involved the re-siting of a waste outfall from a synthetic petrol refinery in north Taranaki to preserve traditional Maori shellfish beds (Cant, 1990).

The settlement over fishing rights distinguished between commercial and non-commercial fishing rights, with the former settled through a NZ$150 million payment, part of which has assisted Maori in acquiring a half share in Sealord Products, a multinational joint-venture company holding approximately one-quarter by volume of the country's commercial fishing quota (Memon and Cullen, 1996). In addition, Maori will receive one-fifth of quotas for all new species as they are introduced to the quota management system and they will have a guaranteed participation on those statutory bodies concerned with fisheries management. Overall, Maori now control one-third of New Zealand's inshore and deepwater fisheries.

The tribunal reports its findings and any recommendations to the Minister of Maori Affairs and to other ministers with an interest in the claim. The Crown then considers the tribunal's report and may enter into negotiations with Maori to settle the claim. The tribunal can only make recommendations, except for four specific circumstances: SOE land or former-SOE land; Crown forestry lands (but not the forestry on it); railway lands; and certain educational lands. In these cases legislation has established a protective mechanism for the return of land to Maori by entering 'memorials' on the land titles. The memorials notify prospective purchasers that the land may be returned to Maori owners upon a binding recommendation of the tribunal. The claimants must have been, or will be, prejudicially affected by:

- legislation or ordinances;
- regulation, orders or proclamations;
- policies or practices adopted or proposed by, or on behalf of, the Crown;
- any other act or a mission by, or on behalf of, the Crown, and which is inconsistent with the principles of the treaty (Sorrenson, 1995; Stokes, 1993).

In the North Island one of the largest claims has been made by the Tainui in respect of rich farmland in the Waikato which was confiscated in the Land Wars. The Tainui claim that this loss of land breached the treaty and that previous commissions and committees of enquiry returned insufficient land and provided too little compensation. A recent example of the tribunal's rulings is the order placed on the government to return over NZ$6 million worth of land confiscated from Maori in the 1960s to create the Turangi hydro village near Lake Taupo. Subsequently, some of this land used to site housing for construction workers on the project has been sold to private owners. The ruling in favour of the Ngati Turangituka sub-tribe includes monetary compensation of at least NZ$1 million to the claimants.

Given that compensation at today's land values would bankrupt the Exchequer, the claims are being settled through a mixture of some monetary compensation, return of some land held by the Crown (such as that in national parks but not leased by the Crown to farmers), and special grants and privileges. However, despite receipt of funds following tribunal rulings, many Maori are still amongst the poorest within New Zealand society (see Chapter 11).

FURTHER READING

The title of this Chapter is taken from Young (1995). Australian Aboriginal pre-history and the impacts of Aborigines on the environment are covered in Dodson (1996), Flannery (1994), Head (1989), Kohen (1995), Lane (1997) and Pyne (1991). Overviews of the society and way of life of Aborigines pre-1788 are contained in Mulvaney and White (1987). The destructive nature of European contact is considered by Gale (1972; 1987; 1990) and Rowley (1970). Land-rights issues are dealt with by Baker (1992), Howitt (1992), Howitt and Jackson (1998), Jacobs (1988), Mercer (1997) and Young (1992; 1993). See also the essays in Connell and Howitt (1991) and Howett *et al.* (1996). Dale (1992) discusses Aboriginal councils, whilst self-government is addressed by Larritt (1995) and Reynolds (1996). The legal background to Native Title is addressed by Attwood (1996) and French (1996). Maher (1998), Taylor (1998) and Taylor and Arthur (1993) have written about Aboriginal urbanisation. See also the provocative essay by Jacobs (1998).

For the pre-European history of New Zealand see Irwin (1992), Clark (1949), Cumberland (1949; 1962) and Crawford (1993). The history of early contacts between Maori and Pakeha and the ensuing conflicts are covered in detail by Belich (1986; 1996) and the compendium edited by Rice (1992). See also the tremendous work on the Maori chief Te Kuiti by Binney (1995).

Further consideration of the Treaty of Waitangi is given by Brookfield (1995), Kawharu (1989; 1992), Martin (1992), Owens (1990) and of the Waitangi Tribunal by Cant (1990), Cant *et al.* (1993), Memon and Cullen (1996), Pawson (1992; 1999), Sorrenson (1995) and Stokes (1993). The urbanisation of the Maori is covered by Kawharu (1968) and Pool (1991). The classic account of the Pakeha 'loss of memory' and its recovery is given by Orange (1990). Current Maori claims for greater independence are considered in Archie (1995), Banham (1998), Durie (1998), Pearson (1994) and Pearson and Ongley (1996). See Douglas (1984) for a discussion on Maori identity and Davey (1993; 1994) for work on contemporary social problems. Several of the plates in McKinnon (1997) cover Maori issues.

6

OUTPOSTS OF EMPIRE

Although various pieces of evidence suggest that Dutch, French, Portuguese, Spanish and even Chinese mariners 'discovered' Australia roughly two centuries before Captain Cook's epic journey in 1769–70 (Cannon, 1987), it was the British who successfully colonised the southern continent and the islands some 1800 km across the Tasman Sea, which were named New Zealand. Indeed, the name of the country itself, the Tasman Sea, the now defunct names for Australia (New Holland), Tasmania (Van Dieman's Land), and the Torres Strait, separating northern Queensland from New Guinea, are amongst the few surviving testaments to those early voyages of discovery. The European colonisation was dominated by the British throughout the nineteenth century despite French flirtations with settlement in both Australia and New Zealand. However, the initial purposes of British colonisation were different in the two countries, establishing some lasting distinctions that can be seen today in the contrasting social and ethnic compositions. In particular, the legacy of the initial penal settlements in New South Wales and Tasmania and the role of settlement associations in New Zealand and South Australia have had lasting impacts. These are assessed in this Chapter, which also considers some of the long term outcomes of the initial economic basis of colonisation, the significance of minerals exploitation in the nineteenth century and the forms of government adopted.

Purposive British utilisation of Australia as a penal settlement commenced over 50 years before settlement associations began sending settlers to New Zealand. This time lag enabled key characteristics of economic development to become established in Australia, and it is these that are dealt with first here. One characteristic was the emergence of pastoralism as a significant provider of income, which was then repeated in a New Zealand context: this forms the second section in this Chapter. It is followed by discussion of the growth of arable farming in the two countries and the impact of major goldrushes in Victoria and South Island, New Zealand. Finally, the evolution of the systems of government is examined, with reference to the creation of two separate countries rather than a single unified state spanning the Tasman.

THE BRITISH COLONISATION OF AUSTRALIA

The mercantile system

The fact that the first European settlements in Australia were essentially gaols for British convicts has been stamped indelibly on both Australian and British history, fostering a particular *frisson* in the relationship between

FIGURE 6.1 A modern 'face-lift' for Constitution Dock, Hobart (photo: G.M. Robinson)

the two countries long after the abolition of penal transportation. The story of the first convicts arriving in 1788 at Sydney Cove has been a vital component in the creation of both historical fact and myth, and also an integral element in the forging of an Australian nation over 100 years later (Hughes, 1987). However, it can be argued that equally as important in the long term impacts upon settlement and economy was the way in which the penal colony at Sydney Cove and others founded during the next 40 years, notably at Newcastle (1803), Hobart (1804) and Brisbane (1826), were all 'limpet' ports, dependent initially upon ships from Britain to supply them with food and all the materials the prisoners and their gaolers would need to survive in the unfamiliar environment. The ships also brought a steady supply of new convicts and military personnel and, over time, free settlers and the beginnings of a vital trading link. Indeed, it was this link that represented one of the other essential aspects of British interest in Australia: the establishment of bases from which to further commercial operations in the South Pacific.

Initially these bases were part of the whaling trade operating to the south of Australia and

New Zealand, and five of the first six ships arriving in Sydney Cove from Britain returned home carrying cargoes of whale oil. This trade in whales and seals reached its peak in the 1820s and 1830s, with Hobart (Figure 6.1), the southern-most port, being the main Australian base for the whaling fleet. Trading opportunities attracted other entrepreneurs from Britain in the early decades of the nineteenth century, this led to the establishment of new coastal settlements not primarily intended as repositories for convicts: in 1829 at the Swan River in western Australia, in 1836 at Adelaide (to develop trade with India and the East Indies), and in 1834 at Melbourne (attracted by fertile land and a sheltered harbour). This concentration on coastal settlement and a maritime system of trade has had a lasting influence: the Australian states are dominated by their prime port cities; 90 per cent of the country's trade is still conducted by sea; and nearly 80 per cent of the population lives within 35 km of the sea.

Britain's Colonial Office policies favoured centralisation of port activities at a few strategic sites around Australia. This permitted control over the coastline and interior whilst providing the necessary level of isolation for convict settlement. Hence, once they had been

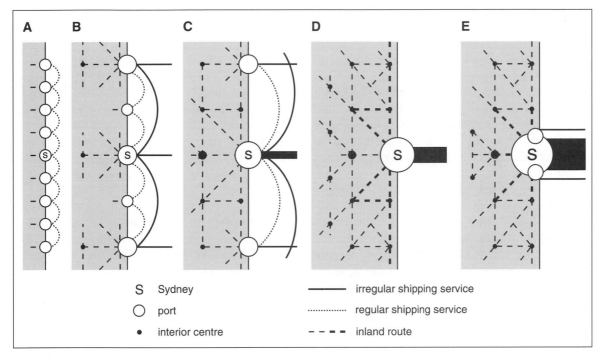

FIGURE 6.2 The process of port development in Australia

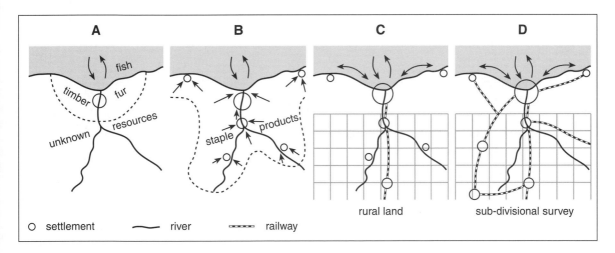

FIGURE 6.3 Vance's mercantile theory

selected, Sydney and the other colonial capitals became unrivalled as ports and as foci of economic growth. Subsequent control by colonial governments over the rural export economy ensured that locational advantages associated with growth processes were tied to these major ports. For example, the state-owned railway networks captured hinterlands and drew trade to the state capitals (Waitt and Hartig, 1997). This process of port development, based on Rimmer's (1967) model, is portrayed in Figure 6.2. It shows how Sydney competed with other small ports in the colony, but captured the larger share of inland trade, effectively condemning some of its rivals to remain small, or to lose port functions. However, the lack of deepwater ports on the New South Wales coast also contributed to Sydney's primacy. The growth of industry associated with port processing hastened the expansion of Sydney's port facilities, leading to its further dominance both as a port and in the urban hierarchy.

The exaggerated role of the major ports in the urban hierarchy can be explained with reference to Vance's mercantile theory as opposed to traditional central place theory (Vance, 1970). The key to the theory is contained in his argument that 'trade spreads outwards from the city, or the mother country, leading to the complex flowering of the rural economy at home as well as in a distant plantation of staple-producing colonists' (p. 11). This emphasises the relationship between the wholesaler and the retailer rather than that between the retailer and the customer as in central place theory. The former relationship encouraged a particular type of urban development in which fewer types of centres developed (see Figure 6.3). The initial urban system was dominated by entrepot cities such as Sydney, Melbourne and Brisbane, which controlled the movement of goods into and out of the country by acting as major ports and controlling transport routes inland.

Two principal types of centre developed in the interior:

(a) workplace towns: points of exploitation of local resources, e.g. Mount Morgan (Qld);
(b) fundamental trading centres, akin to a 'central place' in which small market functions and services operated. These serviced the 'frontier' and were tied to the entrepot by essential lines of communication. In most cases these centres have not grown beyond a population of 10 000, e.g. Dalby (Qld).

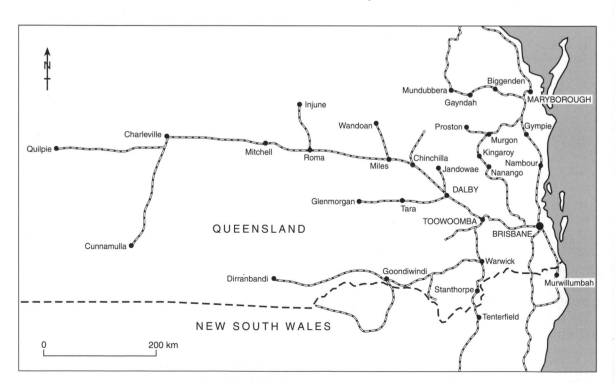

FIGURE 6.4 The 'pastoral railways' of southern Queensland

TABLE 6.1 Population numbers, 1790–1940

| Date | Australia* | | | New Zealand | | | |
	Total 000s	Annual growth %	Net immi-gration 000s	Total 000s	Annual growth %	Maori 000s	Net immi-gration 000s
1790	2						
1800	5	9.6					
1810	12	9.1					
1830	34	11.0					
1840	70	7.5					
1850	190	10.5				20	20
1860	405	7.9		115		56	40
1870	1146	11.0	79.5	300	10.3	47	103
1880	1648	3.7	124.2	534	5.6	46	137
1890	2232	3.1	234.3	669	2.2	44	25
1900	3151	3.5	14.7	816	2.0	46	22
1910	3765	1.8	74.4	1058	2.6	53	86
1920	4425	1.6	17.5	1272	1.9	57	45
1930	6501	2.0	−10.1	1490	1.9	73	64
1940	7078	1.9	5.2	1636	1.9	90	6

*non-Aboriginal
Source: Annual Yearbooks.

The settlement pattern was a dynamic one, evolving as the nature of the frontier altered and as the relationship with the 'mother' country changed, so that serving domestic demand assumed greater importance. This led to the emergence of some manufacturing centres, although manufacturing functions were also taken on by the entrepots. Over time the workplace towns, and especially the fundamental trading centres, developed more substantial central place functions based on consumer–retailer trading.

The development of the railways was vital as a means whereby links between the entrepot and the interior could be maintained efficiently, and for enabling the position of the frontier to be moved further inland as the cost per unit of transporting goods to the entrepot was rendered more affordable. This can be seen clearly in the evolution of the 'pastoral railways' of southern Queensland (Figure 6.4) and the 'wheat railways' of Victoria, South Australia and Western Australia.

Transportees and free settlers

A total of 160 463 convicts were transported to Australia from the UK and the British Empire. Of these 49.4 per cent were shipped to New South Wales and 40.9 per cent to Van Dieman's Land. Males accounted for nearly 84 per cent of the convicts. By the time transportation was ended in 1868, the continent's population was 1.6 million, illustrating how emigration of free settlers and natural growth had become dominant in the process of population growth. Yet only 18 per cent of the 77 000 people who arrived before 1830 were free settlers. There was a substantial influx of free settlers in the next three decades. Between 1835 and 1841, 40 000 migrants received bounties to emigrate under the auspices of private employers. From 1851 the lure of recently discovered gold, especially in the newly established colony of Victoria, led to its population more than doubling in the 1850s: Melbourne, the colony's capital, grew from 23 000 to 140 000 during the decade (Table 6.1).

There was a much higher proportion of women amongst the free settlers than among the convicts, and this helped to stimulate higher levels of natural population increase. This remained above 2 per cent per annum from the late 1850s to the early 1890s, contributing to a reduction in the importance of immigration to population growth: from constituting 75 per cent of overall growth between 1788 and 1861 to 28 per cent between 1861 and 1900 (Burnley, 1976). An additional factor was the placing of restrictions on the immigration of certain ethnic groups. For example, in 1861, 3.5 per cent (40 000) of the population was of Chinese origin. However, strong opposition from goldfields' communities had already led to the Victorian government passing the Chinese Restriction Act in 1854. Chinese miners were hounded out of several areas and they gradually became concentrated in more remote areas or left the country. More measures to exclude their entry followed later in the century, whilst the reduced need to attract immigrants led to the demise of bounty and other assisted passage schemes. By 1891 Australia's population had more people born within the country than born outside.

There were also small communities of Chinese in the mining settlements of New Zealand, but never on the same scale as in Australia. The main source of new migrants to New Zealand throughout the nineteenth century was the UK. This source dominated the 25 000 arriving between 1840 and 1853 and was predominant after 1870 when assisted passage schemes operated. These schemes helped increase female immigration, and were popular amongst farm labourers and their families, and for the first time led to direct migration from Britain outnumbering migrants from Australia (Grey, 1994: 237; Simpson, 1997). These migrants generally found good opportunities for work or purchase of land in a colony in which over half the population lived in rural areas until 1911. In contrast, by 1891 two-thirds of Australia's population were living in urban areas and 40 per cent in the state capitals.

Another important difference between the two countries lay in the composition of the immigrants. Both within the transportees and the free settlers attracted to the Australian goldfields there was a significant representation of Catholic Irish. As a result Catholicism and a degree of anti-British, anti-establishment attitudes were prevalent in Australia in a manner not evident in New Zealand, where minorities were usually Protestants from countries in north-west Europe. In both countries, though, the high birth rates of the 1850s and 1860s gave way to much lower rates by the end of the century. In New Zealand the average number of children per non-Maori household fell from 6.3 in 1870 to 3.8 in 1890. In Australia birth rates per 1000 non-Aboriginals fell from 35 in 1880 to 25 in 1890. In both countries the recession of the 1890s may have influenced this, but the falling birth rates did presage future concerns regarding population and labour force numbers which helped determine the nature of immigration policies.

LIVING OFF THE SHEEP'S BACK

Initial pastoral development

A major problem for the early settlers in Australia was finding sufficient fertile land close enough to supply food at an economic cost or before it perished. The few large coastal valleys assumed importance at an early date for the fertile land they offered and relative ease of transport. Of particular significance were the Hawkesbury, settled from 1794, the Hunter, from 1823, and the Derwent in Tasmania from 1804. The latter became the main source of grain supply for Sydney, some 1000 km away from the Derwent's port at Hobart. However, yields of both wheat and maize were poor and the hazards of grain production in unfamiliar conditions meant that the colony did not become self-sufficient in grains until the 1820s. This bald statement hides over three decades of threats of food shortages and tremendous reliance on the link with the UK as the chief food supply.

The demise of whaling and sealing from the 1830s was accompanied by a corresponding growth of pastoralism as pioneers gradually opened up lands to the west of the Great Dividing Range. One of the earliest major capital investments in the development of the interior was the establishment of the Australian Agriculture Company in 1824, with capital of one million pounds invested to promote production of wool and other agricultural outputs for export. This was part of the first significant pastoral activity in Australia, with official sanctioning of exploration in the 1830s to open up the interior. This revealed the extent of the grasslands on the western slopes of the Great Dividing Range, and they were then settled from gateways in the south-east, south and south-west, so that by the 1860s most of the more well-watered southern grazing lands had been occupied. These were the areas of the winter rains whilst, to the north, the grazing lands were primarily associated with unreliable summer rains. These proved less amenable to sheep and cattle and were not exploited to such a great extent until the 1890s, and then only with the aid of artesian water.

Apart from the physical barrier offered by the mountains, the chief constraint upon the opening-up of the grazing lands was a legal one, as the British government prohibited occupation of the country's interior so that it could more readily maintain control of the growing colony. However, from the 1830s there was widespread illegal occupation by stockmen (known as squatters), and this led to the issuing of temporary grazing permits. At first, tickets of occupation were issued, and then, from 1836, depasturing licences costing £10 regardless of the amount of land used. This was effectively an annual licence for grazing, but without security of tenure. Subsequently, purchase of land was allowed, but the high costs of land determined by the authorities were generally prohibitive. In 1847 'long'

leases of 14 years' duration were introduced by the Imperial Waste Lands Act in New South Wales, for units of land large enough to graze 4000 sheep, and this proved a major stimulus to both sheep and cattle production. The leases created blocks of up to and over 3200 ha on 'unsettled' land. These blocks still exist today and this system of long leases remains in place in many areas where extensive grazing of livestock is still the main land use. Smaller blocks, of eight-year leases of 647 ha for squatters on 'intermediate' land, are less discernible today.

The smaller units of land required for commercial cropping, as opposed to livestock production, meant that land purchase proved more feasible for cropping and there was increased competition for land in the immediate hinterland of the ports. This competition gradually spread inland, forcing some pastoralists to purchase land that they had previously leased and to intensify their production system. However, the potential for such intensification was limited until the 1880s, with the first successful transportation of beef to the UK in a vessel containing refrigeration facilities. This enabled Australian beef to compete on the British market with refrigerated beef arriving from Argentina and the United States. However, sheep remained the mainstay of Australia's pastoral lands, dominating New South Wales, and extending into new, drier, areas from the 1880s with the use of oil-drilling techniques tapping water at depths up to 1000 m in the Great Artesian Basin. This helped to overcome the limitations imposed by previous reliance upon a network of small dams and reservoirs.

The new source of water supply could not eliminate the threat of drought in many parts of the pastoral belt nor competition for pasture from both indigenous species, notably kangaroo, wallaby and emu, and exotics, principally the rabbit. Rabbits were introduced to Victoria in the 1860s, and by the 1890s had spread to the Eastern Rangelands and the southern and northern rims of the desert country. They were not controlled effectively until the introduction of the disease myxomatosis in the 1950s, which wiped out over four-fifths of the rabbit population. Nevertheless, some rabbits developed a resistance to the disease and they continue to pose a major threat to the quality of pasture. Feral pigs, horses (brumbies), goats and camels are other pests affecting pastoralism and native animals. Another threat has been the dingo (*Canidae*), introduced by Aborigines, which will attack and kill sheep, 30 to 40 at a time. One system of control was the erection of a six-foot-high dingo-proof fence extending from South Australia 6000 miles (9600 km) to the north-east into Queensland. This cordons off the south-eastern third of the country.

The merino in Australia

The first Australian wool was exported to the UK in 1807 by which time there were more than 25 000 sheep in New South Wales. Effectively, this entry of Australia into the world wool market marked the end of the first pioneering phase of development, setting in train the drive to open up the interior for wool production based on the Spanish sheep breed, the merino. High prices for wool encouraged the expansion of production, with pioneers moving inland, grazing sheep and making themselves self-sufficient in foodstuffs through general farming. Hence, rural development was predominantly focused on wool and some local cereal production and flour milling. By 1831, 2.5 million lb (1.1 million kg) of wool per annum were exported. The balance between sheep and cereals changed subsquently as cereal exports could also command favourable prices, and the coming of the railways enabled wheat to be transported to the ports more easily.

The merino was first introduced in 1797, with further significant imports in 1804 and the 1820s. Gradually they were grazed in the more arid areas, with such success that by 1850 there were more merinos in Australia than Spain. However, substantial areas of land had to be under one management for both cattle and

sheep rearing to be profitable on the drier pastures. This situation has persisted to the extent that holdings in excess of 400 000 ha for cattle and up to 40 000 ha for sheep are regarded as the optimal size today. This has generally yielded low profit per unit area but relatively high net return on capital investment. The merino remained crucial to the well-being of large areas of rural Australia in the nineteenth century, and wool was the dominant export for eight decades, benefiting from rising demand for wool in the UK through technological advances in the woollen textile industry. Although cereals gradually assumed greater importance, even by the 1860s only 400 000 ha of Australia were under crops, and wheat was still being shipped from Tasmania to the mainland.

The pastoralists dominated the exploration and early settlement of eastern Australia, for example the Bathurst Plains (opened up from 1813) and the upper Murrumbidgee. By the mid-1860s only parts of the Mallee and Gippsland were unoccupied in the south-east, whilst the spread of pastoralism northwards had embraced Queensland's Darling Downs by the 1840s (Butlin, 1962). In 1813 there were 50 000 sheep in Australia; by 1860 there were 21 million and at the end of the century 100 million. Wool exports were 35 million lb (15.8 million kg) in 1860 and 20 times this amount in 1900. Until the 1860s systems of flock management similar to those in Britain were used on large holdings, but with few fences, with the use of out-stations and each shepherd looking after between 400 to 600 sheep (Powell, 1968). Sheep were grazed on natural forage and relied on streams for water. This system of management was altered in the 1860s after the Victorian gold rush had led to an increase in the cost of labour. Gangs of itinerant shearers (as depicted in the 1950s' book *The Sundowners*) became an important component of the labour force instead of relying entirely on full-time labourers. There was also further movement of ranching northwards and towards the arid interior from 1860. Improved breeding techniques led to the recognition of the

Australian merino as a distinct new breed in 1858, better adapted to survive the heat, aridity and more limited forage.

Canterbury wool and lamb

By 1840 some of the British whaling and sealing stations in New Zealand had become permanent settlements, and there had been widespread land speculation with Sydney-based businessmen claiming some substantial purchases. Amongst these claimants was the New Zealand Company (NZC), formed in the mid-1830s. Impatient to promote settlement in New Zealand, the Company 'sold land orders' in London for its proposed settlements before it even knew where they were going to be located. Wellington (originally named Port Nicholson) in 1840 was the first and most successful of the NZC's settlements, the other three being at Wanganui (1840), New Plymouth (1841) and Nelson (1841). However, as land had been sold before settlers arrived in New Zealand, there was a problem of absentee landlords that limited the economic stability of the new settlements. This and difficulties relating to land titles, problems of surveying difficult terrain, failure to appreciate the potential offered by wool and the NZC's hostility to the Treaty of Waitangi contributed to its demise and purchase by the British government in 1850. Nevertheless, the NZC was responsible for sending 57 ships and 19 000 migrants to the new colony.

Before the collapse of the NZC, two related settlement schemes had founded settlements in Christchurch and Dunedin. The former was established by the Canterbury Association in which a prominent part was played by Edward Gibbon Wakefield (1796–1862), brother of the NZC's chief land purchaser (Martin, 1997). Wakefield's ideas on colonisation were pursued in the founding of Adelaide by the South Australia Association, which he left when they fixed a selling price for land at a price that he deemed too low. He also fell out with the NZC, resigned his directorship in

1849, and joined the Church of England-led Canterbury Association. From London he campaigned for self-government for New Zealand. When this was realised in 1852, he emigrated to New Zealand and was elected to the House of Representatives. However, his theory of selling land at a 'sufficient price' was never followed in practice and he did not envisage the growth of pastoralism that occurred.

The first major success in the development of a strong rural economy in New Zealand was the opening-up of the Canterbury Plains and the hill country further south. Although the Canterbury Association's plan for Christchurch and the Plains envisaged the establishment of self-sufficient arable-dominated smallholdings over an area of one million ha, the first settlers in 1850 found that two Scots, the Deans brothers from Lanarkshire, had already settled in the area and had land under a lease negotiated in 1846 with local Maori for £18 per annum. They ran sheep on this land, setting the trend for early settlers also to obtain substantial leases. The Association's granting of a pastoral licence to the Deans on the tussock grasslands of the Plains helped undermine the idea of arable smallholdings, and the Deans and other 'runholders' grazed sheep to develop a thriving economy based on wool production. By 1850 sheep farming had also been established in Marlborough and northern Canterbury, outside the Association's jurisdiction.

In 1851 a wave of new arrivals, known as the shagroons, came from Australia, fleeing drought conditions on grazing lands there, but intent on rearing sheep on the Canterbury Plains. They leased Association land for pasture between the Waipara and Ashburton, contributing to the creation of substantial sheepruns across the Plains. The size of the leases partly reflected the need for sheep to have access to drinking water so that the runs had to front on to watercourses. Virtually all the Plains and High Country had been leased by 1855.

All the most easily utilised land had been occupied by 1860, by which time there were three-quarters of a million sheep in Canterbury. New runholders had to push into more remote areas, known as the High Country, including the writer Samuel Butler who established his estate, called Erewhon, in the Upper Rangitata. Substantial wealth was generated by the export of wool, a surviving testament in modern Christchurch being the Canterbury Club, built in 1861 as the first gentlemen's club in New Zealand, with a membership dominated by the runholders. Other notable surviving buildings from this period are the Provincial Council Buildings (1858) and Christ's College School (1863).

The growth of population helped stimulate grain production and some dairying to supply the expanding town, but at the cost of competition for land between runholders and small-holders (termed 'cockies'), who introduced gorse and hawthorn field boundaries and shelterbelts of Australian eucalypts and Monterey pine (*Pinus radiata*). Between 1850 and 1871, 530 ha of exotic trees were planted. Further impetus to such plantings was then given by the 1871 Forest Trees Planting Encouragement Act which led to a further 1215 ha being planted for shelterbelts, fuel provision and climatic 'improvement' by 1883. Under the Act tree planters were given preferential rights over Crown land.

As in Australia, the coming of the railway enabled the cockies to compete with the graziers and they often purchased land from runholders who did not have pre-emptive right of purchase. To provide more water for both crops and livestock, 'water races' were created to carry water from foothill streams across the Plains, making distinctive straight-line features in the landscape. For example, Ashburton County had over 1600 km of water race by 1894. By 1880 the leases had virtually all been purchased so that there were 400 000 ha of freehold land on the Plains, and a mixture of sheep and grain production. By the 1880s there were 100 000 ha of wheat on the Plains, which also produced 16 million lb (7.2 million kg) of wool (one-quarter of the country's output).

Box 6.1 Staple theory

The emergence of Australasia's reliance upon a few staple exports to the UK in the nineteenth century (Table 6.2) can be considered with respect to staple theory, as applied originally to the Canadian economy (MacKintosh, 1964). This utilises a mild form of geographical and technological determinism to argue that natural endowment, such as the extensive grasslands of Australia, allied to the technical characteristics of production helped shape the whole economic structure through linkage effects: backward linkages into transport investment, and forward into processing and service industries. For both Australia and New Zealand 'the influence of nineteenth-century pastoralism remains firmly imprinted on the modern ... economy' (Schedvin, 1990: 535), in part because of the way in which a reliance upon a few staples encouraged import-replacing industrialisation and thereby determined the nature of economic growth. For both countries, some diversification of production has occurred, but with the export base still heavily dominated by traditional staples. This is especially so for New Zealand with its smaller domestic population and narrower resource base (Gould, 1982). However, both countries have struggled to replace the productive efficiency of nineteenth-century wool and gold production (see below), and pastoral products accounted for about half the value of Australian exports until the 1960s.

The reliance upon wool, for which virtually all the processing was performed in the UK, provided little stimulation for Australasian manufacturing, though there was a need for local financial services, insurance and overseas shipping, so that Australasia had the highest number of banks per capita in the world. Wool provided wealth without a need for sophisticated technology or high levels of investment, but with little manufacturing spin-off. Some proponents of staple theory contend that the export-led economy provided not only high standards of living in general for the white settler economies, but laid the ground for subsequent long term economic growth. Schedvin (1990) contends that this argument is more readily applicable to Canada than to Australasia where the absence of a strong domestic manufacturing industry has led to falling standards of living and economic stagnation coupled with the retention of an essentially nineteenth-century pattern of export staples.

TABLE 6.2 Exports: Australia 1900–1940; New Zealand 1860–1940

(a) Australia

	1900		1920		1940	
	£m	%	£m	%	£m	%
Wool	15	30	48	37.5	58	34
Grain	4	8	34	26.5	9	5
Gold	14	28	4	3.0	9	5
Frozen meats	3	6	6	4.5	14	8
Butter	1.5	3	8	6.0	8	5
Other	12.5	25	28	22.5	71	43
Total	50	100	128	100	169	100

Source: Yearbooks.

(b) New Zealand (percentages)

	1860	1880	1900	1920	1940
Wool	70	48	36	24	22
Grain		3	7		
Gold	6	17	8	1	
Forest products	1	6	10		2
Frozen meats			16	28	30
Dairy produce			7	20	37
Other	23	26	16	28	8

Source: Grey (1994: 21).

A major impetus to sheep farming came when the first lamb was shipped to England from Canterbury in 1882 on a refrigerated vessel. This helped to confirm the basis of the rural economy (Box 6.1): some sheep for wool production; the majority of sheep for lamb for export; cattle for local milk supply, and, from the 1880s, for dairy produce to be exported; and crops for local supply. Variations in the balance between these reflected not only fluctuations in demand and supply, but also changes in government policy. For example, the McKenzie land reforms of the 1890s were aimed at settling more people on the land; this was effected by the Crown purchasing some of the larger estates and converting them to smallholdings (Brooking, 1997), for example as carried out in the Cheviot area, north Canterbury, where dairy holdings were

created. In spatial terms there was a general threefold division in Canterbury:

(a) the High Country: dominated by extensive sheep grazing on holdings in excess of 10 000 ha;
(b) the Hills: with semi-extensive sheep grazing and some cattle; average holding size *c.* 1800 ha;
(c) the Plains: mixed farming including wheat, sheep, some improved pasture for cattle, and some specialist crop production; average holding size 200 ha.

GRAIN AND CANE

Wheat production in Australia

For grain production, the coming of the railways was as significant in Australia as it was in North America (Andrews, 1966). The new transport links enabled farmers to open up the red-brown earths of south-east Australia, gradually extending into the arid limits of the required growing season (Meinig, 1959). When the first expansion of arable farming began and there was some displacement of pastoralism, it was the intention of both federal and state governments to replace large grazing concerns by smaller family farms producing grain. However the Selection Acts in the 1860s created large holdings taken by pastoralists who could enlarge their holdings by taking control of water points. This helped to push grazing further inland, with cereal growers displacing pastoralists closer to the major ports. Some land was subsequently removed from pastoral leases and split into smaller blocks at low prices and with long term credit available. Blocks varied in size from 32 to 130 ha, the most common being 65 ha. Similar legislation to permit this process was passed in the three eastern states between 1850 and 1872, termed 'closer settlement policies'. However, many legal loopholes allowed large estates to survive in many areas. Overall, 3 million square miles (7.8 million sq km) of Crown land was disposed of during the nineteenth century amongst five million people. By 1914 there were at least 146 ways of holding land, including some associated with new closer settlement policies introduced at the turn of the century to create smaller holdings. For example, in South Australia by 1915, 1.3 million ha had been compulsorily purchased by the state government, leading to new settlements, land sub-division, new roads and other infrastructure, including the development of irrigation schemes. This policy was extended after the First World War with the creation nationwide of 31 560 smallholdings for returning soldiers.

The fertile red-brown soils in the immediate hinterland of Adelaide were first to develop a significant trade in wheat. South Australia was settled from 1836 as part of a systematic colonisation based initially on the ideas of Edward Gibbon Wakefield. He argued that problems attracting 'free' settlers in the Australian colonies were caused partly by the system of land sales adopted. He viewed the Swan River settlement, established in 1829, as misguided in its conception. Its mixture of soldiers, convicts and free settlers conflicted with his views that a new colony should represent the best of British life, with a mixture of social classes, but with due social cohesion maintained through an equilibrium between land and capital. Therefore he argued that land must be sold at a price suitable to attract potential farmers who possessed capital whilst not deterring labourers from hoping to purchase land one day. He contended that by fixing the price of land at quite a high level, only men of capital would be able to buy; those who could not would be the suppliers of labour, and the finance generated would then promote investment and development of resources. Land sales were used as a means of financing the settlement scheme and as a means of regulating the new colony. Unlike the Church of England-dominated Canterbury Association a few years later, the South Australian colonists included a strong contingent of dissenters who were regarded as valuing civil and religious liberty whilst upholding the ideals of the new colony. A lasting impact of this recruitment from a cross-section of religions is the large number of churches of different denominations in Adelaide, and hence its title of 'city of churches'.

The proximity of the productive wheat-growing area to port facilities in Adelaide restricted transport costs and helped the competitiveness of the region. By 1861 it accounted for 40 per cent of Australia's wheat production, but its share dropped steadily during the next four decades as the so-called 'fertile crescent', extending from Victoria into New South Wales, was opened up (Figure 6.5). Here the role of the railways was crucial, coupled with both states' new settlement schemes (Robinson, 1976).

Large-scale grain production advanced rapidly from the mid-nineteenth century under the stimulus of falling freight rates, the cheapness of available land, the initial fertility of the virgin soils, rapid mechanisation reflecting a lack of labour, the spread of the rail network and the establishment of an elaborate marketing system. Production in the immediate hinterland of Adelaide was replaced by expansion inland, following the railway. As on the Canadian prairie, production was not economic unless it was within 25 km of the railhead. The first major expansion was into areas of sclerophyllous forest, but subsequently areas of scrub woodland and part of the mallee became wheatlands: drier areas necessitating larger holdings, with sections of 128 ha in New South Wales and 259 ha in Victoria being common. Dry farming techniques were developed (Meinig, 1962) and, from 1902, planting of drought-resistant Federation wheat. The land tenure changes in New South Wales and Victoria in the 1860s were crucial in reducing South Australia's domination (the state had half of all Australian wheat production in 1870), and led to wheat growing in Victoria's Riverina and Wimmera. The current distribution of wheat was apparent by the 1890s (Figure 6.5), with holdings of 120 to 160 ha dominant and the use

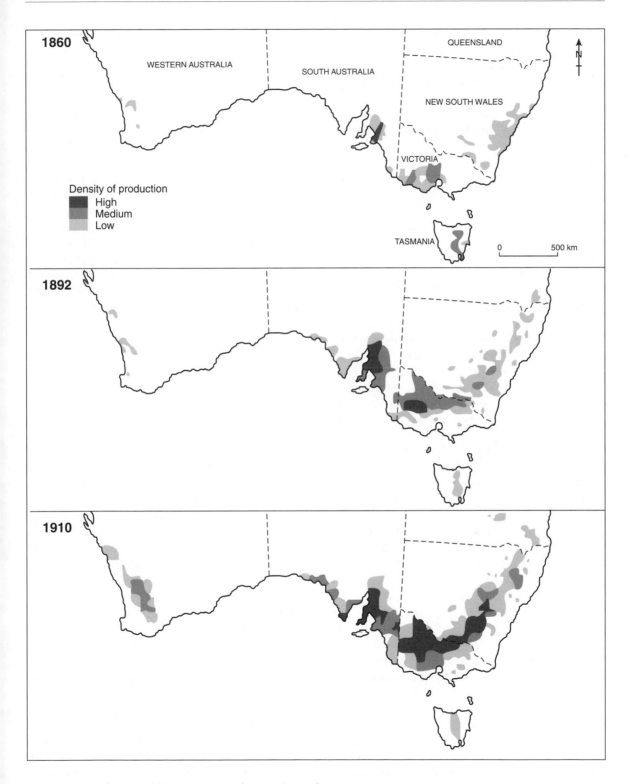

FIGURE 6.5 Wheat production in south-east Australia

of superphosphate to correct the widespread phosphorus deficiency. By this time there had been a gradual merging of wheat and sheep production on the same holdings. A common rotation was wheat for three years, then a four-year ley on which fat lambs were raised. This often produced a land use pattern on a holding that consisted of one-fifth timber, one-quarter crops, one-quarter grass and the rest unimproved land. Local innovations increased productivity, notably through reaping machines from the 1840s, the stump-jump plough (1876), new breeds of sheep, 'long blow' shearing machines, Farrier's improved varieties of wheat and the use of subterranean clover.

Settling the tropics

Sugarcane production and cattle rearing were important components in the settling of tropical Australia from the 1860s onwards, though sugar was only significant in a long, narrow belt along the Queensland coast. Sugarcane was Queensland's most important export crop by the end of the nineteenth century. It was introduced on a very small scale in the 1820s when it was grown at Port Macquarie (NSW). In 1858 the crop was grown in the Clarence River valley, but it was only when it was introduced further north, in areas with frost-free winters, that it became significant commercially. In 1870 there were 4050 ha grown, mainly in Queensland. By 1900 this figure had risen to 53 825 ha, 85 per cent of which was in Queensland, grown on the plantation system using indentured labour. This labour was provided by the Kanakas from Vanuatu and the Solomons. In 1885 there were 11 000 Kanakas working in Queensland, an average of one per 2.4 ha of cane, and in total 57 000 labouring contracts were issued to Kanakas in the last three decades of the century. However, there was strong official disfavour towards the Kanakas and other 'non-white' workers during the 1890s and the reliance upon this labour source was eliminated as the Kanakas were repatriated. This contributed to a different system of management of cane production, in which the plantations were replaced by individually owned smallholdings supplying their cane to a central processing mill. The mill provided a guaranteed market and stimulated increased production through a combination of increased acreage, introduction of higher-yielding canes, and greater use of fertilisers.

Co-operative grower-owned mills were introduced in 1888. The principal processor, the Colonial Sugar Refining Co (CSR), and the millers required growers to sign five to seven year contracts stipulating the types of cane that could be grown. Because cane rapidly deteriorates after cutting and loses over four-fifths of its weight in milling, most cane was grown within 20 km of a mill, an aspect of production that continues today (see Chapter 14).

Throughout much of tropical Australia the only economic activity is extensive grazing. It was this activity that opened the interior of tropical Australia in the late nineteenth century, primarily through the rearing of beef cattle (Courtenay, 1982). Sheep have been important on the margins of the tropics in central Queensland, southern Northern Territory and the Pilbara, though even here cattle have gradually come to the fore because of the better prices for beef, the continuing problems sheep have had with dingoes, and the greater degradation to grasslands as a result of sheep grazing. Given the dry conditions, management of the grassland has to maximise efficient use of available moisture, and hence the annual legume, the Townsville legume, has been planted over large areas. A simple open-range system has been used for beef cattle for over 100 years, with herds averaging from around 1000 in parts of western Queensland to over 15 000 on the Barkly Tableland. There is a similar variation in holding size: from 75 000 ha in western Queensland to over 500 000 ha on the Barkly Tableland and the Vic River. The larger sizes of herds and holdings reflect inaccessibility which has created greater difficulties for marketing and efficient herd management. Traditionally, many cattle and sheep stations have kept labour costs down by employing cheap Aboriginal labour.

The first cattle drive to Cape North was made in 1865, but settlement north of Townsville in the 1870s relied on the lure of gold, as in the Palmer River goldrush. To the west, the harsh environment limited European settlement, and only 1700 Europeans were recorded in the whole of the Northern Territory in the 1911 census. Vast areas provided very low-density cattle grazing, but many leases were taken as part of speculative investment anticipating the discovery of minerals. In tropical Western Australia, in the 1880s long pastoral leases and cheap freehold land were offered in the Kimberley region, enabling the establishment of pastoralism along the Ord River. Elsewhere, pockets of settlement were established in conjunction with mineral discoveries, as at Mount Isa in western Queensland (see Chapter 8).

THE GOLDRUSHES

The Victoria 'rush'

In 1834 settlers from Van Dieman's Land seeking more arable land migrated to the coast of south-west Victoria. The promise of this area led to investigations of the Port Philip area and purchase of land from local Aborigines on either side of the River Yarra. The settlement that developed on this site was named Melbourne after the British prime minister of the day. The fact that this was a 'free' settlement, in contrast to the penal colony established at Sydney, has remained part of the ongoing rivalry between the two cities. It also contributed to the separation of Victoria from New South Wales in 1850 and its development as a self-governing colony, able to prevent any extensions of the penal system of settlement to its territory.

The Victoria goldrush of the 1850s conferred great wealth and prosperity upon Melbourne. The goldfields were located only 100 km away, centred on Ballarat and Bendigo. The discovery of gold attracted immigrants, to the extent that there were half a million new arrivals in the decade, many of whom stayed in the country once the rush had faded. The miners included Chinese, of whom there were 24 000 in Victoria in 1861 even though official prohibition of Chinese immigrants was instituted in 1854, by administrators fearful of further Chinese and Asian colonisation.

When the *Great Britain* left Melbourne in November 1862 bound for the UK, she was carrying one of the richest freights ever. The cargo included over 100 000 ounces of gold (valued at £400 000), 50 000 sovereigns, 732 bales of wool, other colonial produce and 400 passengers. The nature of the cargo and its value symbolised the emergence of Australia (and primarily Victoria) as a key supplier of a highly valuable commodity – gold – on which a prosperous economy had been rapidly developed and which had added to its other staple export, wool. The *Great Britain's* cargo also included gold mined in New Zealand.

Although the easily worked gold deposits were soon exhausted, with peak production occurring very rapidly (1856), the Victorian goldrush yielded gold production of over £100 million, providing a phenomenal stimulus to the local economy. Much of the income went into colonial banks who then gave out loans for the development of new pastoral properties or farms. Therefore capital from mining financed a substantial amount of the spread of rural settlement as Victoria became the UK's richest colony.

The influence of the goldrush was a lasting one, especially on the numbers of free settlers. Australia's non-Aboriginal population had doubled every eight years between 1800 and 1850, but it doubled again in the first five years of the 1850s, reaching one million. The need to supply food and manufactured goods for this population helped advance both domestic agriculture and industry, leading to rapid economic development. This helped attract more migrants.

Much of the 'gold fever' associated with other finds in Australia generated only short-lived, sporadic urbanisation, with settlements being deserted or becoming run-down once the mining boom was over. In a few cases, though, permanence has been assured either by the size of the reserves being sufficient to stimulate lasting commercial development, as in the case of Broken Hill, Kalgoorlie and Mount Isa, or the development of other functions as in the case of Ballarat and Bendigo.

The Otago 'rush'

New Zealand's equivalent of the Victoria goldrush was its own 'rush' based on the Otago goldfields in the 1860s. Gold was first discovered in 1861 near Lawrence, 90 km west of Dunedin. In the next few years further major finds were made elsewhere in central Otago, notably near Clyde and on the Clutha, Shotover and Arrow rivers (close to modern Queenstown), 270 km inland from Dunedin. In six months in 1861 the population of Otago more than doubled, half the immigrants coming from Australia, primarily from the Victorian goldfields, including some who had originally gone to Victoria from the United States, having taken part in the 1849 California goldrush. The peak in-migration was in 1863 when over 24 000 people were working in the Otago goldfields, and the province had over one-third of the country's Pakeha population. In 1864 finds in Westland and the Buller attracted new migrants, and there were smaller 'rushes' in Nelson, Marlborough and Thames-Coromandel. Indeed two of New Zealand's greatest nineteenth-century prime ministers were amongst the goldrush immigrants: Julius Vogel (to Otago in 1861) and Richard Seddon (to Hokitika in 1866). Although most of the Otago goldfields were over 200 km from Dunedin, the city was the source of supplies and finance and was the export port for the gold. The goldrush transformed its economy (Table 6.3), helping to produce grand buildings, both commercial and domestic, making Dunedin the largest and richest city in the country by the 1880s and developing the traditional Scottish skills of banking and engineering.

TABLE 6.3 Production of gold, coal and gas, New Zealand, 1860–1940

	Gold kg	Coal 000 t	Gas $m^3.10^6$ (m)
1860	14		
1870	15 200		
1880	8400	454	6.9
1890	5300	1167	12.1
1900	10 500	1738	22.3
1910	13 400	3386	58.8
1920	6000	3043	96.3
1930	3700	2582[a]	117.3
1940	5300	3817[a]	n.a.

[a]Underground mining only
Source: Grey (1994).

The settlement of Dunedin (originally New Edinburgh) was planned by the Free Church of Scotland which had broken away from the Established Kirk (Church) in 1843 (Matheson, 1994). Under the direction of two of the new church's ministers, a settlement was planned combining Wakefield's ideas on land sales and Scottish presbyterianism in the Otago Association. In 1844 the Association's representatives, in conjunction with the NZC, paid £2400 to local Maori for 162 000 ha alongside Otago Harbour. However, by late 1847 when the first two ships sailed for the new lands only 72 of 2000 lots had been sold, reflecting the controversy over the activities of the NZC. Only 15 of the 344 passengers on the first two migrant ships in 1847 held their own land orders. As a result, the dominance of Scots Presbyterians amongst the early settlers was somewhat diminished as there were land sales made to English and Irish settlers in order to recoup the land purchase costs. Nevertheless, after a slow beginning on an unpromising site, a city with very clear Scottish origins emerged, notably in the Free Church kirks, place and street names (with many echoes of Edinburgh), and buildings (including the law courts and railway station built of Aberdeen granite and the University of Otago's main building reminiscent of Glasgow University, but built on the banks of the Waters of Leith).

GOVERNING THE NEW COLONIES

From colonies to the Australian Commonwealth

The first British administrators drew their colonial borders as straight lines on the map, ignoring the presence of physical features. Hence, the first border of New South Wales was simply a straight line along 135°E bisecting the continent (Figure 6.6). This demarcated an area that was too large to be controlled by a single administration and, with the establishment of new settlements at Swan River, Melbourne and Adelaide, the Colonial Office in London granted official colony status to Western Australia (1829), Victoria (1850) and South Australia (1834) respectively and inserted new boundaries based largely on lines of latitude and longitude. The 1850 Australian Colonies Government Act established Victoria as a separate colony, with its seat of government in Melbourne, and confirmed representative government in both New South Wales and South Australia. Queensland was granted self-government in 1859 and Tasmania (formerly Van Dieman's Land) in 1855. By 1860 all except the sparsely populated Western Australia had self-government. This status was not conferred upon Western Australia until 1890, due to its long retention of transportation of convicts and low population numbers.

The area designated in 1911 as the Northern Territory, under the responsibility of the 10-year-old federal government, was part of New South Wales until 1863, and then under the jurisdiction of South Australia. The first coastal settlement established in the Territory was located at the mouth of the Adelaide River in 1864. However, it was abandoned the following year after heavy flooding, and another site was surveyed and selected for settlement at Port Darwin, which had originally been discovered in 1839 and named after Charles Darwin. The name was changed officially to Darwin in 1911. The isolation of Darwin limited its growth until the construction of the Stuart Highway during the Second World War, linking Darwin with the railhead at Alice Springs, over 1500 km to the south. Since 1931 the Territory has been administered from Darwin, with self-government granted in 1978. On 1 January 2001 it may attain the same status as the existing six states if current government proposals are ratified. First, though, issues relating to Aboriginal land rights, mining royalties and ownership of uranium deposits will have to be settled. Acceptance of the change of status by Aborigines will be crucial as they constitute one-third of the Territory's population and own half the land.

The first meeting representing all the Australian colonies was held in Melbourne in 1890 to debate the motion that 'the best interests and the present and future prosperity of the Australasian colonies will be promoted by an early union under the Crown'. This gave rise to a Constitutional Convention of delegates chosen by the colonial parliaments and, in 1897, an elected convention of 40 delegates. All 40 were federalists!

Federation was championed by a broad coalition of the educated elite in each of the separate colonies on a variety of grounds: defence, a common voice and united action on common problems, economic gains from a national economy and common tariffs, and the greater fostering of the emergent nationalism. Furthermore, federation promised certain advantages as a system of government, partly because it offered an additional tier of government power, with added flexibility in policy development by suggesting a prospect for innovation at state level that did not necessarily have to affect the whole country. The positive examples of the German Customs Union and the federation of the United States, with the retention of strong powers for the individual states of the latter, were often referred to by champions of federalism, alongside a growing expression of fear regarding the possible territorial claims likely to be made by the 'Asian hordes' to the north.

The Australian Commonwealth came into being on 1 January 1901, with Melbourne as

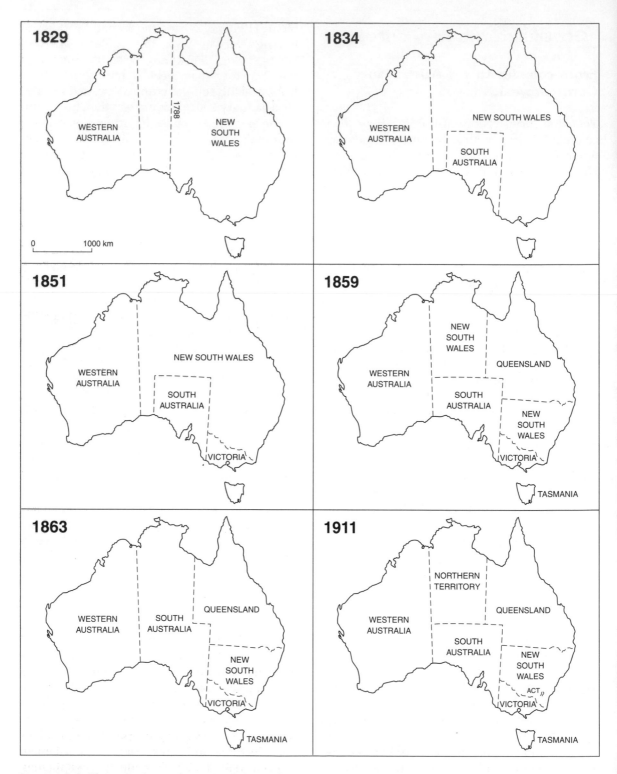

FIGURE 6.6 The colonies and states of Australia

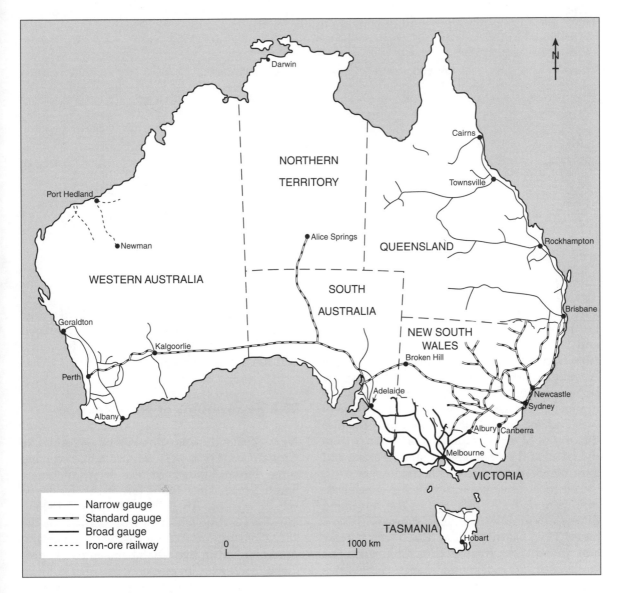

FIGURE 6.7 Australia's railway gauges

the seat of parliament, where it remained until it moved to the purpose-built offices in Canberra in 1927. The census of 1901 revealed a population of 3.8 million, just over one-third of which resided in the state capitals, with two-thirds in New South Wales and Victoria, and just under 5 per cent in each of Western Australia and Tasmania. Both Melbourne and Sydney had populations of around half a million.

The old colonies became the new states with their constitutions intact. Close ties to the UK were maintained via UK-appointed state and federal governors and the UK's power to disallow colonial Bills. Different systems applied in the states over who could vote, whilst Victoria was unique in retaining property qualifications for electors and members of parliament (Powell, 1988: 17–20). Outside the capital cities, disquiet over confederation was expressed in

TABLE 6.4 Railway track per head of population, 1871–1911

Date	Australia		New Zealand		Canada	
	km	km/000 capita	km	km/000 capita	km	km/000 capita
1871	1578	0.96	74	0.21	4337	1.18
1876	2940	1.52	1155	2.52	7731	1.93
1881	6236	2.79	2073	3.88	11 577	2.68
1886	10 325	3.84	2596	4.18	17 337	3.79
1891	14 366	4.56	2964	4.43	22 269	4.61
1896	15 753	4.56	3241	4.36	26 183	5.13
1901	21 808	5.79	3560	4.36	29 192	5.43
1906	22 347	5.46	3874	4.14	34 363	5.46
1911	28 987	6.55	4430	4.19	52 397	7.27

Source: Schedvin (1990: 549) population censuses.

various forms: farmers protesting about higher tariffs on imported machinery; Queenslanders and South Australians fearful of competition from Victoria and New South Wales; Western Australians worried by their isolation from the populous south-east and their own small numbers: although the Swan River settlement was founded in 1829, it was not until 1850 and the arrival of more shipments of convicts from the UK that the small colony started to expand and in 1856 Perth was proclaimed a city.

The most obvious legacy of the nineteenth century's separate colonies was the remaining differences in railway gauges: standard (NSW), broad (Vic), narrow (Qld, Tas, WA), narrow and broad (SA) (Figure 6.7; Table 6.4). The emergence of independent and competitive transport policies in each colony resulted in high inland transport costs which, with the fragmented national market, helped retard manufacturing industry. The railways themselves were built almost entirely of imported steel before 1914, and lines radiated from colonial capitals rather than forming good trading routes between colonies. From 1901 each state retained responsibilities for maintaining its own railways and hence the different system of gauges remained. The Commonwealth built a trans-Australian railway linking Western Australia to the eastern

states in 1917. Adoption of the standard gauge on all major mainland trunk routes occurred from the 1960s, though Adelaide–Melbourne still has a broad-gauge track.

The government of New Zealand

New Zealand was proclaimed a separate Crown Colony in 1841 (Figure 6.8), receiving some initial limited finance from the New South Wales land fund, but thereafter to be self-supporting. An abortive attempt in 1846 to introduce self-government was followed by the 1852 Constitution Act which created a General Assembly with two chambers, and six provinces each with an elected provincial council.

The first capital selected in 1840 by Governor Hobson had been Auckland, deemed more central than Russell (in the Bay of Islands, Northland), and favoured by fertile soil, a good port on Waitemata Harbour and with a network of internal waterways based on Kaipara and Manukau Harbours and the Waikato and Waipa rivers. Unlike the other main urban centres in the country, it was not a planned settlement based on the activities of a settlement association. Although it started to receive immigrant ships from 1842 and had a military garrison, Auckland suffered from

FIGURE 6.8 The Treaty House at Waitangi in the Bay of Islands. One year after the Treaty of Waitangi was first signed here in 1840, New Zealand became a Crown Colony (photo: G.W. Martin)

economic depression after the outbreak of the Land Wars in 1860, the competing attraction of the Otago goldfields and the decision to relocate the capital to Wellington in 1865, reflecting that city's own centrality as the economic focus of the country also moved south. It was not until the 1890s that Auckland's longer-term economic dominance became apparent, based on its port, the rich agricultural lands in its southern hinterland and, for a short time, the goldfields of Thames-Coromandel (Stone, 1987).

The earliest towns dominated their confined hinterlands, and retained a position of strength by virtue of their favourable locations and conversion of initial advantages into lasting growth. This dominance has only recently been challenged by later-established towns, notably Hamilton and Palmerston North. Initially the six main settlements (Auckland, Christchurch, Dunedin, Nelson, Wanganui and Wellington) developed as growth poles linked only by coastal shipping routes. Hence it was proposed by the future Prime Minister, William Fox, in 1851 that a variation of the American federal structure should be adopted. This was implemented the following year in the New Zealand Constitution Act establishing provinces based on the six main settlements. As Grey (1994: 166–8) points out, even these were too large for the meagre communications of the time, and this contributed to four subsequent subdivisions (Figure 6.9). The provinces tended to take control of pastoral licensing and making provisions for 'waste' land, though they did not have authority to grant municipalities powers for effective local government. Canterbury, Otago and Wellington derived significant revenues from land sales and loans, and they refused to surrender land to the 'federal' government for use as land reserves. Bickering over this and provincial opposition to other government policies was one of the causes of the abolition of the provinces in 1876. In their place 63 counties were created, each with its own council 'responsible for making by-laws, constructing public works, subsidising roads, and aiding charitable institutions, museums, and libraries' (Dalziel, 1992: 104).

Original provinces
Provinces created post-1853

Auckland

Taranaki

Wellington   Hawke's Bay (1859)

Marlborough (1859)

Nelson

Westland (1873)   Canterbury

Otago

Southland (1861–70)

0   200 km

FIGURE 6.9 New Zealand's provinces

New Zealand representatives attended federal conferences in Australia in 1890 and 1891, but they were absent from subsequent meetings. This was a reflection of the separate development and identity of the two countries. Despite common ties with Britain that Australia and New Zealand shared, most Pakeha and Maori New Zealanders did not identify themselves as Australian and they had no desire to be part of a single Australasian state. Hence New Zealand stayed outside the Australian Commonwealth and sought a separate identity and development. The 1907 Colonial Conference declared that the UK's self-governing colonies should be termed 'Dominions' to distinguish them from other dependencies. Hence the Dominion of New Zealand was created, with a Governor-General rather than a Governor and with development of its own armed forces. This further reduced the limited powers of the UK's Colonial Office over New Zealand affairs.

The following two Chapters develop two of the themes pursued above: the long term reliance by New Zealand upon a relatively narrow export base of agro-commodities (Chapter 7) and the partial substitution of mineral exports for agro-commodities in the Australian economy. Aspects of regional and local government will be discussed in Chapter 15.

FURTHER READING

Attracting much attention at the time of the bicentenary of the landing at Sydney Cove, Hughes (1987) is a compelling account of the years of transportation to Australia. A solid geographical account of this period and the subsequent social and economic development of Australia is Powell (1988); a comparable recent text on New Zealand is Grey (1994). Compare these with 'standard' historical works, e.g. Blainey (1966), Clark (1981), Martin (1978), Rice (1992), Sinclair (1990). Various aspects of exploration and settlement of the Australian interior are covered in Cannon (1987), Davidson (1965), Heathcote (1965), Meinig (1962), Robinson (1976) and Williams (1974). Pioneering settlement in New Zealand is considered in the 'classic' by Clark (1949); see also Cumberland (1981). The impacts of land reforms in late nineteenth-century New Zealand are examined by Brooking (1997). Recent work on Edward Gibbon Wakefield includes Martin (1997) and Simpson (1997). Schedvin (1990) is an excellent introduction to the nature of the nineteenth-century rural economies. The growth of Australian manufacturing industry is covered by Linge (1979). A recent celebration of Australian federalism is given by Daumanting (1997).

7

FROM 'BRITAIN'S FARM IN THE SOUTH PACIFIC' TO FARMING WITHOUT SUBSIDIES

RELIANCE ON THE UK MARKET

Supplying lamb, beef, dairy produce and wool

The fundamental character of New Zealand's agriculture was shaped in the nineteenth century when reliance upon the staple exports of wool, lamb and butter was developed. This was heavily dependent on refrigerated ships, cheap production methods and a high degree of efficiency at both farm level and in processing prior to export. The trading ties to the UK were confirmed by improvements in shipping, the opening of the Suez Canal in 1867 and of the Panama Canal in 1914. By the 1920s the cost of shipping butter from New Zealand to London or New York was only marginally greater than the cost of shipping it by rail from Wisconsin to New York (Grey, 1994: 294). The legacy of the focus upon export-based production is clear today: New Zealand remains unique within the OECD in that most of its produce is exported (Figure 7.1), and New Zealand still accounts for 46 per cent of the international trade in sheepmeat and 6 per cent of the beef trade (Le Heron and Pawson,

1996: 142–3). The New Zealand Dairy Board (NZDB) exports 700 000 tonnes of manufactured dairy products to over 100 countries annually, providing 20 per cent of the country's export income.

The quantity of wool exported increased more than three-fold between 1880 and the Second World War, but its share of the total value of exports fell from 48 per cent to 20 per cent during this period (Table 7.1). Meanwhile exports of foodstuffs rose dramatically: from 3.5 per cent to 65 per cent, virtually all to the UK. Butter, cheese and sheepmeat were the prime components of these exports, reflecting the demands of the UK market and New Zealand's ability to supply the produce at affordable prices. The result was that nineteenth and early twentieth-century forest clearance replaced timber with pasture (for sheep and dairy cattle) or grain growing, and extensive pastoralism for wool gave way to dairying or meat production. A network of dairy factories and meat freezing works was developed as processing outlets. Expansion of the road, rail and coastal shipping systems improved transport efficiency, and there was a stream of technological improvements both on-farm and to processing factories that

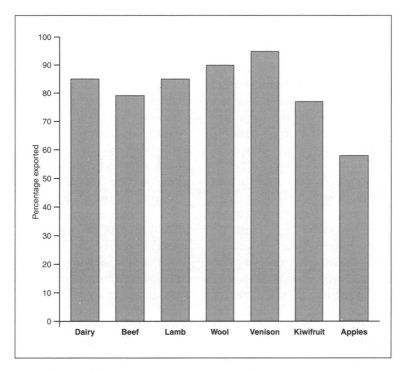

Figure 7.1 Percent of New Zealand production exported (based on Journeaux, 1996: 2)

Table 7.1 New Zealand's principal export items by percentage of value

	1885	1910	1930	1950	1970	1990	1995
Wool	54.8	41.0	17.3	40.9	19.3	8.6	4.7
Meat	6.4	19.0	24.6	15.7	34.5	15.2	14.4
Dairy produce	2.4	14.8	42.0	29.5	17.8	13.6	17.5
Casein			0.4	0.4	2.5	2.3	2.9
Hides/skins	3.6	5.6	5.0	5.5	4.6	3.3	3.3
Apples		<0.1	1.5	0.4	0.9	1.4	1.7
Kiwifruit					<0.1	3.1	1.9
Vegetables					0.5	1.0	1.4
Fish				0.2	2.3	5.5	3.5
Grass/clover seed	0.6	0.6	0.4	1.2	0.4	0.2	0.2
Timber	2.7	2.0	<0.1	0.2	3.2	3.0	5.7
Wood pulp/newsprint					9.3	4.9	3.6
Minerals	19.5	14.3	3.3	2.3	<0.1		
Manufactured goods/ machinery/transport equipment/others	10.0	2.7	5.4	4.7	4.7	37.9	39.2

Source: Statistics New Zealand, *New Zealand Overseas Trade Statistics.*

maintained competitiveness (Grey, 1994: 348–9). From the First World War the New Zealand government took measures to help regulate production and marketing (see details of producer boards below), establishing buffer accounts to counter poor returns in certain years. However, this tended to reinforce the existing pattern of production, thereby making it harder to break away from the reliance on the UK market.

Between 1885 and 1935 the area under agricultural production rose from 2.65 million ha to 7.9 million ha, with 89 per cent under sown grass (today 81.5 per cent of the agricultural area of 16.6 million ha is under grass). Sheep numbers increased by 63.5 per cent to 29 million, with 11 million sheep carcasses exported annually, of which 72 per cent by weight were classified as lamb. During this period cattle numbers rose four-fold to 4.3 million, of which 42.6 per cent were dairy cows. Factory-based butter and cheese production rose from just under one million kg to over 255 million kg, with regional concentrations in Waikato and Taranaki.

The owners of the large sheepruns were a dominant force in the country until the 1890s when the Liberal government encouraged the development of smaller family farms. Some of the large landed estates were broken up under new legislation, and a further 2.64 million ha were purchased from Maori between 1892 and 1929, but the changing economics of farming also contributed to sub-divisions and sales (Fairweather, 1982). Notably in the form of the 1892 Land Act and the 1894 Advances to Settlers, the legislation promoted land settlement and further development of agricultural land by European settlers. Similar legislation continued through the first half of the twentieth century, even into the 1950s with the 1952 Land Settlement Promotion and Land Acquisition Act.

Spatial patterns of production changed as a result of the expansion of output and cultivation of new lands. For example, from 1900 Auckland and North Auckland experienced a very rapid growth of dairy production, so that by 1939 they had just over half the country's dairy herd and produced two-thirds of the butter and one-fifth of the cheese. The North Island had two-thirds of the cultivated land and 70 per cent of the sown grassland. In 1906 Wellington superseded Canterbury as the district with the most sheep (though today Canterbury has 7.3 million sheep as does Wellington–Manawatu–Wairarapa). In the early years of the century the Romney Marsh breed, with excellent meat-producing qualities, displaced the Lincoln and the merino.

Increased production on farms was developed by harnessing the latest machinery and improvements in plant and animal breeding. By the Second World War only 14 per cent of the dairy herd were not milked by machine, contributing to a very hygienic milk-handling system to give a saleable, uniform product (Grey, 1994: 359). Milk processing using on-farm cream separators was developed at the turn of the century and helped shift the location of butter factories into more remote areas, the factories housing machinery for pasteurising and processing the farm product. However, the fortunes of both the dairy and meat factories fluctuated greatly in response to new technological developments and changing demand for their products. The key difference between the two sectors was that co-operatives came to dominate dairy processing whilst meat factories were in private hands, being largely owned by British firms such as Borthwicks, Swifts and Vesteys.

The establishment of small co-operatively owned dairy factories in the late nineteenth century was attractive to many farmers because relatively little capital was required, an assured market for the farmers' milk was guaranteed, and any liability faced by the co-operative could be easily countered by deductions from the monthly milk cheque paid to each supplier or shareholder. In the 1880s it usually cost under NZ$1000 to set up a dairy factory (Yerex, 1989: 66). Subsequently, government support has assisted the dominance of co-operative dairy manufacture.

Both butter and cheese factories needed to be located close to the source of milk supply,

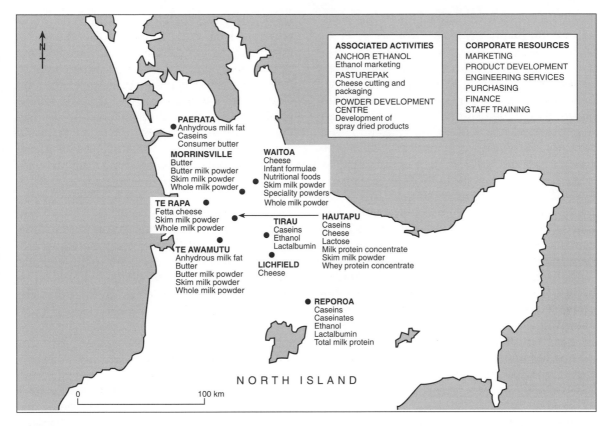

FIGURE 7.2 The New Zealand Dairy Group (NZOG) (based on Begg and Begg, 1997: 169)

especially in the case of cheese production which used whole milk. Butter production used cream and was usually manufactured in larger-scale factories. In 1945 the country had 230 cheese factories and 116 butter factories, but subsequently substantial rationalisation has occurred. This has been driven by improvements in the transfer of milk from farm to factory (e.g. bulk tankers) and scale economies at factories. Small factories have also been unable to meet the costs of introducing sophisticated equipment. Hence there have been several waves of closures: in the Waikato the 43 factories controlled by eight companies in 1971 have been reduced to 10 factories and two companies (Begg and Begg, 1997: 167). Concentration of ownership has produced substantial enterprises, with the largest being the New Zealand Dairy Group (NZDG), which had assets valued at $NZ992.5 million in 1995.

Rationalisation was also promoted by major changes in manufacturing processes, which have enabled factories to produce several products according to demand. The complexity of the organisation of this production is illustrated in Figure 7.2 by the pattern of horizontal integration employed by the NZDG. The contrast with nineteenth-century reliance essentially upon just two products is obvious.

From the 1920s dairy farming has become more focused on environmentally advantaged areas, a process encouraged from the 1950s by co-operative dairies employing differential transport pricing policies to discourage suppliers at the spatial margins (Moran and Nason, 1981). In response, dairy farmers outside these favoured areas have formed small specialist companies producing particular products for the 'top end' of the market (Moran, 1987). This contrasts with the general trend towards fewer but larger integrated processing plants.

In contrast to dairy manufacturing, meat processing has remained largely in private hands. Even in the early 1960s foreign-owned factories processed over half the sheep and lambs slaughtered compared with 27 per cent for New Zealand-owned companies and 7 per cent for producer co-operatives (Le Heron, 1987). From the late 1960s a major change has been the reduction in foreign ownership: to less than 11 per cent of sheep slaughtered and processed for export in 1994 compared with 60 per cent for producer co-operatives and 29 per cent for other New Zealand-owned firms. A similar trend has affected processed beef, but with a higher proportion remaining in foreign hands (15 per cent) and under co-operative ownership (40 per cent) (Le Heron, 1992; Le Heron and Pawson, 1996: 142). In part these changes reflect a response to the need to seek diversified markets from the early 1960s, with the prospect of the loss of ready access to the UK market. This prospect became a reality in 1973 when the UK became a member of the European Economic Community (EEC), later renamed the European Union (EU), and quotas

were placed on dairy produce and meat imported from New Zealand. One outcome of this has been the promotion of diversification of output by product, the upgrading of some freezing works, opening of new plants and closure of others, e.g. closures at Petone (Wellington), Southdown (Auckland) and Patea (Taranaki); and new plants at Oringi and Takapu (Wellington). The latter are on green-field sites outside major centres but use the latest technology and a small labour force to process meat obtained from a large catchment area. Increased reliance of factories upon contracts for special types of output, such as boneless lamb joints or meat produced according to Halal slaughter for the Moslem markets, has fostered a greater concentration of factories on North Island green-field sites (Robinson, 1988) (see Figure 7.3).

The producer boards

From the early 1900s more government control over rural production systems and marketing was introduced, for example via the 1908 Land Drainage Act, helping landowners to drain

FIGURE 7.3 An integrated dairy processing complex in the Manawatu, North Island (photo G.M. Robinson)

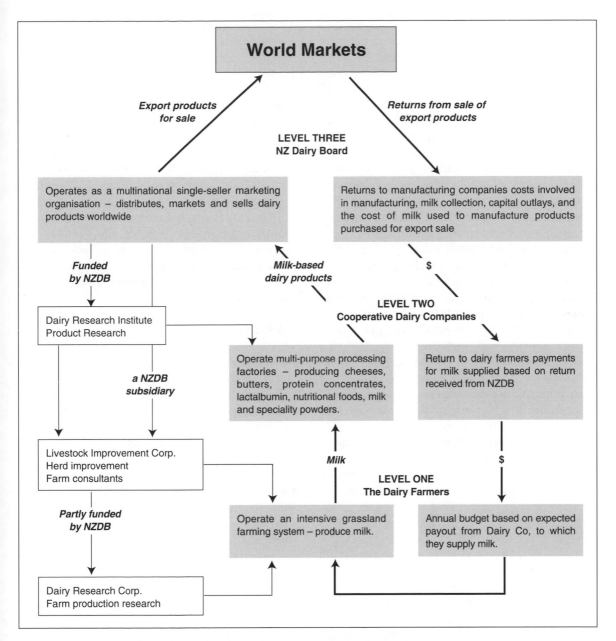

FIGURE 7.4 The vertically-integrated dairy industry (Begg and Begg, 1997: 160)

land otherwise unsuitable for pastoralism. The co-operative structure of dairying was strengthened in the same year by the Dairy Industry Act which ensured that all dairy processing would be performed by co-operatives. This also introduced production stand-ards that ensured a reliable quality for all the country's dairy produce, thus enhancing its attractiveness in its prime export market, the UK. At the farmers' behest, further controls were introduced during the depression that followed the First World War, through the creation of control boards by the Reform government to regulate collection of farm output and its transportation to the UK. This re-established regulations first utilised during the war when the UK had commandeered all New Zealand's meat, wool and dairy exports. The producer-controlled boards had statutory powers to buy, sell and market produce in order to stabilise agricultural prices and ensure adequate returns to farmers, and they could purchase surplus products at prices generating a basic income to farmers. However, their chief function was to promote agricultural exports for meat, wool and dairy produce.

One of the first boards established (in 1923) was the Dairy Board (NZDB), created to arrange shipping contracts, insurance cover and the distribution of dairy produce overseas. Marketing was excluded from the board's initial remit, but greater government control of export selling was introduced by the first Labour government in 1935 in conjunction with a system of guaranteed prices to farmers for their milk sold for manufacture. Marketing and price support were amalgamated as responsibilities of the board in 1961. The board's members were elected by directors of dairy companies who in turn were elected by dairy farmers. Its chief function was to establish the price paid to dairy companies for export produce and to market that produce overseas. In return for marketing milk products overseas the NZDB returns to the manufacturing companies the costs involved in manufacturing, milk collection and capital outlays (Figure 7.4).

The NZDB provides financial stimuli to processors to vary the quantity of individual products and to undertake product development. It co-ordinates output by acting as the sole purchaser of dairy products for export, and has made substantial investment since the early 1980s in developing speciality cheese, milk powders and related products. It has become the largest milk marketing organisation in the world, operating as a transnational enterprise in which a principal aim has been to add value to its basic raw material, milk. As part of this strategy, it has utilised own-brand product

'placement' (e.g. Anchor and Fernleaf), contracts with fast-food chains (e.g. McDonalds), and established regional development centres in the form of eight holding companies in different parts of the world. These centres have enabled specialisation to be pursued in terms of different products being supplied to different markets in response to demand. This has helped the NZDB to increase its leading national share of merchandise export earnings. Its 18 per cent share (NZ$3500 million) compares with 7 per cent (NZ$1425 million) for Air New Zealand, the next largest company (Le Heron and Pawson, 1996: 66). The NZDB is the second-largest trading company in the country, with a turnover in excess of NZ$5 billion and a market value of NZ$3.5 billion (Le Heron and Pawson, 1996: 294–5).

A key subsidiary of the NZDB is the Livestock Improvement Corporation (LIC), established in 1939 to carry out the activities of the Herd Improvement Plan: to help dairy farmers improve their profitability through the main activities of herd testing, artificial breeding, farm advisory services, research, service development and herd recording. Subsequently, production per cow has increased by over 50 per cent, production per labour unit has trebled, and production per farm has quadrupled (Smith, 1997: 177). Until 1972 herd improvement was part-funded by government, but now almost all funding comes from the dairy industry itself.

The producer boards helped secure a significant part of the UK market between the world wars, with a series of controls established in the 1930s and post-1945. In particular, these reinforced producer-board dominance of agricultural marketing, with an emphasis upon the export sector. Through the boards and co-operative processing establishments, farmers and pastoralists have been able to exert a high degree of control over the processing and marketing of their produce (Moran *et al.*, 1996). This continues today, as when economic deregulation was introduced on a large scale from 1984 the marketing boards for export produce remained in place, whilst similar structures were deregulated in industries that served only the local market (e.g. town milk, wheat, tobacco, eggs and poultry) (Roche *et al.*, 1992). The boards' dominant status, controlling around 80 per cent of all agricultural and horticultural exports, was anomalous, given the Treasury's calls for market competition between exporters to be encouraged. So there have been some changes to their remit and they have focused more upon facilitating quality enhancement and fostering research and development. Access by the boards to concessional finance was removed and there was pressure to review their ownership structures, lift their statutory monopoly, open them to competition and remove their power to make compulsory levies. In recent years further moves to restrict the role and market dominance of the boards have come from the New Zealand Business Round Table and from potential competitors. Nevertheless, despite attacks from within and outside government, the boards have survived, their record in obtaining new markets securing them a continuing highly significant role in the New Zealand economy.

The range of powers of the boards varies widely between the different industries that they represent. For example, the NZDB, the Apple and Pear Marketing Board (NZAPMB) and the Kiwifruit Marketing Board (established in 1988) have near monopolies on export marketing and own substantial assets whereas the Wool and Meat Boards play only a small role in marketing and have limited assets.

THE LOSS OF THE ASSURED MAKET

The era of substantial state support

EEC quotas on imports of butter from New Zealand adversely affected dairying in the 1970s, and the number of dairy cattle fell below 3 million for the first time since the 1920s. A reorganisation of dairy processing was a further outcome of the threat to an assured market, with economies of scale contributing to

the closure of small factories and a reliance on fewer but larger units located in the main dairy producing regions. Some dairy farmers turned to beef production (Le Heron, 1989). Meanwhile, exports of frozen lamb were directed to new markets in Arab countries and Japan, encouraging increased stocking of sheep: from 55 million to 70 million in the 1970s.

The effects of rising farm input costs were also readily observable in the 1970s when the cost of some inputs to the livestock sector rose four-fold. High costs reflected costs of expensive home-produced farm machinery and the dramatic impacts of world oil price rises upon manufacturing industry. Thus, at the same time, farmers faced both rising costs and the loss of their assured overseas market. Under Sir Robert Muldoon's National government, the solution from 1975 to 1984 was to develop an elaborate price-support system and incentives so that by 1984 approximately 30 per cent of revenue for sheep and beef enterprises was provided by government-funded support structures. This built upon subsidies introduced in 1963 for phosphate fertilisers and pesticides and tax allowances on farm investment. In the year ending March 1984 the government spent around NZ$2 billion or 6 per cent of national income on assistance to agriculture (Hopkins, 1991).

Farm price supports were raised substantially in 1978 and peaked in 1984 at NZ$505 million, of which four-fifths were supplementary minimum prices (SMP), representing the difference between the price set by government at the start of the year and the lower price that the farmers could obtain on the world market. In 1981 wool and meat received NZ$52 million in subsidies; by 1984 this had increased over seven times. However, dairy production remained largely unsubsidised and reliant on its high efficiency, which started with the economies of large herd sizes. The average dairy herd in New Zealand had 141 cows in the early 1980s; the corresponding figures for the UK and the USA were 58 cows and 39 cows respectively. One farm labourer in New Zealand looked after 140 cows compared with just 16 cows by his or her European counterpart (*The Economist*, 1 June 1985, p. 17).

Despite the rapidly escalating subsidies, farmers' profits fell, primarily because the cost of inputs was at least double the cost of subsidies. In part this reflected the fact that the costs of many of the home-produced inputs were artificially high because domestic manufactured goods were also heavily protected and uncompetitive in world terms (Le Heron, 1988). One solution for farmers was to diversify their enterprise, and new sources of income began to emerge, for example deer farming on South Island pastures (over 4000 farms now stock deer, with half a million deer in total). Two more internationally well-known developments were the growing of grapes and of kiwifruit (see Table 7.1). The area under grapes trebled in the 1970s (to 4500 ha) and by 1985 covered 5500 ha (primarily in Hawke's Bay, Gisborne and Marlborough). In 1984 exports of kiwifruit totalled NZ$125 million, equivalent to 31 per cent of the value of all fruit and vegetable exports, from 17 000 ha, 60 per cent of which was in the Bay of Plenty. However, animals and animal products continued to account for 62 per cent of all merchandise exports, and the 'kiwifruit revolution' waned somewhat in the 1990s as international competition grew (especially from Chile and South Africa). The area under kiwifruit in 1996 was nearly 8000 ha less than the corresponding area a decade earlier. Amongst other 'exotics' marketed from the mid-1980s have been the Asian pear, with the appearance of a golden apple, which had the marketing slogan of 'so moist and juicy you need to wear a wet suit to eat it!'; the fuji, a red apple from Japan; and the kiwano, also known as the African horned cucumber, a type of melon, of which one million were exported to Japan in 1986.

The removal of farm subsidies

New Zealand is distinctive in the developed world in that the long term alliance between the state and farmers was broken dramatically in

TABLE 7.2 The Crown Research Institutes

- *AgResearch*: research into pastoral agriculture
- *HortResearch*: research into permanent horticultural crops
- *Crop and Food Research*: research into annual crops – arable, vegetables and flowers
- *Landcare Research*: research into the environmental management of land resources
- *National Institute of Water and Atmospheric Research (NIWA)*: research into marine and water systems
- *Institute of Environmental Science and Research (ESR)*: research in the area of public health, environmental and forensic sciences
- *Institute of Geological and Nuclear Sciences (IGNS)*: research into geological processes and nuclear science
- *Forest Research Institute (FRI)*: research into forestry

Source: Journeaux (1996:5).

the mid-1980s as part of the economic reforms of the fourth Labour government (Robinson, 1997b; Sandrey and Reynolds, 1990). The elimination of various farm supports created a set of circumstances largely unfamiliar to farmers in the rest of the developed world, namely a lack of government price or farm income maintenance. In 1984 SMPs were eliminated and there was a progressive increase of Rural Bank lending rates to market levels. Other key reforms included phasing out fertiliser subsidies, a lowering of irrigation and water supply subsidies, elimination of a noxious weeds subsidy, the reintroduction of cost recovery for product inspection, and the conversion of the livestock standard value system to a similar basis to general business taxation. The free government advisory service to farmers was replaced by a user-pays regime and then a fully commercial business, Agriculture New Zealand, which remained part of the Ministry of Agriculture and Fisheries (MAF) until sold to estate and land agents Wrightson in 1995. Agricultural research has become part of the

Crown Research Institutes (CRIs) (see Table 7.2), funded via the Foundation for Research and Science Technology (FRST). However, the producer boards, and especially the NZDB, retained substantial control over marketing and a limited ability to regulate prices (Cloke, 1989), though in 1986 the NZDB lost its access to cheap credit from the Reserve Bank, and in 1988 it was made responsible for its own borrowing, investment and the setting of milk prices. The board continues to fund research and development and provides loans to dairy companies for their capital works. Through the Closer Economic Relations (CER) agreement with Australia it has acquired Australia's largest dairy processing company, thereby obtaining a dominant position in the Australian market.

New Zealand's overall level of producer subsidy in 1984 was around 18 per cent of farm income compared with just 3 per cent today. This contrasts with equivalent figures of 40 per cent in 1984 and 49 per cent today in the EU (Journeaux, 1996: 4), though, as Table 7.3 shows, there are other forms of government support for agriculture in New Zealand that have been retained. The immediate consequences of the removal of many producer subsidies were a reduction of farm incomes in real terms over a 15–month period to just one-third of their level in 1975, or for dairy farmers about a 40 per cent fall. Of the country's 76 000 farms, 5000 (or 10 per cent of the country's farmland) were put on the market at prices between one-third and one-half of their 1982 value. Farm indebtedness rose to over NZ$8.5 billion, reflecting the high levels of borrowing that had occurred during the 1970s when farm prices were high under the influence of inflation, the high cost of imported oil and the introduction of farm subsidies. Nearly two-thirds of all New Zealand farmers were significantly in debt in 1986, though some relief measures were introduced in the 1986 budget to permit the restructuring of debt repayments.

The devaluation of the New Zealand dollar in 1984 (from NZ$2.16 to the £ in early 1984 to NZ$2.70 to the £ later that year) helped render

TABLE 7.3 Assistance to pastoral agriculture (NZ$m), 1980–90

	1980	1985	1990
Output assistance[a]	136	630	35
Input assistance[b]	79	57	18
Assistance to value-added factors[c]	189	348	153
Total assistance	405	1035	206
Total value of output	2621	4577	6148

[a]NZDB stabilisation; meat industry stabilisation; supplementary minimum prices; inspection, grading and hygiene, town milk subsidy.
[b]Fertiliser subsidies; livestock incentive scheme and land development encouragement bonus; agricultural pest control.
[c]Advisory services, labour, research and extension, animal health and quarantine, interest concessions, taxation concessions, agricultural organisations, Rural Bank debt write-off, climate relief grants.
Source: Le Heron (1992: 152); Tyler and Lattimore (1990: 72–3).

produce more competitive on overseas markets. However, it adversely affected the cost of imported inputs, adding to the cost–price squeeze experienced by farmers. Its effects were most pronounced on the costs of fertilisers, transport and processing costs. The loss of subsidies also had a dramatic impact upon the prices of some agro-commodities, notably of fat lambs whose price was more than halved before recovering at a level 20 per cent below that of the late 1970s. The impact was exacerbated by the continuing reductions in access to European markets. For example, between 1973 and 1986 the quota for New Zealand exports of cheese to the EEC fell from 68 000 tonnes to 9500 tonnes whilst the comparable figures for butter were a reduction from 165 000 tonnes to 79 000 tonnes. The latter helped to account for the development of a 60 000 tonne stockpile.

In protest at the reduction in government support for farmers, in May 1986, 10 000 farmers staged a protest rally at the Parliamentary Grounds in Wellington, the first large-scale march on parliament in the country's history. Whilst seven crop-sprayers 'buzzed' parliament, an effigy of the Prime Minister was hanged and one of the Treasurer was burnt. Over a decade later there are sharply opposing views on just how severe the impacts of the reforms were upon farm incomes. For example, a former leader of Federated Farmers, Brian Chamberlin (1996: 51), asserts that there was relatively little 'fallout' from the removal of subsidies because:

(a) many farmers had little debt and did not have to cope with high interest rates;
(b) many of those who were substantially indebted had their debt restructured in 1986 when the Rural Bank was allowed to offer farmers reduced levels of liability. Many also started working off-farm to bring in money to reduce debt and make them less vulnerable;
(c) management of most farm businesses was of a high standard;
(d) farmers worked collectively through their organisations to bring about the changes required for long-term viability;
(e) a huge improvement was made in efficiency of the whole farm service sector.

This presents a positive view of the loss of subsidies, based on the reasoning that it was not the state's business to protect farmers from the economic realities of the world market. It is also based on the observation of a slow recovery in farm profits from 1988 whilst acknowledging that heavy protectionism for farming in the EU and North America has restricted New Zealand's ability to compete in certain countries, and hence the strong support given by successive New Zealand governments to principles of free trade. Despite protectionism in many foreign markets, New Zealand's agricultural exports doubled in value between 1984 and 1996, though as a proportion of all merchandise exports, agriculture's contribution fell from 70 per cent to 52 per cent. The latter reflected concerted moves towards a

more diversified economy (see Chapters 13 and 14), paralleling diversification within agriculture itself, which has been one of the key strategies adopted by farmers to meet the challenge of the new economic context in which farming operates (Cloke *et al.*, 1990).

FARMING WITHOUT SUBSIDIES

The international farm crisis

'The image of the family farm that resides in the Kiwi psyche – say, 200–300 ha won from the bush in pioneering days, dotted with sheep and cattle and sustaining a farming couple and their children – simply isn't workable these days'.

'The family farm is either getting bigger, to gain greater economies of scale and productivity, or smaller, to accommodate the lifestylers and boutique farmers. Alternatively, it is being converted to forestry, or maybe deer. Perhaps it is operating as a farm-stay to earn a few extra dollars, or one spouse is working off-farm to service the family's debt' (Watkin, 1997: 20).

These quotes symbolise what has been referred to as the 'international farm crisis', affecting not only New Zealand but also other countries in the developed world (Goodman and Redclift, 1991). It is a crisis in which some of the traditional certainties of farm life are being challenged, resulting in new farming practices and a greater polarisation between large commercial operations and smaller-scale 'family' farmers. Whilst there are particular local dimensions to the 'crisis' in New Zealand, there are also factors common to the developments elsewhere. In particular, the crisis reflects the twin impacts of rising costs of farm inputs and falling returns from the sales of farm produce, as processors, wholesalers and retailers seek to maintain or expand their own profit margins. In most developed countries the crisis has arisen against a background of continuing strong state support for agriculture in the form of subsidies and tariff protection.

Box 7.1 Wine production

In 1960 grape production and wine-making were concentrated in West Auckland and Hawke's Bay. In the former, vineyards were operated to a substantial degree by Dalmatians with a strong cultural tradition in wine-making, e.g. the Nobilo family. Production was dominated by small family farms, using family labour, with the growing, processing and selling of wine all as one operation. In contrast, the Hawke's Bay production was in the hands of a few individuals, and vineyards were largely company-owned, using machinery, wage labour and large-scale specialist commercial production. Workman and Moran (1993) recognise four stages of expansion from this base, stimulated initially by increased wine consumption in New Zealand itself. The basic geographical changes are shown in Table 7.4, demonstrating substantial dispersion of the industry, including an extension of wine production further south than elsewhere in the world.

(a) Stage One
During the 1960s the largest area of new plantings was around Auckland, with new areas exploited away from the traditional grape-growing lands around Henderson. Multiple-site operations developed and there were expansions to the south of Auckland by firms that have considerable reputations today, e.g. Villa Maria, Montana and Cooks, which planted its first vineyard in 1969. Dalmatian family wineries were still dominant, but with inter-generational transfer of control. Larger, private companies were formed which was indicative of the new commercial era.

(b) Stage Two
The larger commercial Auckland-based wineries became inter-regional operations in the 1970s, in particular contracting grape-growers in Gisborne and Hawke's Bay. As a result, Gisborne became the largest grape-growing area by the early 1990s. The expansion was led by Corbans and Montana and followed by other Auckland wineries, e.g. Nobilos and Villa Maria. Gisborne was a popular area of expansion as it produced good crops of the Muller Thurgau variety, which companies were keen to grow instead of the hybrids that dominated existing vineyards. Its high yields made it a profitable variety to be grown on contract and assisted mass production with mechanical harvesting facilitated by the flat terrain. The companies also favoured contract production as they could not afford to purchase new land for vineyards.

(c) Stage Three
In 1973 Montana expanded into the South Island, helping to develop Marlborough as a major wine-producing region: the country's largest today. In contrast, the importance of Auckland for grape-growing has declined. Montana's move coincided with its original owners, the Yukich family, selling a 40 per cent share to the international wines and spirits company, Seagrams, and publicly floating another 20 per cent. This raised income for purchase of an initial 1600 ha in Marlborough, where land was cheaper than in the North Island and possessing both suitable soils and the longest annual hours of sunshine in the country. In 1977 Montana built a winery near Blenheim, though bottling is still perfomed in Auckland which accounts for nearly half their market. Montana has been limited by the land acquisition court in its ability to buy more land for production in Marlborough and so has used contract growers elsewhere. In successfully marketing their Marlborough Sauvignon Blanc internationally, Montana have stimulated other producers to develop in the region. In 1979 Corbans, by then subsumed by the international conglomerate, Rothmans, purchased land there. Other Auckland-based firms have either established their own vineyards in Marlborough or have utilised contract production. In addition, in the last ten years several small 'boutique' wineries have been established in the region.

(d) Stage Four
From the mid-1970s the wine industry has spread over a much wider area, with the accompanying growth in the number of small wine-makers. New areas of grape-growing have emerged in Martinborough, Nelson, Canterbury and, in the 1990s, central Otago (Fitzharris and Endlicher, 1996) (see Figure 7.7). These smaller concerns have brought experimentation and innovation to the industry, often drawing solely upon household labour and capital accumulated in other sectors to develop wine production. Some of this activity has been part of hobby or part-time farming, or as part of developing an attractive lifestyle. Frequently, the smaller entrants to the industry have combined grape-growing, wine-making and retailing (by cellar door sales or mail order) so that profits (and losses) from each stage all accrue to the family enterprise.

The area under vines increased by 25 per cent between 1992 and 1997, with over 57 million litres of wine produced in 1996. Of this amount, 38 per cent was exported, netting NZ$75.9 million in 1997, though this should be compared with a total value of NZ$1302.5 million for exports of horticultural produce (dominated by fresh fruit). The UK is the dominant export market for New Zealand's wine (62 per cent of exports), followed by Australia (12.5 per cent).

From 1984 this has not been the case in New Zealand, although as Table 7.3 shows there are still certain types of state support for farming in existence.

In New Zealand the crisis has not destroyed the fundamental core of beef, lamb and dairy produce, but there is now a more diverse mix of products and markets, and important changes to some production systems. There has also been a diversification in the agricultural export base, involving wine (Box 7.1), timber, fish, pipfruit and kiwifruit amongst others, and a significant growth in the recreational use of farmland. This diversification is both part of the farm crisis and, possibly, evidence for a new food regime (Le Heron and Roche, 1995; Moran *et al.*, 1996).

The concept of food regimes has linked international relations of food production and consumption to forms of accumulation and regulation under capitalist systems since the 1870s (Friedmann and McMichael, 1989; Robinson, 1997a). Its proponents generally recognise the existence of three regimes, the first of which was determined by white settler agriculture in the late nineteenth century as part of a new international order. For Australasia this meant the supply of unprocessed and semi-processed foods and materials to the industrial 'core' of Europe and North America. The second regime took shape post-1918, based on the development of agro-industrial complexes focusing on production of grain-fed livestock, fats and durable foods. It was associated with the growth of consumerism and various forward and backward linkages (to markets and suppliers respectively). However, in Australasia this did not supplant the previous pattern of production for the core, and neither has the incipient third regime, which is associated with global restructuring and financial circuits linking production and consumption (Moran *et al.*, 1996). For New Zealand this new regime may involve a re-orientation from the traditional primary exports to a more important role in supplying fruit and vegetables to the global market (Le Heron and Roche, 1995). This global dimension also applies to

both the successes in marketing the traditional staples to new customers (especially in southeast Asia and the Middle East), and to new sources of investment (Le Heron, 1991a), e.g. the takeover of the New Zealand firm Watties, first by the Australian-based Goodman Fielder and then by H.J. Heinz, an American-based multinational corporation (MNC). The latter may encourage more export-based production of fruit and vegetables, as part of 'the global sourcing of produce by footloose transnational food companies' (Coombes, 1997: 37) under the sponsorship of the World Trade Organisation (WTO).

It remains to be seen whether fundamental shifts occur in the character of New Zealand's agricultural production as part of a possible new food regime, and also whether farm ownership by external capital will increase under this new regime, despite the growth of farm linkages with agribusiness for credit, inputs and processing (Lawrence, 1996: 59). However, it is clear that the need to diversify markets and the impact of reduced government price supports has opened New Zealand's food and fibre sector to the impacts of the global economy. The demands of the new challenges are being met by strategies that add value to farm produce 'through further processing, product differentiation, branding and income management' (Le Heron and Pawson, 1996: 121). Diversification of products and markets means that New Zealand meat is now exported to around 80 different countries, and even medium-sized processors can fulfil up to 1500 individual specifications sought by overseas clients.

The government's wish to avoid further involvement in agricultural price support activity means that New Zealand's farmers have been rendered more susceptible to international trends in commodity prices and developments in international food and monetary systems. The most obvious consequences of this have been declining farm incomes, especially for smaller farmers and for certain sectors (notably sheep and beef cattle), increased indebtedness despite debt restruc-

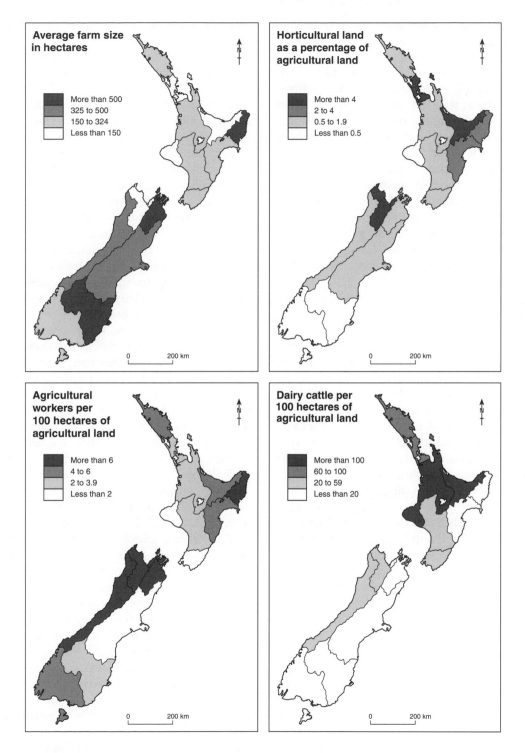

FIGURE 7.5 Agricultural distributions in New Zealand, 1997

turing measures, and various adjustments to farm practice. The latter have included measures to reduce expenditure and increase efficiency (e.g. O. Wilson, 1994; 1995). For example, spending on farm inputs and maintenance has been reduced, as has expenditure on labour and machinery, whilst off-farm wage-earning by farmers and the farm household has increased.

Changing practices have resulted in lower stock numbers and diversification into new enterprises, including low-input activities such as sphagnum moss collection in Westland, direct replacements for 'traditional' staples, e.g. deer for sheep, and new enterprises directed at evolving new and niche markets, e.g. viticulture, pipfruit, organic production (Campbell, 1996) and olives. Yet some of these emerging enterprises may be short-lived or rely too heavily upon a volatile world market. This is typified by recent experiences in the pipfruit industry where 'a world oversupply of fruit has shaped poor returns to growers for the third straight season [1996–8]' (McKenna, 1998: 38). This has contributed to the export agency of NZAPMB introducing a 'clawback' on payments made to exporters of red delicious apples, and also proposals for reorganising the structure of the relationship between the NZAPMB and growers.

The impress of overseas market requirements is being felt through the need to meet new environmental and food safety regulations in these markets. Thus, as the UK takes nearly 30 per cent of New Zealand pipfruit exports, new UK regulations are encouraging the advance of integrated fruit production (IFP) in three of the main growing areas (Hawke's Bay, Nelson and Central Otago), involving 'reduced use of organo-phosphate pesticides; selective and targeted chemical use after monitoring for actionable pest levels or disease presence . . . an overall commitment to more "environmentally produced" fruit' (McKenna, 1998: 41) and the use of new varieties, e.g. Pacific Rose, Southern Rose, Pink Lady and GS 494. In contrast, current USA standards do not admit IFP fruit and hence this may restrict IFP in Nelson,

where growers have the closest export links to the USA.

One of the most significant changes post-1984 has been the expansion of dairy farming (Figure 7.5). The numbers of dairy cattle have risen by 37.8 per cent (+1.1 million, 1980–95) and the amount of land devoted to dairying by 500 000 ha, comparable with the expansion in forestry. Income per ha from dairying has been three times greater than that from sheep or beef cattle in the 1990s and hence dairying has extended into areas previously dominated by livestock rearing, e.g. the Waitoto in the Waikato (Begg and Begg, 1997: 173).

One feature of this development has been the maintenance of opportunities for entry to farm ownership or tenancy by a new generation of farmers. In dairying a pathway has been developed from the early years of the twentieth century by sharemilking, where an enterprise is operated on behalf of the owner by the sharemilker who receives a share of the profits (Blunden *et al.*, 1997). Over one-quarter of the 15 000 dairy farmers in New Zealand is a sharemilker who can enter dairying this way without needing a large initial capital investment. Regulations preventing exploitation by owners have been on the statute book since 1937 (Begg, 1997).

Other changes include substantial increases in the numbers of deer (+1.1 million 1980–95) (Figure 7.6), and farms >40 ha (from 27 per cent to 45 per cent of all farms, 1972–92). The latter reflects a strong increase in part-time farming as off-farm income has become well established as an important means of supporting both the farm business and the farm family. This has been most marked in conjunction with kiwifruit and pipfruit enterprises, and also in the arable sector (Moran, 1997).

One of the ironies of these developments is that they have many similarities to those occurring throughout the rest of the developed world, where there has been a much greater amount of direct government support for farmers. However, a common dimension to the international farm crisis has been the way in which farmers have been affected by the

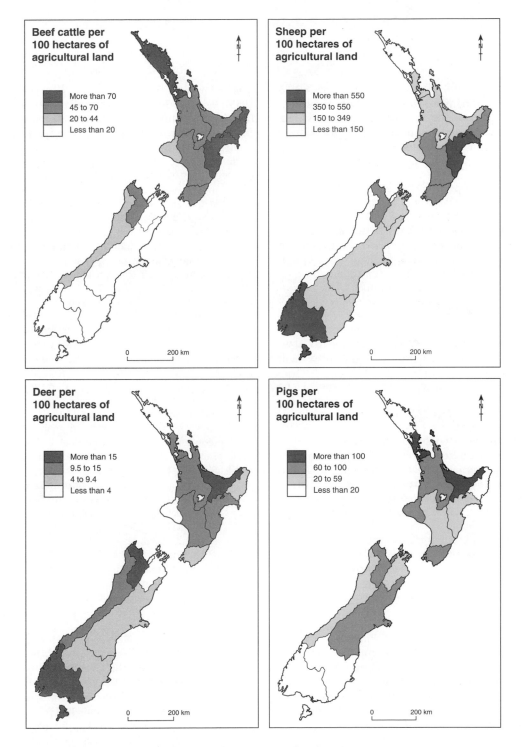

FIGURE 7.6 Livestock distributions in New Zealand, 1997

Box 7.2 Pluriactivity

Olivia Wilson's (1994; 1995) survey of sheep and beef farmers in Southland found that only those farmers with high levels of debt at the start of the downturn in the mid-1980s experienced severe financial difficulty. These were mainly farmers who had purchased land in the early 1980s when land prices were high. Nevertheless nearly all farmers experienced reduced incomes and increased indebtedness. They followed various survival strategies to cope with this, often involving reduced inputs of labour, machinery and purchased inputs. Minimisation of risk was part of these strategies as was greater reliance on family labour on the farm (often unpaid) and off-farm income (Le Heron et al., 1994). This has emphasised the importance of the role of farmers' wives in farm labour and in contributing to farm income through off-farm work.

Pluriactivity by farmers has not simply been a response to the 'crisis' caused by deregulation. Rather it existed before 1984 as one part of a spectrum of farm survival strategies, but has now emerged as a more important component of capital accumulation in rural New Zealand. For example, in the mid-1980s Pomeroy (1986) recorded 17 per cent of farm households interviewed in King Country, central North Island, with a male household member engaged in off-farm work. Approximately one-third of spouses had employment off-farm, often aided by the higher level of qualifications held by farm women when compared with their husbands. These levels of pluriactivity were observed to have increased by the end of the 1980s in work by Le Heron et al. (1994) on dairying, livestock rearing and apple-growing in selected areas of the North Island. The proportions of enterprises with pluriactivity recorded in their survey were: two-thirds for dairying (Manawatu), one-third for livestock rearing (Dannevirke) and one-half for apple-growing (Heretaunga Plains).

Coombes and Campbell (1996) recorded three predominant actions taken by farmers in Ashburton County, Mid-Canterbury, between 1984 and 1989 to offset falling incomes. First, there was restructuring of production away from commodities falling in price towards new crops or stock types. Moves into non-traditional activities were often blocked by banks or stock firm managers who were controlling farm debt. However, in general, throughout the country diversification of the farm enterprise has been a common strategy. For example, in Southland this has included moves into deer farming, timber production, horticulture and dairying (Wilson, 1995). Second, there was widespread reduction of farm inputs and personal spending, and, finally, family labour activities were intensified and paid labour was removed.

Pluriactivity in the form of off-farm work was not seen primarily as a survival strategy but as part of a standard means of enhancing family income. For men this almost invariably meant agriculture-related work, but this was not the case for women, for whom jobs in teaching and nursing were common. In the Caitlins, Southland, it was mainly women who played the leading role in work related to farm-based tourism which developed rapidly from the mid-1980s. Some of these tourism ventures were part of a need to increase income to meet 'the crisis' but few were associated with attempts to relieve substantial indebtedness (Coombes and Campbell, 1996: 15–16).

Martin and McLeay's (1998) nationwide survey of 1384 sheep and beef farmers recognises five different groups of farmers in terms of their risk management practices. These are:

(a) risk managers (27 per cent): utilising a wide range of on-farm activities to generate income, but carrying large debt burdens, and practising agricultural diversification;

(b) income spreaders (19 per cent): also pursuing enterprise diversification but typically focusing on beef cattle which gives them flexibility to keep animals longer or sell them earlier, depending on market conditions and feed availability;

(c) production managers (19 per cent): paying the most attention to pest and disease management and feed management. They emphasised high productivity and had low levels of debt, but often with relatively poor gross income levels;

(d) capital managers (18 per cent): protecting their capital base, in part by off-farm investment generating a steady cash flow with little indebtedness. There was a tendency towards larger numbers of beef cattle and low lambing percentages;

(e) part-timers (17 per cent): emphasising off-farm work, usually with smaller holdings than the other groups, and more intensive management of sheep enterprises.

TABLE 7.4 The location of grape plantings in New Zealand (ha)

Date	Auckland	Gisborne	Hawke's Bay	Marl-borough	Martin-borough	Canter-bury	Otago	Nelson	Waikato/ Bay of Plenty
1960	163	13	149						
1970	658	278	327	2	1				
1982	455	1922	1891	1214	7	49			
1992	232	1498	1577	1902	160	161			
1995	248	1514	2276	3233	271	325	152	137	137

Sources: Workman and Moran (1993); Le Heron and Pawson (1996: 302–5).

cost–price squeeze, whereby the rising cost of farm inputs has not been matched by returns from sales. The desire of supermarkets and processors to maintain food prices at a level attractive to their consumers, whilst making healthy profits, has prevented farmers from readily passing on the rising costs of farm machinery, seeds, fertilisers and pesticides. One consequence has been the development of cost minimisation strategies in farming, notably expressed by shedding farm labour, but seen also in pursuit of economies of scale and by increased integration of production functions. The latter has fostered new marketing and purchasing arrangements, blurring the distinctions between farm supply, production, processing and marketing. Farmers have also been forced to become pluriactive (Box 7.2), diversifying their sources of income, either by adopting new agricultural enterprises or by generating income from off-farm sources.

Meat and dairy processing in the 1990s

Dramatic changes post-1984 have meant that the meat processing industry has now become completely New Zealand-owned, with farmer co-operatives dominant. However, processing capacity has fallen by 25 per cent and employment by 40 per cent. The entry of the Meat Board into meat marketing in 1982 produced a rapid increase in the amount of lamb and mutton being processed into cut and boneless products (Le Heron, 1990). The return to private marketing three years later encouraged more rationalisation of processing in order to maintain profits (Le Heron, 1991b; Robinson, 1988: 72–3), but subsequently some long-established firms have left the industry and have been replaced by nearly three times as many, mainly smaller companies. In 1986, 11 processor and exporting businesses operated 45 plants; by 1995 there were 32 businesses and 61 plants (Le Heron and Pawson, 1996: 142). Although economies of scale have helped produce some very large processing complexes, smaller 'niche' producers have also emerged. The long term major players have disappeared, including Borthwicks and Weddel (a subsidiary of Vesteys International which as recently as 1994 was the second-largest beef and fourth-largest sheep processor in the country), in part because of weaknesses in their operations outside New Zealand (Roche, 1996).

The pattern of closures and openings is illustrated by the demise in 1994 of Fortex, a Christchurch-based meat processing company. Founded originally in 1971 as Cattle Services Ltd, a farmer-owned operation, it expanded in the early 1980s under government export incentives, with several meat-related ventures combining as the Fortex Group, operating modern factories near Ashburton (mid-Canterbury) and Mosgiel (near Dunedin). However, the inability to develop sufficient profit margins brought in the receiver in 1994

FIGURE 7.7 Viticulture in central Otago, showing frost protection covers for the vines (photo: G.M. Robinson)

and redundancy for a workforce of 1800. Subsequently, the Ashburton plant was bought by a joint venture of other meat processors, reopening with a workforce of 600. The plant at Mosgiel has also been sold and reopened as a meat processing enterprise (Le Heron and Pawson, 1996: 147–8).

The demise of the Fortex Group and the purchase of another large processor, Weddel, by its rivals who then closed some of its factories, have enabled the remaining producers to purchase animals more cheaply from the supplying farmers whilst also selling at lower prices. This has contributed to falling incomes for sheep and beef enterprises (e.g. Perry and Morad, 1997), and hence a move by some graziers into deer or dairy production.

More than most sectors of food processing, the manufacture of dairy produce has moved from a series of small specialist producers to the generation of a range of products from large integrated plants. The latter have been better able to switch from one product to another according to market requirements whilst also benefiting from scale economies. Although

processing in the dairy industry is dominated by ownership on a co-operative system by farmers who supply the milk to factories, there are wide variations in the size of operations. However, faced with diminishing profit margins, co-operatives have increased output, closed branch plants and maximised economies of scale by complex inter-factory transfers of milk, cream, skim milk, buttermilk and whey. The accompanying plant rationalisation has reduced the number of dairy factories, for example in Taranaki from 44 in 1969 to six in 1983 (Willis, 1984). The three leading companies process over 70 per cent of milkfat: New Zealand Dairy Group (NZDG) (42.3 per cent), Kiwi Co-operative Dairies (18.4 per cent) and Northland Co-operative Dairy Company (9.9 per cent) (NZDB, 1994).

New Zealand's largest co-operative dairy company, with its well-known Anchor trademark and 6000 suppliers of milk, is the NZDG. Its headquarters are in Hamilton in the Waikato, the region containing two-fifths of the country's dairy cows. The NZDG controls nine of the ten remaining dairy factories in the Waikato, though proposed rationalisation will reduce its factories there to just four by 2015.

FURTHER READING

Grey (1994) provides a thorough account of key changes in the evolution of the rural economy prior to 1939. The roles of the various producer boards are examined in Begg and Begg (1997), Le Heron (1992) and McKenna (1998). The era of state support in the late 1970s and early 1980s is analysed by Le Heron (1988: 1989; 1991a). Several studies have dealt with the consequences of the removal of farm subsidies in 1984; they include Cloke (1989), Cloke *et al.* (1990), Robinson (1997b), Roche *et al.*

(1992) and Sandrey and Reynolds (1990). Contrasting views are presented from within the farming industry by Chamberlin (1996) and Yerex (1992). The position of New Zealand's rural economy within a possible third food regime is considered by Le Heron and Roche (1995) and Roche (1996). There are several studies focusing on growing farm diversification in the 1990s, including O. Wilson (1994; 1995), Le Heron *et al.* (1994), Coombes and Campbell (1996) and Martin and McLeay (1998). Specific sectors are analysed in work by Robinson (1988), Le Heron (1991b; 1992), Le Heron and Pawson (1996, Ch. 5), and Workman and Moran (1993). Journeaux (1996) and Moran (1997) provide straightforward descriptions of recent change. Useful collections are those edited by Alston (1991), Burch *et al.* (1996) and Burch *et al.* (1998).

8

BOOM AND BUST IN THE AUSTRALIAN MINERALS AND ENERGY INDUSTRIES

If New Zealand's long term economic development has been inextricably linked to the fortunes of its farmers, graziers and the processing of primary produce, then Australia's has been reliant on a mixture of both 'farm' and 'quarry'. Whilst the Australian economy has developed a diversity of manufacturing and service industries in response to post-1945 population growth and expansion of its domestic market, the primary sector, both in agriculture and mining, has continued to be important. This Chapter examines the significance of the mining, minerals and energy sector, noting its particular influence upon the nature of economic development and emphasising its role in the generation of export revenues. Other aspects of the Australian economy are discussed in Chapters 13 (focusing on manufacturing) and 14 (services and agriculture).

MINING'S ROLE IN THE AUSTRALIAN ECONOMY

The mining 'boom' of the 1960s

Despite the fact that mining accounts for one-third of Australia's export income, with two-thirds of the output being exported, it employs less than 2 per cent of the workforce and produces only just over 6 per cent of GDP. Its significance lies in its contribution to exports and the extent to which it is controlled by overseas capital, with the Australian government having relatively little ability to regulate prices. Its export orientation means that this sector of the economy is heavily dependent upon fluctuating world market prices, which in turn are closely related to changing demands and supply in various non-Australian areas of production and consumption. Scarcity of, or rising demand for, a given mineral at a particular time can lead to high returns for producers, as evidenced periodically for gold mining. However, oversupply can quickly turn mining 'boom' to 'bust', as seen in the closure of uranium mines in Queensland and the Northern Territory in the 1960s. Such sharp swings can be prompted by a number of factors, frequently beyond the control of the producing country, thereby injecting a strong element of economic instability into those regions or countries heavily dependent upon this sector.

In Australia's case the instability of the minerals sector has had significant impacts upon the value of both mineral exports and the

TABLE 8.1 Production of selected minerals in Australia

	Coal 000 tonnes	Gold kg	Copper 000 tonnes	Lignite 000 tonnes	Crude oil 000 litres	Iron ore 000 tonnes
1860	362	759 574	5			
1880	1561	365 089	10			
1900	6488	994 265	23			
1920	13 011	293 509	27	0.1		
1940	11 914	511 341	22	4.6		
1960	22 931	338 004	111	16.3		
1980	93 406	16 805	206	32.9	23 647	4141
1990	160 400	214 400	555	40.1	27 583	111 475
1995	243 020	298 697	365	48.3	33 910	137 525

Source: Duncan (1987); *Australia Yearbooks.*

Australian dollar, the domestic rate of inflation, the levels of exploration and production of minerals, and rates of capital investment. The caveat to this picture of an uncontrollable fluctuating and inherently unstable sector is that the Australian mining industry is broadly based, so that boom and bust for different minerals are not necessarily in phase. For example, depressed coal and iron ore prices in the late 1980s were counter-balanced by soaring gold prices (Walmsley and Sorensen, 1993: 104–5). Nevertheless, the history of the mining industry and minerals development in Australia is one characterised by boom and bust in which there have been certain key periods when mining has played a significant role in overall economic development.

Three periods have been of particular importance:

(a) the 1850s and 1860s, when gold discoveries in Victoria and New South Wales helped trigger dramatic economic, social and political changes;

(b) the 1880s and 1890s, when discoveries of a wide range of minerals led to the growth of settlement in more remote parts of the country, e.g. Broken Hill (NSW), Kalgoorlie (WA) and Zeehan and Mt Lyell (Tas);

(c) the 1960s, when new discoveries of huge reserves of coal, iron ore, natural gas and other minerals contributed to mining's

share of export revenue tripling between 1964 and 1975. The other key factor at this time was the expansion of the Japanese economy which dramatically increased its demand for minerals and energy.

In the 1960s the chief benefits from the rapid growth of the mining industry were the substantial generation of export revenues, its attraction of few related imports and initial benefits to the Australian dollar (Powell, 1988: 252). Much of the 1960s' 'boom' was financed by foreign capital, with international mining concerns heavily involved in both the initial exploration phase and subsequent development. This reflected the relatively small size of the Australian capital market and long term balance of payments deficits that restricted the availability of domestic capital for investment.

Although the 1960s' mining boom was followed in the mid-1970s by an economic recession associated with the 1973 'oil crisis', subsequent efforts to substitute coal for oil in many parts of the developed world benefited Australia, which also prospered through exploration for new domestic sources of oil. As a result, coal production rose by 50 per cent between 1979 and 1984, and exports by 75 per cent.

This growth in both production and exports of coal was part of a broader expansion in the Australian mining industry, with gross output quadrupling between 1970 and 1992. This rise in output contrasts with the much weaker growth in the manufacturing and agricultural sectors and was part of significant long term structural change in the economy in which a very heavy reliance has been placed on mineral exports and on the nature of overseas markets. From the late 1960s onwards coal and metal ores have accounted for at least 20 per cent of exports by value; by the early 1990s exports of mineral resources accounted for nearly half of Australia's export earnings (Table 8.1).

The new mineral discoveries of the 1960s, coupled with improved mining technology, led to sharp rises in the export of unprocessed or semi-processed iron ore, bauxite, coal and nickel, especially to Japan. Indeed, the reliance on the Japanese market has been commented on unfavourably by many politicians, and is symbolised by the saying, 'if the Japanese economy sneezes, Australia catches cold'. For example, the economic crisis in Japan and south-east Asia in 1997/8 contributed to a 20 per cent fall in the value of the Australian dollar, emphasising Australia's vulnerability to falls in demand in this particular market.

Despite commanding a large proportion of world trade in certain minerals, Australia has not been able to maintain high prices for these exports, largely because individual purchasers, and especially the Japanese, have been able to resist this by utilising, or threatening to utilise, alternative supplies. As a result there has been a steady fall in the price of mineral exports from the early 1980s, contributing to the depreciation of the Australian dollar. Japan's intention not to commit itself too much to a single supplier for key raw materials, coupled with reforms to its economy since the 1970s, has restricted opportunities for Australia to expand its already sizeable share of the Japanese market. However, the securing of new markets in south-east Asia, especially in China, South Korea and Taiwan, has meant that the recent regional economic downturn has impacted strongly on the Australian economy.

Nevertheless the export-led boom in minerals development has also been aided by sales to more distant markets, notably in Europe and North America, through improvements in handling and in the long-distance transport of bulky minerals such as coal. Other important technical developments have included new exploration techniques such as satellite remote sensing, improved geophysical equipment and advances in mining, abrasion resistance and blasting techniques. Oil and pipeline technology has enabled oil and gas reserves underneath the continental shelf in the Bass Strait and the North-West Shelf to be exploited. Here the changing economics of oil production after the early 1970s' 'oil crisis' has been among several factors in the energy industry favouring Australia, so that since the early 1980s the energy sector has accounted for over half the value added in the country's resource industries.

In addition to the revenue derived from exports, successive federal governments have also obtained a high tax yield from some parts of the minerals sector. In order to maintain high levels of investment, certain tax concessions have been operated, notably in the oil industry. However, state governments have tended to charge royalties and to levy rail-freight charges for minerals and oil transport on the state rail systems at higher rates than for other users. On the negative side, mining has generated relatively few jobs and has relied substantially upon involvement by transnational corporations (TNCs).

Foreign control and overseas-based processing

The mining industry has had the highest level of foreign control within the Australian economy (Wheelwright, 1984), though the rapid rise of Japanese investment in the tourist sector may now be challenging this. The perceived loss of control over Australia's resources has been of periodic concern to politicians, but underlying the rhetoric related to ownership and nationalism is a more fundamental problem associated with economic management. Foreign investment in mining has brought about an influx of capital in the initial stages of operation, but subsequently there have been outflows in profits, interest, royalties and loan repayments (McColl, 1984). This is difficult to manage and, given the size of the industry, has had an effect upon the currency and the government's policies in other areas. Periodically there has been concern shown by federal government about the high levels of foreign ownership, with temporary restrictions on its amount by the Labor government in the early 1970s. However, the continuing need to develop minerals as a central component of the economy has helped maintain the need for overseas investment so that this sector retains a reliance on TNCs and capital from the world's leading economies.

The mining industry consists of several different stages of development, from exploration to extraction and processing in which there is a modification or upgrading of the uncovered material to prepare it for market. This processing can take various forms, some of which can take place at the site of mining activity (e.g. crushing, sorting, sizing, grinding, classifying and dewatering). However, when processing is more readily identified with a manufacturing stage, the use of ancillary raw materials such as energy sources and chemical inputs may necessitate the use of alternative locations. For example, iron ore from the Pilbara is shipped by rail to Dampier or Port Hedland where it is blended, sized and classified prior to shipping to Japan for further processing that will use the ore in the manufacture of electro-technical goods.

Processing within the mining sector is often classified into initial and final stages, the former including activities such as ore dressing, beneficiating and concentrating, and the latter involving a manufacturing capacity as in the case of refining, smelting and alloying. These manufacturing processes may then be linked to further crucial value-added manufacture of products beyond the mining industry, ready for purchase by individual consumers from wholesalers or retailers. Thus raw materials that have been smelted or refined may be incorporated into electro-technical goods in which the biggest share of the final retail price accrues to the manufacturing stages beyond the control of the mining industry. Therefore it is the Japanese refiners and manufacturers of electro-technical goods who derive the greatest profits from the ore, despite the ore's source being in Australia. As the initial stages of prospecting, extraction and processing may be in the hands of non-Australian-based companies, the overall returns of income to Australia from these activities may be limited, depending upon the taxation yield and additional benefits such as those derived from provision of infrastructure and employment. This calls into question the substantial reliance upon TNCs in the Australian mining industry and also the relatively limited development of the final processing and downstream manufacturing sectors. Nevertheless, the character of the mining industry is now firmly established and has been championed by successive Australian governments.

THE RESOURCE BASE

The foundations for the growth of a substantial export-led mining industry are the significant reserves of several key minerals. On the basis of current production rates Australia has reserves of iron ore for another 350 years, black coal for 280 years, and bauxite and manganese ore for over 100 years. Yet the magnitude of the reserves of coal, iron ore, oil and natural gas has only been apparent since major discoveries that started in the late 1950s. These have transformed the export base of Australian minerals, adding considerably to the growth in foreign exchange developed from minerals exports. These developments of new reserves over the past 40 years have tended to focus on single, low-value, bulky minerals mined on a large scale. It has been the scale of the operations that has made minerals' exploitation profitable

FIGURE 8.1 Metalliferrous provinces

despite the remote location of many of the reserves.

One of the most important characteristics of the resource base is the wide distribution of metalliferous provinces, with rocks and geological conditions suitable for yielding iron ore, copper, lead, zinc and precious metals (Figure 8.1). In addition there are a number of sedimentary basins containing coal, oil and natural gas, and prospects of further discoveries at depth. The location of some of the major finds, well away from the main areas of population, has helped to promote settlement in the interior, but the nature of the nineteenth century discoveries and initial development by individual prospectors working small leases has been superceded by corporate enterprises establishing a substantial infrastructure in spite of the remoteness. An important characteristic of the metalliferous provinces has been the presence of a range of minerals in the same area, most notably copper, gold, lead, manganese, silver and zinc, and this has offered mining companies a degree of flexibility, in which production can be switched to reflect the most favourable current market price. Apart from minerals utilised for energy production, the development of three key ones with significantly different development histories is considered below, namely iron ore, gold and uranium.

Iron ore

The slow development of Australia's steel industry before 1914 reflected the low cost of iron and steel imports from the UK before protective tariffs were introduced in 1907, and the limited supplies of domestic ore before extensive reserves of high-grade iron ore were discovered in the 1960s. Prior to the Second World War there had been relatively small finds in the Eastern Highland Zone in conjunction with coal, but the magnitude of these deposits has been dwarfed by new reserves in remote locations and new techniques permitting their exploitation. In 1937 reserves were estimated at 522 million tonnes, of which only 260 million tonnes were deemed to be exploitable (mainly because of the distance from the coast of the other reserves). At the 1930s' rate of consumption this gave only 50 years' worth of supply. This was one reason for an embargo being placed on the export of ore. This was not lifted until 1960 when the usable reserves were reassessed at 368 million tonnes. Just five years later they were re-estimated at 16 000 million tonnes, of which 1254 million tonnes were usable immediately. This marked the discovery of massive finds in Western Australia, which then prompted further exploration for other minerals there and elsewhere in the country, yielding substantial amounts of nickel (at Kambalda, WA), phosphates (Qld), natural gas (WA), coal (Qld) and oil (WA).

Today Australia is the world's second-largest exporter of iron ore (112 million tonnes per annum) and there are reserves of over 15 000 million tonnes of high-grade ore.

The huge increase in iron ore reserves owes much to the discovery of ore in the Pilbara region in Western Australia (Figure 8.2; Box 8.1). The story of the find has entered legend, with the entrepreneur Lang Hancock credited with an 'accidental' discovery of a 'mountain of ore' when his plane was forced to fly low in a storm. Subsequently he experienced difficulties in launching the initial phases of development of the new orefield, but with the help of foreign investment from Rio Tinto Zinc (RTZ), mining in the Pilbara got underway in 1962 in an area 300 km from the coast. Major mining operations were established at Mt Newman and Mt Tom Price, with railways constructed to link the mines to new port facilities at Dampier and Port Hedland. The exploitation of the ore owed much to input from just a few companies, notably Hammersley Iron, who built the facilities at Dampier for the export of ore to Japan for processing, and the inland towns of Karratha, Tom Price and Paraburdoo, which were founded at a cost of A$300 million. At the latter two locations mining has been open-cast, with mined ore then shipped by rail in giant 180-car loads.

Tom Price has a population of 4000 of whom 1200 are employed in mining operations, on high salaries and with a substantial tax discount for working so far from a metropolitan centre. That centre is Perth, 1100 km away (equivalent to the distance from Edinburgh to Bordeaux or New York to Atlanta, Georgia), but which is still the centre for mining companies' headquarters and sourcing of supplies.

Gold

The first reported find of gold in the colony of Victoria was in July 1851 at Clunes, 35 km north of Ballarat. However, it proved difficult to transport supplies to this location and attention was quickly diverted to discoveries at

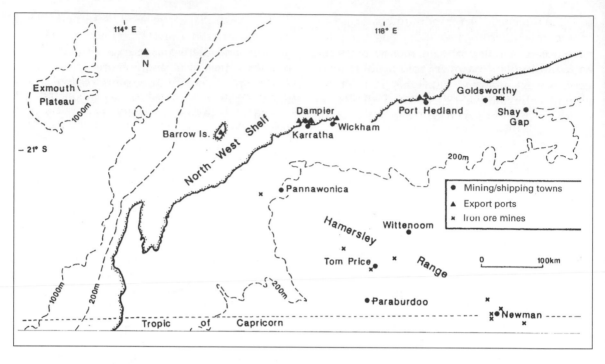

FIGURE 8.2 Distributions of selected minerals

Ballarat (Figure 8.3), Buninyong (10 km to the south) and Bendigo. Within three months of the first discovery there were 8000 inhabitants in the area between Ballarat and Buninyong. The numbers had swelled to 30 000 early in 1852 and to 100 000 four years later. Initial workings extracted gold from near-surface deposits, with individual claims predominant. These were soon displaced by deep-shaft mining financed by companies who also operated steam-powered stamping and crushing processes at the surface. Testament to the wealth generated between 1860 and 1890 can be seen in the towns' civic buildings, churches, hotels and residential houses, not to mention the wealth generated within Victoria and its capital, Melbourne, as described in Chapter 5. However, most of the mining activity had disappeared by the beginning of twentieth century, with the exception of deep-reef developments such as the Central Deborah Goldmine in Bendigo. Sunk in 1909 and closed in 1954, it was 411 m deep and had 17 mining

levels. It has been restored as a tourist attraction. In Ballarat an early gold-mining settlement has been re-created for tourists at Sovereign Hill, a reminder of the origins of the town which now, like Bendigo, functions primarily as a market centre with a prosperous agricultural hinterland.

One of the few goldrush towns still surviving as a major gold-producing centre is Kalgoorlie–Boulder, the second-largest settlement in Western Australia after Perth–Freemantle, with a population of 21 000. The discovery of gold led to rapid development in the 1890s to exploit the alluvial gold which occurs at depth in the Kalgoorlie and Coolgardie region, with deep mines being sunk, some over 1000 m deep. The aridity of the local area led to the development of the technique of 'dry blowing' or winnowing of powdered ore being used until a 550-km pipeline bringing water from Mundaring Weir, near Perth, was laid in 1896 at a cost of £3 million. Kalgoorlie–Boulder lays claim to

Box 8.1 Mining and the Western Australia economy

The impact of mining on the economies of Perth and the state of Western Australia itself has been profound. For over a century and a half since the founding of the Swan River Colony, Western Australia had been a 'poor relation' compared with the eastern states, and was heavily reliant upon federal aid before the Second World War. It was remote from the centre of government and there was a strong feeling amongst Western Australians that they were ignored by politicians in far-away Canberra. Indeed, in the early 1930s the Western Australian parliament actually voted to leave the Australian federation, before intervention from London and the rest of Australia prevented secession. The iron ore and minerals boom has helped transform Perth–Freemantle, aided by the additional boost provided by tourism, which has expanded greatly since the stimulus provided by the hosting of the America's Cup by the Royal Perth Yacht Club in 1987, after the Alan Bond-led challenge had been successful in 1983. Bond was one of several Perth-based entrepreneurs who came to public attention in the 1980s. For example, business empires led by Kevin Parry and Robert Holmes a'Court helped contribute to the statistic that, in 1990, 29 of the top 200 richest families in the country were from Western Australia. Several of these were based on income from minerals developments, which during the last two decades have also included exploitation of natural gas from underneath the continental shelf. Development of the North-West Shelf, allied with the iron ore from the Pilbara, now supports 10 new towns, four new railways and three new deepwater ports.

Western Australia was a major beneficiary of free-market policies pursued by the federal government from 1983, and, from its rich mineral base, it became the country's premier exporting state (by value of merchandise exported) by the early 1990s. It also has the highest employment in mining, the highest turnover and more than half of the country's recent mining investment (Table 8.2). Perth's population grew by 25 per cent during the 1980s, with new industries emerging to serve this population, including the new shipbuilding complex at Henderson, south of Freemantle, employing 2000 workers producing luxury yachts, high-tech catamaran ferries and military patrol boats for export.

Western Australia has 55 per cent of the Australian mineral exploration expenditure, worth A$500 million per annum. In 1996 it had A$5.8 billion in projects under construction, A$4.3 billion in 22 projects committed and A$12.2 billion in 51 projects under consideration. Its minerals sector output totalled A$14.6 million (gold 3.1; heavy mineral sands 3.0; crude petroleum 2.0; LNG 1.4; natural gas 0.42; iron ore 3.0; bauxite/alumina 1.8). Increasingly, though, new mining developments in the state involve a 'long-distance commuting' (LDC) operation or 'fly-in/fly-out' (FIFO) in which a temporary work camp is established at the site, at which food and accommodation are provided and to which workers commute for a fixed number of days of work followed by a fixed number of days at home (Jackson, 1987; Storey, 1998). Between 1984 and 1998 40 such commute-mines went into production in the state, producing significant capital savings, e.g. the Argyle diamond mine has reportedly saved A$50 million in this way, but incurred higher staff turnover (Dillon, 1991).

having the richest few hectares of gold-bearing land in the world, the Golden Mile, and this has possibly more unworked gold than that removed over the last 100 years. However, the fortunes of the town have been heavily dependent on the price of gold on the world market. As a result of fluctuations in this price, the town has gone through periods of boom and bust, the former often characterised by conspicuous public buildings and works, e.g. the

Olympic-size swimming pool built in 1937. Nearby Coolgardie became one of the best known 'ghost towns' in the country. Alluvial gold was discovered here in 1892, and within 10 years the settlement had grown into a town of 15 000 people, two hotels, six banks and two stock exchanges. However, the source of gold was limited and was soon overshadowed by the substantial reserves at Kalgoorlie. In 1985 only 700 people lived in Coolgardie, but there

TABLE 8.2 Mining production by state, 1996

	NSW	Vic	Qld	SA	WA	Tas	NT	Total
Employment	14 869	2036	14 756	2123	18 736	1086	1742	55 348
Employment (% of total)	26.9	3.7	26.7	3.8	33.9	1.9	3.1	100
Turnover (A$million)	4589	3435	6669	940	11 817	371	1115	28 936
Turnover (% of total)	15.9	11.9	23.0	3.2	40.8	1.3	3.9	100
Net capital expenditure (A$million)	620	566	677	68	2613	45	110	4698
Net capital expenditure (% of total)	13.3	12.0	14.4	1.4	55.6	1.0	2.3	100

Source: Australia Yearbook.

has been something of a renaissance recently through tourist development.

Kalgoorlie–Boulder produces 70 per cent of the gold mined in Australia. Open-cut mining is now replacing underground mines and extending operations north of Kalgoorlie to towns where traditional mining activity ceased earlier in the century, e.g. Kanowna and Broad Arrow. To the south, metal discoveries in 1966 have revived the fortunes of Kambalda.

In the 1860s gold was responsible for over 90 per cent of mining's contribution to GDP. This proportion had fallen to 20 per cent in the 1920s, but by 1939 stood at 43 per cent. By this time coal's contribution had risen to 24 per cent, with silver, tin, copper, zinc and iron ore together representing 20 per cent. In 1985 Australia's gold sector was valued at A$700 million; by 1996 this had risen to A$18 000 million, with gold becoming the third-largest merchandise export, and Australia was the fourth-highest gold producer in the world (260 tonnes of pure gold p.a.). Gold accounted for 7.2 per cent by value of exports (Table 8.3).

FIGURE 8.3 Sovereign Hill, a re-creation of a gold-mining settlement in Victoria in the 1850s (photo: G.M. Robinson)

In the early 1990s it looked unlikely that this level of output of gold could be sustained as there had been few significant discoveries in recent years. Furthermore, the cost and time involved in resolving native title land claims by Aboriginal groups were starting to raise production costs. For example, RTZ spent A$200 million in settling a title claim on its Century Zinc mine. However, a surprise discovery in early 1997 by the Helix company, in the largely unexplored Gawler Grafton region of South Australia, brought a surge of new investment by exploration companies and has raised hopes of the maintenance of high levels of output.

Uranium

Uranium mining in Australia symbolises some of the problems associated with reliance upon an oversupplied or contracting world market, of mining in remote locations, of conflicts over land rights with Aboriginal groups (see Chapter 5), and political factors relating to a mineral that is one of the key elements in nuclear power generation and nuclear weapons manufacture. Australia has 30 per cent of the uranium in the developed world, recoverable at a general cost threshold agreed

by the International Atomic Energy Agency, and huge potential reserves that have not been fully explored (Figure 8.4). Exploration in the 1940s and 1950s revealed at least a dozen locations with substantial reserves, especially in northern parts of the Northern Territory and at Mary Kathleen, 50 km east of Mt Isa (Qld). Today there is an annual production of 4000 tonnes of U_3O_8 ('yellowcake').

Uranium mining has been subject to the extremes of boom and bust, as the mine at Rum Jungle (NT) exemplifies. Uranium was discovered at the site, 100 km south of Darwin, in 1949 and a mine was opened. Its development was strongly promoted, with visits by the Prime Minister, the Governor-General and the Duke of Edinburgh, as Australia hosted the testing of British nuclear weapons based on uranium, primarily near the rocket-testing base at Woomera (SA), 490 km north-west of Adelaide. However, the mining, milling and sale of uranium from Rum Jungle ended in 1963 when world prices were depressed by oversupply. Some copper mining was continued there until 1971, but then all mining operations were suspended, leaving an environmental abscess polluted by radium and heavy metals. A recent A$16 million federal allocation for restoration has led to clean-up activities.

A similar story applies to uranium mining at Mary Kathleen (Qld). A mine and treatment plant was opened here in 1958, only for operations to cease in 1963. They were operational again from 1976 to 1980 but had completely ceased two years later. By the end of the following year the miners' houses had been sold and removed from the site. The area has since been returned to its natural state, leaving little visible trace of its former inhabitants. This uneven production record reflects both strong political opposition and international competition. The most concerted protest over the possibility of Australian uranium being utilised in the manufacture of nuclear weapons has come from the Australian Council of Trade Unions (ACTU). Their opposition to nuclear weapons has been influential in shaping Labor Party policy on this issue, to the extent that since 1977

TABLE 8.3 Exports of minerals from Australia, 1995/6

	A$ million	Percentage of all merchandise exports
Metal ores	7600	11.3
Coal	6939	10.4
Gold	4820	7.2
Non-ferrous metals	4509	6.7
Petroleum products	2951	4.4
Gas	1355	2.0

Source: *Australia Yearbook.*

FIGURE 8.4 The Pilbara and The North-West Shelf

it has been official party policy to bring all uranium mining and related operations to a halt in Australia (Duncan, 1987: 331). However, when Labor came to power in 1983 Australia's two existing uranium producers were exporting 4431 tonnes of uranium oxide worth A$345 million. Furthermore, a deep-seated copper–uranium ore body had been discovered at Roxby Downs (SA), with reserves equal to those of the Northern Territory mines. The new government permitted development at Roxby Downs and continuation of the existing uranium operations, but vetoed all but two other small contracts pending an inquiry by the Australian Science and Technology Council into Australia's role in the nuclear fuel cycle.

This inquiry recommended that uranium exports should be subject to stringent controls on supply, designed to strengthen the nuclear non-proliferation regime. In accepting this and recommendations for Australia to process uranium beyond the basic mining and milling stages, the government withdrew its previous commitment to phase out the uranium industry. However, it has maintained controls in export agreements relating to the nature of the use of uranium and dumping of radioactive waste. For example, France has been excluded as a purchaser because of its use of French Polynesia as a testing ground for nuclear weapons. At Roxby Downs there is now a purpose-built 2000–population township to accommodate the employees of the Olympic Dam mining project.

Other minerals

Australia is a leading producer of a wide range of other minerals (Figure 8.4). These include bauxite, copper, manganese, lead, zinc, nickel, silver and the mineral sands, rutile, ilmenite, zircon and monazite, which are used in the manufacture of titanium dioxide pigment and as refractories used in the aerospace industry, and for the manufacture of advanced ceramics, lasers and fibre optics. Australia is also the world's leading producer of diamonds and opals.

Bauxite and manganese were amongst the ores exploited on a large scale from the 1960s in tropical Australia. Substantial deposits of bauxite at Weipa and Gove (Qld) were developed from 1966 by the Consolidated Zinc Corporation Ltd, building facilities to ship 9.5 million tonnes per annum and employing over 1250 workers. Australia is currently the world's largest producer of bauxite and alumina, exporting three-quarters of its output and accounting for half of the world's trade. It is also the fourth-largest producer of aluminium.

Manganese deposits on Groote Eylandt (NT), used in manufacturing batteries and for

increasing the strength of steel, have been worked since 1966 by a subsidiary of BHP. Much of the annual two million tonnes production is shipped to Tasmania for smelting and export to Japan. Australia is one of only three exporting nations.

Australia is the world's leading producer and exporter of lead and zinc (often found in close proximity) and the third leading producer of nickel (mainly from nickel sulphide ore). Port Pirie (SA) is the world's largest lead smelter. Risdon (Tas) is one of the world's largest zinc refineries. Two lead and zinc complexes in particular have given rise to major settlements in Australia, at Mount Isa (Qld) and Broken Hill (NSW) (see Box 8.2).

Mount Isa, with excellent amenities and facilities, is an oasis of civilisation in the midst of thousands of square kilometres of arid spinifex and large cattle stations. Located 900 km west of Townsville, it owes its existence to the discovery of a rich silver–lead deposit in 1923, since when it has been the company town of Mount Isa Mines (MIM), operating one of the largest silver–lead mines in the world, as well as copper and zinc. The silver smelter stack is the country's tallest freestanding structure. Despite its isolation, the town has a population of 25 000 and is the commercial and administrative centre of north-west Queensland.

The initial mining developments in the north-west of Queensland occurred around Cloncurry where the first copper lodes were discovered in 1867. The rail link from Cloncurry to Townsville in 1908 helped bring about a copper 'boom' here during the First World War, when it was the largest source of copper in the country, with four smelters operating. But the discoveries at Mount Isa led to the focus of mining operations moving there following initial injections of capital from Britain and the USA. In 1930 the American Smelting and Refining Co. bought a controlling interest in MIM. In the 1970s rare, exceptionally pure 22–carat gold was discovered near Cloncurry, leading to some renewed development, with the gold being mined for the jewellery industry.

Box 8.2 Broken Hill: an iconic example of mining in the outback

Of all the towns in Australia still recognised as a quintessential mining town, Broken Hill is pre-eminent. With a population of 25 000 it is significantly bigger than virtually all other mining settlements, with the exception of Kalgoorlie–Boulder, and its history symbolises the development of mining and industrial labour relations in Australia. With its record of strikes and union power it has been regarded as a symbol of both good and bad in the union movement and, with an inability to diversify its economy, it has also been symbolic of problems associated with reliance upon a narrow industrial base. As featured in the film *Mad Max II*, its reputation worldwide has been increased, as a remote and unusual place in a vast sunburnt, sparsely populated region. Located 1170 km from Sydney, but only 480 km from Adelaide, its inhabitants have frequently felt neglected by the New South Wales government, complaining that it has extracted royalties from the mines but has given insufficient in return.

The town was established in the 1880s following the discovery of silver chloride, lead and zinc in 1883. The first 'rush' was associated with shallow and open-cast mining, soon replaced by shaft mining. The lack of sufficient supplies of water locally meant that little smelting of ore took place there. Instead, a railway was constructed to Port Pirie on the coast, opening in 1889 when a smelter was built at the coastal terminus. Acquired by Broken Hill Associated Smelters (BHAS) in 1915, the smelter employs over 1000 people or around one-quarter of Port Pirie's workforce. Problems of water supply to the mines and other industry in Broken Hill were not solved until the construction of a pipeline in 1952 bringing water from Menindee on the Darling River. This has helped Broken Hill to become an island of green in the middle of the semi-arid plains of western New South Wales (Solomon, 1959), in the heart of extensive wool-producing country, with 90 per cent of wool from the stations west of the Darling going to Broken Hill for consignment to Adelaide or Sydney.

For many years Broken Hill was a classic 'one-horse' mining town, with a reputation for poor housing, the dominance of mining leases and a stark landscape of winding gear and processing plants. This contrasts sharply with the 'green oasis' view that can be obtained from a distance by someone approaching by air. It was also a company town, long dominated by the mining firm Broken Hill Proprietary (BHP) which originated as a silver-mining enterprise, but subsequently diversified into a range of other mining, processing and manufacturing activities including iron ore mining and iron and steel manufacture.

The reputation for poor labour relations is tied most closely to events occurring during and just after the First World War. In 1916, 1.25 million working days were lost to strikes and walkouts in Australia: in 1920 this number rose to 3.5 million, and Broken Hill was known as one of the places most closely associated with strikes. Its mines were closed from April 1919 to November 1920 by a strike of miners asking for better working conditions and compensation for industrial diseases such as neumoconiosis. This lengthy strike is regarded by many as being pivotal in securing a 35–hour working week for underground workers, abolition of night-shift working, holiday pay, higher wages and pensions for workers laid off through ill-health. In addition, in 1923 the Barrier Industrial Council (BIC) was established as a 'parliament' of the trade unions in Broken Hill and subsequently helped make a contribution to a range of improvements in the town. The collaboration between workers' representatives on the council was the forerunner of the ACTU, founded in 1927 and marking the end of the era of major disruptive strikes. That is not to deny the continuation of a high level of union-backed strike action in the country, but there has been a tendency for strikes to be shorter and perhaps more ritualised as a means of achieving better living standards for workers rather than disrupting the existing pattern of labour relations (Kitay and Lansbury, 1997).

Broken Hill's mining and processing industries today have 11 trade unions, all affiliated to the BIC, and union membership is compulsory for workers. The control of the council has been absolute as indicated by the restrictions placed upon the conditions of work in industry in the town. To work in the mines, a miner must be born in Broken Hill or have been resident there for at least eight years. In any of the town's industries outside labour is only employed if it possesses skills not available amongst locals seeking employment. Single women are given strong preference over married women competing for employment.

   Throughout the post-war period there has been a strong fear of decline in Broken Hill as the heavy reliance upon the mining base and the poor physical environment have not encouraged economic growth. Doubts about the continued viability of mining produced fears of wholesale closure in 1972, but mining has survived and efforts to promote economic diversification have grown. These efforts have had relatively little success despite the town's discovery by film-makers, with Broken Hill used as a backdrop in *Mad Max II*, *Sunday Too Far Away*, *Wake in Fright*, *Razorback*, *Outward Bound* and the TV series *A Town Like Alice*, and the growth of the 'Brushmen of the Bush' group of artists that includes Pro Hart and Jack Absalom. However, state and federal funding for a A$63 million link of the mines to the NSW electricity grid in 1986 cut operating costs and has helped maintain production.

   Today Broken Hill's mines produce two million tonnes of ore per annum, or two-thirds of NSW's metallic minerals. Mining and related operations still account for nearly 30 per cent of the town's workforce. Although the richest and most accessible ores have been worked out, four large mining companies have remained, though BHP itself ceased operations there in 1939 (Rich, 1987: 149–53):

(a) The Zinc Corporation and New Broken Hill Consolidated (NBHC), both wholly owned by Australian Mining and Smelting, a subsidiary of Conzinc Rio Tinto of Australia (CRA), itself controlled by Rio Tinto Zinc (UK). They operate adjacent mines under common management, sending lead concentrate by rail to the Port Pirie smelter which is operated by BHAS, jointly owned by CRA and NBHC;

(b) North Broken Hill Holdings (NBHH), reformed in 1912 and again in 1976 from a company whose original lease dates to 1883. NBHH is working ore bodies discovered in the 1980s, but at great depth;

(c) Mineral Mining and Metallurgy Ltd, which is reworking residual surface deposits and old mine-waste, though it has cut new mines since 1984.

Mount Isa contributes over one-third of Queensland's mineral production by value. Nearly 40 per cent of the town's employment is in mining, with a range of minerals extracted, including copper, lead, zinc, silver, uranium and gold. The main outlet for the area has been through Townsville.

DIVERSIFYING THE ENERGY BASE

There is spatial variation in the reliance upon different forms of energy supply in Australia. New South Wales, Western Australia and Queensland rely most on black coal; Victoria on brown coal and natural gas; South Australia on natural gas; and Tasmania on hydroelectric power (HEP). However, this pattern also reflects some concerted actions post-1945 to exploit reserves of a variety of different forms of energy, diversifying the domestic energy base and also producing revenue from export of some of these raw materials. Key discoveries of coal, oil and natural gas have been the basis for this diversification.

Oil and natural gas

Natural gas supplies in Australia come from a number of sources (Figure 8.5). Melbourne uses gas from the Bass Strait field whilst Adelaide, Canberra and Sydney are supplied from the Moomba fields in central Australia, and Darwin from the Palm Valley and Mereenie fields also in central Australia. The Surat Basin fields supply Brisbane and Gladstone whilst the Perth Basin and North Rankin field supply Perth. The Copper Basin fields in Queensland are being linked to the Moomba pipeline, thereby enabling gas from Queensland to supply South

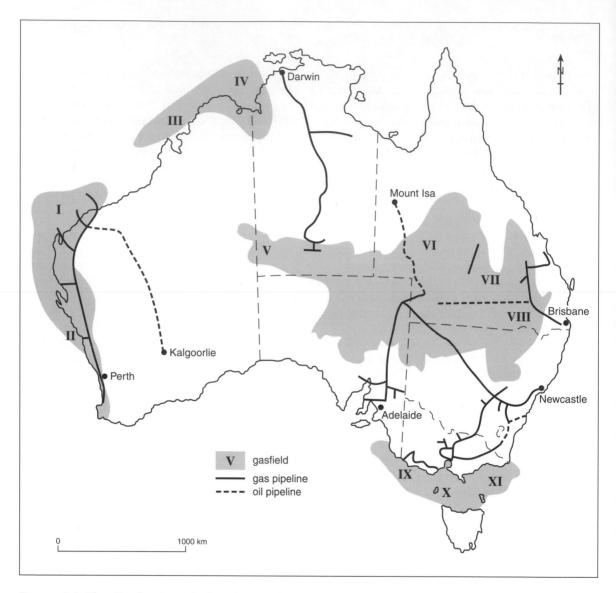

FIGURE 8.5 The distribution of oil and natural gas: I – Caernarvon Basin (49 per cent); II – Perth Basin (<1 per cent); III – Browse Basin (23 per cent); IV – Bonaparte Basin (9 per cent); V – Amadeus Basin (1 per cent); VI – Cooper/Eromanga Basin (5 per cent); VII – Adarelala Basin (<1 per cent); VIII – Bowen/Surat Basin (<1 per cent); IX – Otway Basin (<1 per cent); X – Bass Basin (<1 per cent); XI – Gippsland Basin (12 per cent)

Australia. However, the position has been altered significantly by the tapping of undersea reserves in the Bass Strait and off the coast of Western Australia.

The North-West Shelf project, exploiting reserves of natural gas underneath the Indian Ocean and costing A\$12 billion, supplies gas by pipeline to the State Energy Commission of Western Australia. The pipeline system has been extended in recent years, for example through the opening of the Goldfields Gas Pipeline in October 1996, linking the North-

West Shelf to Kalgoorlie via the Pilbara, costing A$450 million. Also, from 1989 liquified natural gas (LNG) has been shipped to large Japanese power and gas companies, and also to South Korea and elsewhere in Asia as stored energy for domestic or industrial use. Overall, the offshore gas basins of Western Australia have four-fifths of the country's reserves of natural gas, the majority of the remainder coming from offshore supplies under the Bass Strait which were first discovered in the 1960s. Drilling rigs using technology developed in the Gulf of Mexico and the North Sea have successfully exploited these reserves, though heavier crudes are still imported as the crude oil that has been tapped is so light.

Domestic crude oil production currently accounts for 60 per cent of the country's consumption, but there is potential for this proportion to be increased as new discoveries are made. The federal government has continued to encourage exploration under the 1987 Petroleum Resource Rent Tax Assessment Acts which have helped create an attractive fiscal regime for oil companies. The North-West Shelf currently generates over A$2000 million per annum in exports from an investment of A$12 000 million, 70 000 workers and payment of A$200 million annually in royalties to the state government. However, at present the fastest-growing region in the state is the far south-east around Mandurah and Bunbury, in part related to extraction of coal and bauxite and the production of aluminium worth A$2000 million per annum.

Hydro-electric power (HEP)

Melbourne and Sydney have always made the most demands upon power generation in Australia, and they were both fortunate initially in having coal supplies relatively close from which they could generate electricity or other forms of energy. For the last three decades a vital supplement has come from hydroelectric power (HEP) which has also had the effect of restricting oil imports and which now accounts

for 11 per cent of the country's electricity production. The HEP has been supplied from the only two areas which have sufficiently large volumes of water and suitable sites for installation of HEP generators. These are the Snowy Mountains watershed on the New South Wales–Victoria border and western Tasmania.

The Snowy Mountain scheme was initiated in 1949 with the intention of generating 15 per cent of all the country's electricity needs, though in some respects the Tasmanian schemes were more important because of their role in the state's economy, powering plants processing zinc and calcium carbide, paper, textiles and food. The Snowy Mountains scheme is one of the largest HEP and irrigation schemes in the world. It took 25 years to construct, finally being completed in 1974, diverting water inland from coastal watersheds, for HEP and to supplement irrigation waters from the Murray and Murrumbidgee. It includes seven power stations, a pumping station, 16 dams, 144 km of tunnels and 80 km of aqueducts. The scheme was also a major destination for southern European immigrants in the 1950s, creating one of the few inland areas with a highly diverse ethnic population.

In diverting the Snowy River westwards through tunnels to the Murrumbidgee and Murray Rivers, the waters drop by 800 m, thereby providing power to seven power stations, with a total generating capacity of 3.74 million Kw and 2.38 million megalitres per annum of irrigation water. The system is geared to water conservation by means of diverting water to storage when river flows are in excess of power station requirements. Power from oil- or coal-fired thermal stations is used to pump water back to a storage area, as the water loses elevation in response to peak power demands. The scheme is now operated as a joint Commonwealth, NSW and Victoria venture.

Conservation versus HEP generation in Tasmania

Since the 1960s conflicts between, on the one hand, the Hydro-Electric Commission (HEC) of

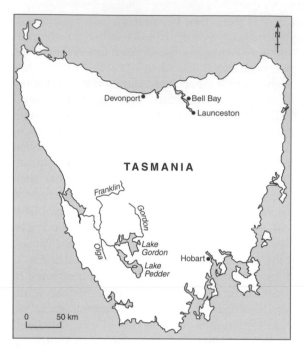

FIGURE 8.6 The Franklin–Gordon HEP project, Tasmania

Tasmania (a semi-autonomous public corporation enjoying a large degree of independence) and other resource development interests, notably the timber and forest products industries and, on the other hand, conservationists and some sections of local communities, have been a prominent feature of economic development in Tasmania (Dargavel, 1991; Davis, 1980; Kellow, 1986). Ongoing opposition to logging and other HEP schemes has been fierce and has resulted in federal protection measures being adopted for wild tracts of the Franklin River (Figure 8.6). Initially, conflict over proposals by the HEC to utilise Lake Pedder for HEP generation attracted interest from outside the state. This development went ahead in 1972, but it had alerted an international audience to the multipronged threats to the vast wilderness in the south-west of the state. The Tasmanian government and the HEC were criticised in a review undertaken by the Whitlam federal government, with proposals

that, in future, major schemes should be considered by planning guidelines containing environmental impact assessments. This was then incorporated in the 1974 Environmental Protection (Impact of Proposals) Act (Davis, 1986; Mercer, 1985). There were also a series of state enquiries into resource management issues, but the state government largely rejected guidelines for project evaluation and public participation, leaving the HEC free to act 'as a proxy planning authority for the state, with all other authorities and the parliament itself reacting to Commission demands, often in a hasty and ill-considered manner' (Davis, 1986: 217).

During the 1970s arguments between a well-organised conservation lobby and the HEC took place against the background of economic recession, high unemployment, reduced federal government expenditure, heavy borrowing by the state government to prop up local manufacturing, and the high costs of proposals for additional HEP exploitation. Two-thirds of electricity generated was sold to energy-intensive industries, principally pulp and paper production and metallurgical processing (aluminium, ferromanganese and zinc). These industries had been seen as a means of employment creation since 1917 when a zinc smelting operation and carbide works were established under state ownership of the related electricity generation.

Championing the cause of these industries and their perceived power needs was a consistent policy of the state's Labor government, which was in power almost continuously until 1982. The HEC's plans were based on financial encouragement of further hydro-industrialisation to supply large industrial customers such as Comalco (smelting bauxite shipped to Tasmania from Weipa in northern Queensland), though alternative processing sites for Comalco at Gladstone (Qld) and in New Zealand weakened the argument.

In 1979 the HEC recommended a A$1360 million development on the King, Franklin and Lower Gordon Rivers which would have inundated some of the state's most spectacular wild river scenery. The state's own National Parks and Wildlife Service countered by suggesting the creation of a Wild Rivers national park in the middle of the area scheduled for the hydro scheme. Amidst international media coverage, a Wild Rivers national park was gazetted in March 1981, aiming to protect the Franklin and Lower Gordon Rivers, though excluding some areas of high conservation value.

In the ensuing state referendum, 40 per cent of the compulsory vote inserted a 'No Dams' option on their ballot as urged by the conservation lobby. However, the original HEC proposal received half the votes, and the following year the newly incumbent Liberal state government gave approval to the HEC scheme. This prompted a blockade of the main construction site by conservationists who involved overseas media personalities, such as the naturalist David Bellamy, in this action. Meanwhile UNESCO nominated a substantial part of south-west Tasmania as a world heritage site.

Resolution of the conflict came via the election in 1983 of the Hawke federal Labor government who were committed to 'Save the Franklin'. They passed a World Heritage Properties Conservation Act, thereby overriding the wishes of the Tasmanian government. This was viewed by some as a dangerous interference by the Commonwealth into state affairs, but conservationists argued that the nature of the issue transcended the interests of a particular state and generated concern that extended beyond state or federal boundaries.

This argument was accepted by the high court in 1983 when it ruled that the federal government did have the power to override a state government on this issue. Subsequently the World Heritage Register has been used as a device whereby the federal government has been able to assume authority over the states in controlling land-use management (Kellow, 1989; Young, 1996: 125). This has led to several conflicts with mining interests, e.g. in the Lake Eyre Basin (with the Roxby Downs mine), the Ruddell River in Western Australia where there is a major uranium deposit, and Kakadu

national park (see Chapter 5). The Tasmanian government received A\$270 million compensation for its loss of the Gordon-below-Franklin power scheme.

The conflict exposed some of the tensions underlying relationships between federal and state governments and the weakly developed nature of strategies for national resource management. The concern of successive Tasmanian state governments to champion destructive development schemes by the hydro and forest products industries has been termed a 'cargo-cult mentality' (Sanders and Bell, 1980) based on perceived short-term electoral advantage and the primary argument that 'hydro-industrialisation' produced employment in a state where high unemployment has been a chronic long term problem (Hood, 1987). Despite the Lake Pedder and Franklin Dam cases, there is little evidence that basic attitudes held by many Tasmanian politicians have changed, and the HEC and the small number of large forest products companies remain very powerful.

By the early 1990s there was such a surplus of electricity in the state that there was an overwhelming case for reform of the HEC, turning it from being primarily a dam-building company into one focusing on electricity production and distribution (Kellow, 1996: 39). Its workforce was cut by 44 per cent between 1985 and 1992 as part of a commercialisation process which has subsequently included a further 20 per cent cut. In 1996 the HEC became the Hydro Electric Corporation, a government business enterprise, but this may not mean an end to the rationalisation process.

In 1997 a new deal was signed with Comalco, which consumes one-quarter of the state's electricity, to provide 256 mW per annum until 2014. However, Comalco has announced its intention to close its aluminium smelter at Bell Bay (on the Tamar River, 45 km north of Launceston) in the year 2000 in favour of expansion at Gladstone and its Tiwai Point smelter in New Zealand (at Bluff, using bauxite from Weipa).

Today Tasmania has 27 HEP stations, two diesel-powered stations and an oil-powered

FIGURE 8.7 Coalfields

station at Bell Bay as back-up. Future expansion of HEP is unlikely, though, as there are no sites suitable for HEP generation that are likely to receive federal government sanction. The state government may look to the Basslink cable to import electricity from Victoria to meet future demands that exceed the capacity of the state's own power stations. Export of minerals still accounts for 40 per cent of Tasmanian export income and 2.9 per cent of state GDP, though the fortunes of some of the principal mines in the state have fluctuated greatly during the last decade. For example, the Mt Lyell copper mine was closed at the end of 1993 and then subsequently re-opened to produce 3.5 million tonnes per annum. It faces heavy competition as there is surplus production worldwide and falling prices.

Coal

Australia is the world's largest exporter of coal (around 140 million tonnes), accounting for 30 per cent of the world's sea-borne trade. Over half the production of black coal is exported (50 per cent of which goes to Japan), one-third is used for electricity generation and 12 per cent in the production of iron and steel. It is the world's sixth-largest producer and has 51 billion tonnes of recoverable coal (Figure 8.7).

Discoveries of coal in the Lower Hunter Valley at the end of the eighteenth century led to the establishment of settlement near the mouth of the river, primarily to supply coal, lime and timber to Sydney by sea. Convict labour was used in adit mines, though later deeper mines were sunk to exploit high-quality steam coal at greater depth. This coal was used for powering steam ships and steam trains, and there was also gas coal for lighting and heating. A series of mining villages developed on the concealed coalfield in the Hunter, with Newcastle as the service centre, exporting coal and wheat and importing manufactured goods. In addition to this entrepot function, Newcastle also developed as a processing centre. In 1915 it was the location for the first major Australian iron and steel plant, operated by Broken Hill Proprietary (BHP) alongside the state dockyard, itself opened in the previous year. This plant was not surpassed in size until the opening of the vertically integrated steelworks at Port Kembla in 1950.

In the 1950s some of the small Lower Hunter mining villages were swallowed by urban sprawl and their mining function declined as the demand for gas coal fell. With its reliance on heavy industry, Newcastle and the Lower Hunter were recognised as a 'depressed area' by the New South Wales government, and from 1958 were eligible for aid in its decentralisation scheme. This helped to establish a wider industrial base, although it also encouraged further development of heavy industry on reclaimed land in the river estuary, including aluminium smelting and a pulp and paper complex. From the early 1970s open-cast mining has also been used to extend coal mining from the Lower to the Upper Hunter, exploiting the anticlinal structure between Singleton and Muswellbrook that brings coal seams close to the surface (Figures 8.8 and 8.9). As a result, in addition to several new large mines, there have been new power stations, dams, inter-basin transfers of water, expansion of an existing aluminium smelter and construction of two new ones to increase aluminium production in the area eight-fold.

Whilst Newcastle was able to utilise coastal transport for distributing its coal, elsewhere the development of coal resources has depended on more complex arrangements, usually reliant on the railways. For example, rail transport of briquettes (blocks of brown coal) to Melbourne was crucial to the opening-up of brown coal reserves in the Latrobe Valley, south-east of Melbourne. For the coalfield centred on Leigh Creek, 480 km north of Adelaide, the railway to the power station on the coast at Port Augusta was vital.

The coalfields of central Queensland have been developed largely as a result of the strong demand for coal by the Japanese from the 1960s. Coal in the Bowen and Styx Basins was discovered in the nineteenth century, but attempts to develop a coal mining industry in the region failed and a proposed steelworks at Bowen was aborted. The difficulty of attaining sufficient coal from surrounding 'country' rock to make operations profitable was not overcome until the combination of new technology and increased demand occurred in the 1960s. Even so, much of the coal has to be carefully washed to produce a consistent saleable product. Further exploration of the Bowen Basin has revealed substantial reserves extending over a region of 75 000 km^2 to the west of the Dividing Range. One estimate is of 23 600 million tonnes of coal reserves, though measured reserves are a more conservative but still substantial 6600 million tonnes, of which 59 per cent is coking coal. One-fifth of this coking coal can be mined by open-cast means. This has led to the development of some of the world's largest open-cast mines, the growth of new

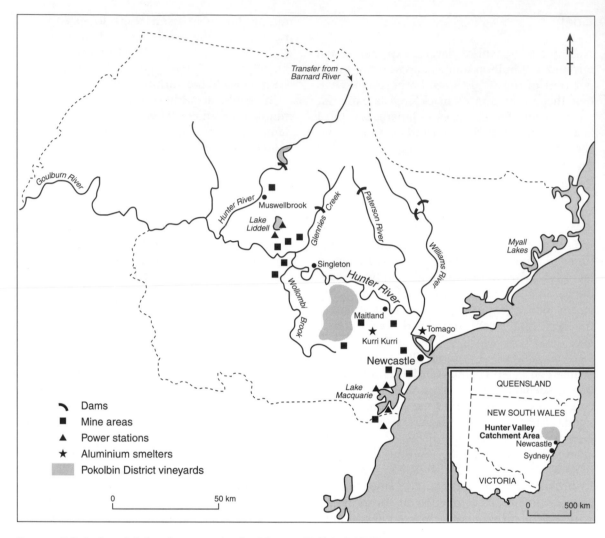

FIGURE 8.8 Industrial development in the Hunter Valley (NSW)

towns linked to the mining activities, e.g. Moranbah and Peak Downs mines, and major port complexes at Hay Point and Gladstone. This coal has been a major provider of domestic energy, supplying Queensland's largest power station at Gladstone which produces 30 per cent of the state's electricity needs. Nearby are the state's only alumina refinery (a 'halfway' stage towards aluminium) and the Boyne Island aluminium smelter.

One of the most significant companies involved in developing Queensland's coal has been Conzinc Rio Tinto of Australia (CRA), largely owned by UK-based RTZ which made substantial profits through its subsidiary, the Zinc Corporation, in Broken Hill in the 1890s, but moved its headquarters to London just before the First World War to obtain more funds. CRA illustrates the interconnected nature of the mining and energy sectors in Australia and the extent of foreign control. For example, at various times CRA and subsidiaries have been involved in lead–silver–zinc mining in Broken Hill, iron ore mining in

FIGURE 8.9 Open-cut mining in the Hunter Valley (NSW) (photo: A.I. Loughran)

the Pilbara (as a major shareholder in Hammersley Holdings), bauxite mining at Weipa, Qld (through shares in Comalco), copper mining at Cobar, smelting at Boolaroo, near Newcastle (through its controlling shares in the Sulphide Corporation), nickel prospecting in Kalgoorlie, and uranium mining (at Mary Kathleen). Through companies like CRA, UK capital owned as much of the mineral production of Australia as all other foreign countries combined in the mid-1960s (Dyster and Meredith, 1990: 247).

Subsequently, more American and Japanese money has entered the Australian mining sector, e.g. the Utah Construction and Mining Co. of San Francisco which has invested in the Queensland coal mining industry since the late 1960s, exporting to Japan, and (with Consolidated Goldfields of the UK and Cyprus Mines of Los Angeles) exporting iron ore to Japan from Port Hedland since 1970.

Australia's role as the largest supplier of the Japanese market has been maintained in part by the lower delivered price for Australian coal relative to other major supplying nations (Colley, 1997). This price disparity has been explained in academic research by applying

bilateral monopoly theory, which postulates that an increased surplus available from the trade, for example due to savings in freight costs because of proximity, is shared between Japanese buyers and Australian sellers (Bowen and Gourlay, 1993). However, Colley's (1998) examination of the trade during the 1980s shows an unequal distribution of the surplus consistently favouring the Japanese buyers. This inequality reflects the greater power wielded by the Japanese purchasers who are characterised by large-scale companies with interlocking shareholdings, notably the eight major steel mills, nine major power companies and specialist trading companies which are the main source of Japanese direct investment in the Australian coal industry and are the nominal coal buyers in most cases (Parker, 1992). The ownership structure has created a strong strategic orientation on behalf of the purchasers, favouring the national interest. This has not been the case amongst the coal sellers despite the fact that four companies (BHP, Rio Tinto, Cyprus Amax and MIM) account for two-thirds of coal exports and 10 account for 85 per cent (Colley, 1998: 66–7). Approximately 40 per cent of production is owned by foreign interests (Europe = 16 per cent; Japan = 13 per cent; USA = 10 per cent; other Asian = 1 per cent), encouraging a concentration of operations into a small number of large mines. However, this understates the level of foreign control, exercised through a controlling interest or substantial influence where a company does not have majority control. These foreign owners also have an increasing number of investors with links to Japanese coal-consuming and trading companies. Various consortia and joint ventures have given Japanese interests a significant influence over investment decisions and policies followed by the Australian coal sellers. This contributes to an unequal bargaining power existing between the producers/sellers and purchasers which contributes to the Japanese enjoying a competitive advantage from the coal trade with Australia.

It is concern over this type of potentially adverse relationship affecting the exploitation of Australian resources that has led to periodic calls for closer constraints upon foreign investment in the minerals and energy sector. It has contributed to political jibes that Australia was becoming a 'banana republic' and has raised concerns over the erosion of controls over national economic decision-making (Goodman and Pauly, 1993). However, it has not prevented the pursuit of economic reforms during the past two decades that have enabled greater ease of foreign investment in Australia. Ironically, whilst these reforms have aimed to transform Australia into a competitive producer and exporter of high-value-added manufacturing products and services, they have helped contribute to increased foreign involvement in the export-orientated primary sector (Yu *et al.*, 1997).

FURTHER READING

Background to the development of mining in Australia, and especially the 1960s' mining boom can be found in Duncan (1987), and Walmsley and Sorensen (1993: 102–14). There are useful references to the mining industry and the development of Broken Hill in Rich (1987). Storey (1998) provides a good overview of FIFO mining. A detailed example of one such development in Western Australia is covered in Dixon and Dillon (1990). Conflicts between mining and Aboriginal groups are considered in two sets of collected essays edited by Connell and Howitt (1991) and Howitt, Connell and Hirsch (1996). Pellow (1996) has an excellent Chapter on the battle between conservationists and champions of HEP development in Tasmania (see also Davies, 1986). Colley (1997; 1998) examines the nature of the relationship between Australia and Japan regarding the coal trade. Restructuring in the Australian mining industry is discussed by Howitt (1991b).

9

POPULATION CHARACTERISTICS

A New York resident speaking to an Australian visitor to his city was recently overheard saying: 'So, you're from Australia? I know someone who just moved there. She lives in Auckland.' It is somewhat understandable that people might confuse parts of New Zealand and Australia. Not only are they located close to each other, but they are both 'new world' countries that have developed rapidly during the nineteenth and twentieth centuries. Similarities in population characteristics are evident in: population growth rates (at least in the 1990s); falling mortality rates and birth rates, leading to an increase in the proportion of elderly in the population; changes in family structures, including increases in the proportion of single-parents; a more rapid rise in the number of households than the number of people; high proportions living in urban areas; and increasing ethnic diversity, with growing proportions of the population from Asian countries. Also, in both countries the characteristics of indigenous populations differ dramatically from the non-indigenous populations. In many ways, the population characteristics of indigenous Australians (Aboriginal and Torres Strait Islander peoples) are very similar to the New Zealand Maori (notwithstanding the fact that within each of these groups there is considerable diversity and distinct cultural traditions).

There are also some important differences between the two countries. These are most noticeable in: the history of immigration and its relative importance for population growth; the level of ethnic diversity and the average age of ethnic minorities; the size of the indigenous population; and overall population densities. However, despite such differences, Australia and New Zealand appear very similar when seen in terms of international comparisons. This Chapter provides an overview of important population characteristics, many of which will be discussed in detail in the following three chapters.

The chapter begins with a discussion of population size and growth rates. This is followed by a discussion of the related issues of fertility trends, life expectancy and age composition. Attention is then focused on family characteristics, which themselves are related to population growth. The distribution of the population within Australia and New Zealand is then examined, including a brief explanation of current internal migration trends. Finally, ethnicity and religious affiliations are discussed. The general population characteristics are illustrated in three

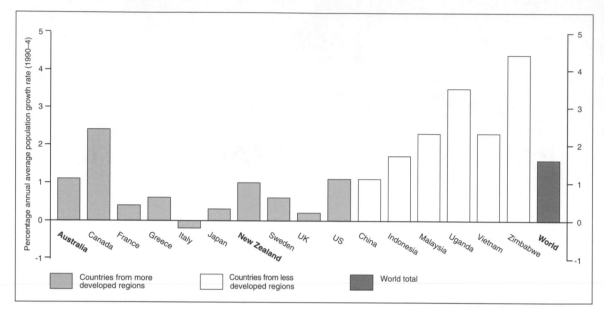

FIGURE 9.1 International comparisons of population growth rates (ABS, 1997b: 5)

case studies: the Gold Coast, the 'greying' of the population, and Chinese in New Zealand.

POPULATION SIZE AND GROWTH RATES

Compared to other countries in the world Australia and New Zealand's total populations are minuscule. In 1994 Australia comprised only 0.3 per cent of the world's population while New Zealand was less than 0.1 per cent. At the same time the United Kingdom had about 1 per cent, the United States of America about 5 per cent and China had a massive 21 per cent (ABS, 1997b: 5). The population growth rates of Australia and New Zealand in the 1990s were very similar, and lower than the world population growth rate of 1.6 per cent. In comparison to many developed countries, Australian and New Zealand growth rates are relatively high, largely due to their younger population and high levels of immigration in the 1990s. Figure 9.1 provides some international comparisons of annual growth rates from 1990 to 1994.

At the beginning of the twentieth century Australia's population was 3.8 million, virtually the same as New Zealand's population near the end of this century. By the mid-1990s Australia had reached 18.3 million. Population growth in Australia surged after the Second World War due to the strong natural increase from the baby boom of the 1950s as well as greatly increased overseas migration. For virtually the entire century natural population growth (excess of births over deaths) has been the main contributor to population increase, providing about two-thirds of the total population increase from 1901 to 1996 (ABS, 1998d). From the 1950s to the 1990s natural increase has not varied greatly in its contribution to population.

In Australia immigration has accounted for about one third of the total growth in population during the twentieth century, with the bulk of this contribution occurring since 1949, when migration from European countries surged in response to increased demand for labour in growing manufacturing industries (see Chapter 10). In the first half of the twentieth century, immigration contributed relatively

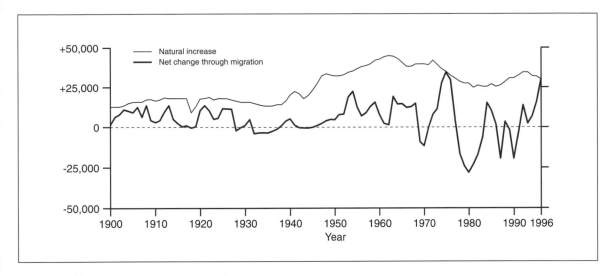

FIGURE 9.2 Natural increase and net change through migration in New Zealand, 1900–96 (SNZ, 1997a: 120)

little to overall growth, and there were times of significant negative net migration (e.g. during the First World War and during the depression years of the early 1930s). Although there has been noticeable variation in the level of immigration, since 1947 it has contributed positively to population growth.

The population of New Zealand has grown at an average rate of 1.4 per cent per annum, from about 800 000 in 1900 to 3.7 million in 1996. More than 80 per cent of this growth was due to natural increase. New Zealand's growth rate over the twentieth century has shown more variability than Australia's. This has been largely due to dramatic fluctuations in immigration rates, particularly since the late 1960s. In contrast to Australia, there have been a number of periods of negative net migration since the Second World War II (most noticeably in the late 1970s), caused by New Zealanders leaving the country during adverse economic times. These dramatic fluctuations in the size and direction of external migration are illustrated in Figure 9.2.

Since the 1950s Australia's net migration levels (Figure 9.3) have appeared to be relatively stable compared to New Zealand's. After a period of population loss from migration in the late 1980s in New Zealand, changes

to immigration policies led to positive net migration in the 1990s. From 1991 to 1996 net immigration accounted for almost 40 per cent of the total population growth. Diverse source countries for immigrants in the 1980s and 1990s have contributed to the increasingly multi-cultural make-up of New Zealand society (see Chapter 10).

In the twenty-first century, the populations of Australia and New Zealand are likely to continue rising, but at progressively lower rates. According to the United Nations' *World Population Prospects*, during the period 2020–50, when world population growth rates are set to fall to 0.7 per cent per annum, Australia and New Zealand will have growth rates of around 0.3 per cent (ABS, 1997b: 7). By this time, many developed countries, including Japan, Hong Kong and Italy, are expected to be well into a negative growth scenario.

Fertility trends and life expectancy

Australia and New Zealand have had a broadly similar pattern of changes in fertility over the twentieth century. Fertility, the actual reproduction performance of a population, is measured by

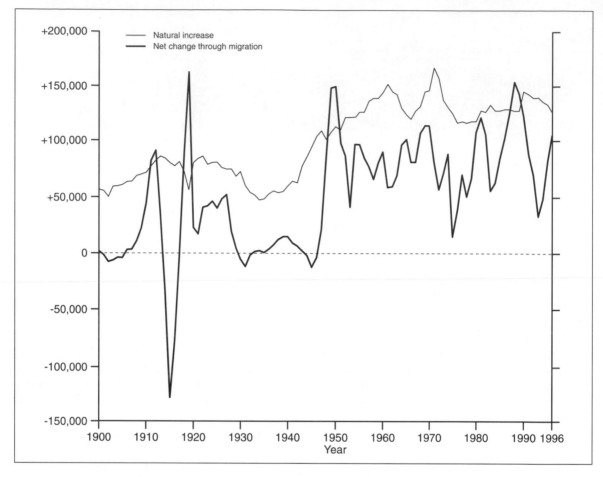

FIGURE 9.3 Natural increase and net change through migration in Australia, 1900–96 (ABS, 1992a: 7; 1998a: 2)

live births per 1000 women aged between 15 and 49 years. Both countries showed a general decline in fertility apart from the baby boom years (1946–65) which reshaped the age structure of the population by producing a large bulge in the population born during this time. This bulge has caused problems over subsequent decades for planners and policy-makers. In both countries the boom ended in the early 1960s, when the oral contraceptive pill become widely available. However, as Cook (1997) suggests, the cause and effect relationship is unclear. Reasons for the decline in fertility are complex, and may include other factors, including an increased percentage of women in the workforce (see Chapter 11),

changing patterns of marriage, rising divorce rates and general economic conditions.

Fertility rates in New Zealand tend to be slightly higher than in Australia. In 1995, the total fertility rates for Australia and New Zealand were 1.82 and 2.04 respectively. These rates are similar to the United Kingdom (1.8 in 1993) and the United States (2.1 in 1991), but much lower than neighbouring countries such as Indonesia (2.9 in 1990–5). Compared to some developed nations (such as Italy, Greece and the Netherlands which have all recorded rates as low as 1.5), fertility rates in Australia and New Zealand are still relatively high (ABS, 1992a; 1997b; SNZ, 1997a).

TABLE 9.1 Historical life expectancies: international comparisons of selected countries

Country	Year/ period	Males	Females
New Zealand	1901	58.1	60.6
Australia	1901–10	55.2	58.8
Denmark	1895–1900	50.2	53.2
Japan	1899–1903	37.8	38.2
Norway	1891–1900	50.4	54.1
United Kingdom	1906	48.0	51.6
New Zealand	1950–2	67.2	71.3
Australia	1953–5	67.1	72.8
Denmark	1951–60	70.1	73.2
Japan	1950–2	59.6	63.0
Norway	1951–5	71.1	74.7
United Kingdom	1951	66.2	71.2
New Zealand	1993–5	73.7	79.1
Australia	1994	75.0	80.9
Denmark	1993–4	72.6	77.9
Japan	1994	76.6	83.0
Norway	1994	74.9	80.6
United Kingdom	1992	73.5	79.1

Source: SNZ (1997a: 118).

In terms of life expectancy, Australia and New Zealand were once the most advantaged nations in the world. Until the Second World War, New Zealand had the lowest mortality rate of any country. However, the relative advantages of Australia and New Zealand changed over the twentieth century (see Table 9.1). Although death rates in both countries have fallen fairly consistently since 1900, there are now a number of developed countries in which life expectancies are higher. By the 1950s, Norway and Denmark had achieved very high life expectancies compared to Australia and New Zealand. By the 1990s, when Australians had a noticeably higher life expectancy than New Zealanders, Japan was well ahead of either country. In the 1990s, the gap between life expectancy in the developed and developing countries was still very evident. Australians could expect to live, on average, 15 years longer than Indonesians, and 22 years longer than residents of Papua New Guinea (ABS, 1998a: 193).

These figures on life expectancy mask significant variations within society for both Australia and New Zealand. In Australia, while most states and territories have very similar life expectancies, the exception is the Northern Territory, where males can expect to live 6.5 years less than average for Australians, and females 6.8 years less. These much lower rates are related to the high percentage of indigenous people there. The small proportion of indigenous peoples in Australia means that the distinctive demographic and health characteristics of this group (see below) have little influence on the health status of Australia as a whole. However, the Northern Territory stands out from the rest of the country due to its larger indigenous population (29 per cent of the total population in 1996, compared to a national figure of 2 per cent).

In Australia, the discussion of differences in life expectancy between indigenous and non-indigenous people is somewhat limited by data availability. Reliable national statistics on deaths of indigenous people are not available, due to incomplete recording of indigenous status in death records for some areas. However, available information indicates that the life expectancy of indigenous people is about 15–20 years less than non-indigenous people (ABS, 1996: 55). There is also evidence that life expectancy for indigenous females has actually fallen in three Australian states in the early 1990s (ABS/AIHW, 1997: 92).

In the 1990s the life expectancies of indigenous Australians in northern, central and Western Australia were lower than any other indigenous minority in a first world country (ABS/AIHW, 1997: 5). This reflects appalling living standards and a long history of discrimination as well as the failure of government policy to eliminate extreme disadvantage among indigenous Australians (see Chapter 11).

In New Zealand, differentials in life expectancy between Maori and non-Maori are not as marked as between indigenous and non-indigenous Australians. Indeed, these differen-

TABLE 9.2 Maori and non-Maori life expectancy (years)

Period	Non-Maori	Maori	Difference
Male			
1950–52	68.3	54.0	14.3
1970–72	69.1	61.0	8.1
1990–92	73.4	68.0	5.4
Female			
1950–52	72.4	55.9	16.5
1970–72	75.2	65.0	10.2
1990–92	79.2	73.0	6.2

Source: Cook (1997: 15).

TABLE 9.3 Age structure of population (as percentage): 1986 and 1996

Age group (years)	Australia		New Zealand	
	1986	**1996**	**1986**	**1996**
Under 15	23	21	25	23
15–64	66	67	65	66
65 and older	11	12	10	12

Source: ABS (1997b: 2); Cook (1997: 23).

tials have shown remarkable reductions since the 1950s (see Table 9.2). However, even by the early 1990s, a newborn non-Maori boy could expect to outlive a Maori boy by 5.4 years. For girls, the difference is even greater, at 6.2 years (Cook, 1997; SNZ, 1994). The differences in life expectancy are related to high levels of economic and social disadvantage for Maori (see Chapter 11). Differences in smoking rates may be partly to blame. Smoking rates among Maori are about twice those among non-Maori.

Age composition

As a result of low birth and death rates, as in other economically developed countries, both Australia and New Zealand have low numbers of children and high numbers of elderly people compared to less developed countries, which in turn slows population growth prospects. The proportion of children aged 0–14 has been steadily declining in the last few decades of the twentieth century, while the proportion of elderly has been increasing (see Box 9.1). The changes in age structure from 1986 to 1996 are shown in Table 9.3. Although the two countries are very similar, New Zealand has a slightly younger structure. In stark contrast with Australia and New Zealand, two of Australia's near neighbours, Indonesia and Papua New

Guinea, had 33 per cent and 40 per cent respectively in the age group under 15 in the 1990s.

Although the age structure of Australia and New Zealand is similar to other developed countries such as the United States and Canada, there are population sub-groups with very distinctive age structures. Some ethnic groups in Australia, particularly those from European countries such as Poland, Italy and Greece, are characterised by an aged population. The more rapid aging of these groups is a result of the post-war immigration policies of Australian governments. Large numbers of these European migrants arrived in the 1950s, but the rate of immigration slowed, and without a continued inflow of immigrants the age structure matured (Graycar and Jamrozik, 1993: 25). The Italian-born community, for example, is now aging and declining rapidly, with a median age of 57 years, compared with an Australian average of 34 in 1996 (ABS, 1998a).

Other population sub-groups have more youthful age structures. This applies to the Australian indigenous population as well as to Maori and Pacific Islander populations in New Zealand. The age structures of these groups are completely unlike the age structures for Europeans, being more characteristic of less developed countries.

In the Aboriginal and Torres Strait Islander population the proportion of children aged less than 15 years (40 per cent) is almost double the proportion that these groups represent in the non-indigenous population. The age structure of the indigenous population is similar in all Australian states and has changed little since

---

Box 9.1 The 'greying' of the population

At the start of the twentieth century the age structures in both Australia and New Zealand were similar to the present age structures of Indonesia or Malaysia. In 1900, only 4 per cent of the population in Australia and New Zealand were elderly (aged 65 or over). By the 1990s, the figure was 12 per cent, and by 2051 it is expected to have surged to over 23 per cent in Australia and about 25 per cent in New Zealand. By the same year, the percentage of the population aged under 15 years will have fallen significantly to about 16 per cent in New Zealand and about 17 per cent in Australia (ABS, 1998a; Cook, 1997). These figures represent what is commonly referred to as 'population aging' or the 'greying of the population', a demographic trend with alarming policy implications.

Though certainly not unique to Australasia, governments in both countries must face the increasing costs that this 'greying' will impose through social security, superannuation benefits and geriatric health care. The elderly comprise an important dependent group, especially those in the very elderly category (aged 85 and over). In Australia, this category will increase from 1.1 per cent in 1996 to an estimated 3.5 per cent in 2041. In New Zealand, the number in this group will jump from 39 000 in 1996 to 260 000 in 2051 (a six-fold increase) (Cook, 1997).

Rapid advances in medical technology have proved very effective in keeping people alive longer, but at considerable cost. In the 1990s, people having heart attacks are much more likely to survive than in the 1960s, due to medical advances such as heart bypass surgery. Keeping these people alive usually imposes considerable further medical costs (such as more heart bypasses and hip and knee replacements). Advances in medicine have also been partly responsible for the lowering of birth rates in Australia and New Zealand, through the availability of oral contraception. The result is a rising index of aging (the ratio of elderly to children).

The greying of the population will also create implications for other people, especially the families of the elderly providing physical and social support. In particular, middle-aged women will be expected to do much of the unpaid care. However, the 'supply of such unpaid women workers will become scarce with the increasing participation of women in the paid workforce' (Zodgekar, 1994: 313–14). Also, increasing life expectancy has resulted in two-generation 'geriatric families' where people enter old age and have surviving parents.

Not only will the greying of the population impose particular burdens on society, but it may also lead to a new group of elderly in poverty. Those without adequate retirement funds may find themselves increasingly unable to survive in countries with a 'winding back' of the welfare state (see Chapter 11). However, as the elderly become a larger proportion of the population, they will become more politically powerful. Evidence of this already exists in New Zealand, with an influential 'Grey Power' movement (Mansvelt in Le Heron and Pawson, 1996).

1980 (ABS, 1994: 7). The extreme contrasts in age structure between the indigenous and total Australian population are starkly illustrated in Figure 9.4.

Within New Zealand the Maori and Pacific Islander ethnic groups have age structures which are more like those of Australian indigenous people than the rest of the New Zealand population. While New Zealand as a whole has a median age of 32 years, the median age for Maori and Pacific Island people is about 21 years. These ethnic groups contain approximately twice the percentage of children under

15 years as non-Maori, non-Polynesian people (SNZ, 1997a: 121). There are a number of reasons for this. Maori and Pacific Islander people have lower life expectancies than Pakeha as well as a history of higher fertility rates. Women in these groups also tend to have their children at a younger age. Maori teenagers are about three times more likely than non-Maori teenagers to have children. Teenage pregnancy may be linked to high unemployment rates amongst Maori women, as well as to a 'sense of self-worth and useful-ness otherwise denied them by unemployment

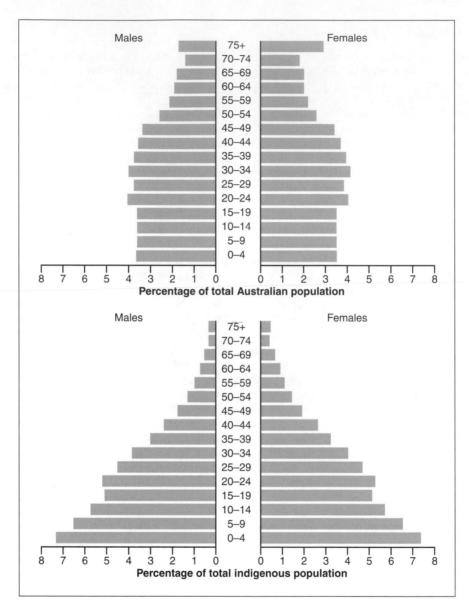

FIGURE 9.4 Age–sex distribution for indigenous and total Australian populations (Anderson *et al.*, 1994: 9)

and discrimination' (SNZ, 1994: 16–17). The age structure of the Pacific Island population also reflects their relatively recent settlement in New Zealand.

Another ethnic group in New Zealand, broadly described as Asian, though containing considerable diversity (see below in pp. 171–5), also has a distinctive age structure, unlike any other ethnic group. Because of their relatively recent arrival in New Zealand the population pyramid for the Asian ethnic group shows concentrations in the 15 to 19 and 30 to 39 (mid-working age) groups. The median age for the Asian ethnic groups is lower than for the European ethnic group, but higher than for Maori (SNZ, 1997b).

Family characteristics

When statistics at the national level are examined, New Zealand and Australia have remarkably similar twentieth-century histories in terms of changes in the characteristics of families. While the family is still an important institution in both countries, there have been considerable changes in the make-up of families. Divorce rates have risen, the average age of marriage has increased and the number of children per family has fallen. Also, the proportions of different family types have changed, with substantial increases in single-parent households and one-person households, particularly over the last 10 years. One noticeable effect of this is that over the last few decades, the number of households has been rising more rapidly than the population, placing increasing pressure on housing. Two key variables relating to families will be examined in this section: single-parent families; and age at first marriage.

Increases in the proportion of single-parent families have important implications for the well-being of children. Single-parent families are much more at risk of poverty than couple families, largely due to the difficulty of undertaking paid employment at the same time as bringing up children. The number of such families has increased with the number of divorces. However, there are also increases in the number of lone parents who have never been married.

A 'feminisation of poverty', or the 'impoverishment of women', can be associated with the growth in single-parent families, as well as the aged (Marcuse, 1996). Females represent a majority in each of these groups, which are rapidly growing low-income groups in Australia and New Zealand (Cheyne *et al.*, 1997: 108).

In Australia, the proportion of all families with dependent children who were single-parent families increased from 14 to 19 per cent between 1986 to 1996. In 1996, 87 per cent of all lone parents were lone mothers. This figure had changed little over the last 10 years (ABS, 1992a; 1997b: 34). In New Zealand, a very similar story emerges for single-parent families. However, care must be taken in the comparison of data on such families. In Australia, dependent children are classified as those who are less than 15 (as well as others aged 15–24 who are in full-time study), while in New Zealand dependent children are defined as children under 18 and not in full-time employment. Nevertheless, in New Zealand in 1996 approximately one in four families with dependent children were single-parent families. This is higher than most other Western countries (Cook, 1997: 26).

Within New Zealand there are particular groups with much higher proportions of dependent children living in single-parent families. Once again, ethnicity has a remarkable effect. Among Maori, almost two out of every five children (39.4 per cent) lived in a single-parent family in 1996. This represented a considerable increase from the 1986 figure of 28 per cent. Pacific Islander children are also more likely to live in such families. For the European ethnic group, only one in five children lived in these families (see Table 9.4). These differences compound the contrasts in socio-economic status between these groups in New Zealand (see Chapter 11).

Another important variable influencing family characteristics is the age at first marriage. The changes in the median age at first marriage exhibit a remarkably similar trend for Australia and New Zealand, falling

TABLE 9.4 Dependent children in New Zealand in single-parent families, by ethnicity of child, 1996 (%)

	Maori	Pacific Islander	European	Total
Single-parent family	39.4	31.3	20.3	23.7

Source: Cook (1997: 27).

TABLE 9.5 Median ages at first marriage

Australia			New Zealand		
Year	Males	Females	Year	Males	Females
1975	23.4	21.0	1976	24.8	22.2
1995	27.3	25.3	1995	28.9	26.9

Source: ABS (1992a); ABS (1997b: 24); SNZ (1997a: 137).

from the 1940s to the early 1970s, and then rising rapidly. Since the mid-1970s, in both countries the age at first marriage has been increasing at the rate of over two years per decade (see Table 9.5). This is largely a reflection of: increasing numbers of people remaining single; cohabitation before marriage; increasing numbers of women participating in post-compulsory education and work; increasing availability of contraception; rising unemployment levels; and the growing proportions of people living in *de facto* unions (ABS, 1997b; SNZ, 1997a).

DISTRIBUTION OF POPULATION

By world standards, overall population densities in Australia and New Zealand are extremely low. Australia has a population density of a mere two persons per square kilometre, compared with 13 for New Zealand, 28 for the US, 240 for the UK, 377 for the Netherlands and 5638 for Hong Kong (SNZ, 1997b). The populations of both Australia and New Zealand are heavily concentrated in coastal areas (see Figures 9.5 and 9.6). In the 1990s almost 60 per cent of the Australian population lived in the south-eastern states of New South Wales and Victoria. Within these states the population is remarkably concentrated in the capital cities of Sydney and Melbourne (see Table 9.6).

In New Zealand the population is highly concentrated in the North Island, and this concentration is likely to continue despite changes in internal mobility patterns in the 1990s. The North Island, in particular

FIGURE 9.5 All of Australia's state capital cities are located on the coast. Australia's largest city, Sydney, is located around the spectacular Port Jackson. On this occasion, the harbour is crowded with smaller vessels welcoming the *Queen Elizabeth 2* (photo: P. Tranter)

FIGURE 9.6 Most of New Zealand's cities are coastal cities. Wellington, the capital, is located on a deep harbour on the south of the North Island (photo: P. Tranter)

Auckland, is by far the most common destination for new overseas migrants. Since 1896 the North Island has consistently experienced faster population growth than the South Island. By the late 1990s 75 per cent of the population was living in the North Island (Cook, 1997).

The process of urbanisation has had a very similar history in Australia and New Zealand. In New Zealand by 1911 census data indicated for the first time that more than 50 per cent of the population were living in urban centres. In Australia at the same time 57 per cent lived in urban areas. Since then the urban proportion of the population has risen steadily (with a more rapid burst in the 1950s in Australia related to increased immigration) so that in the 1990s both countries had about an 85 per cent urban population. Their percentages of urban population have both fluctuated only very slightly since the mid-1970s, making Australia and New Zealand among the most urbanised countries in the world (see Figure 9.7). Despite this, non-urban characters have dominated the history of national myths and identity of both countries. Australian identity is often linked with the 'outback' and rugged settlers in the bush (Bishop, 1996). In New Zealand, the

TABLE 9.6 Percentage of Australian population in each state and percentage of each state's population in a capital city

	NSW	Vic	Qld	SA	WA	Tas	NT	ACT
Percentage in each state	33.9	24.9	18.2	8.0	9.6	2.6	0.9	1.6
Percentage in capital city	61.7	71.5	45.5	73.5	72.8	31.2	45.5	99.9

Source: ABS (1997b).

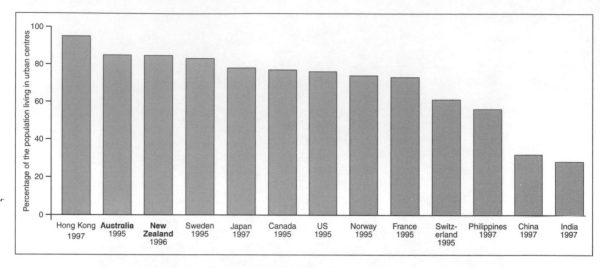

FIGURE 9.7 International comparisons of urbanisation

backblocks settlers, the gold miners and the shepherds in the High Country are celebrated (Damer, 1995; Law, 1997).

One of the distinctive features of urbanisation in New Zealand is the increasing concentration of population in Auckland. In the 1920s less than 15 per cent of New Zealand's population lived in Auckland; by the 1990s almost 30 per cent of the population lived there. The increasing dominance of Auckland in New Zealand contrasts with the Australian situation, where Sydney's share of total population has been declining in the 1980s and 1990s.

The urbanisation of indigenous peoples has been very different from that of non-indigenous peoples in both Australia and New Zealand. Both Maori and Australian Aboriginal peoples have moved to cities at a later stage and in smaller proportions than non-indigenous peoples. As late as 1945 Maori were still largely non-urban dwellers, with 75 per cent living in rural areas. However, in the next two decades, in response to the rapid growth in the need for labour in New Zealand's industries, Maori migrated to cities in unprecedented numbers. This rapid urbanisation led to widespread urban poverty for Maori in the 1980s and 1990s, as they were disproportionately affected by job losses in

manufacturing (see Chapters 10 and 11). By 1975, 75 per cent lived in urban areas, and by 1981 this had increased to 80 per cent, a level at which it stabilised, slightly lower than the national figure. The level of urbanisation for indigenous Australians contrasts even more markedly. Even in the 1990s approximately one in three indigenous Australians lived in rural areas, compared to one in seven for non-indigenous people.

Internal migration

In both Australia and New Zealand there has been a significant level of internal migration, resulting in some clearly discernible net migration patterns (Cook, 1997; Newton and Bell, 1996a). In Australia in the 1980s and 1990s, there was an accelerating 'drift north' of the population. This contrasts with the New Zealand situation, where a 'drift south' was evident by the mid-1980s.

Internal migration patterns in Australia are characterised in the 1990s by moves northwards up the east coast as well as westwards to Western Australia. Most states in Australia lost population from internal migration between 1981 and 1996. The only two states to

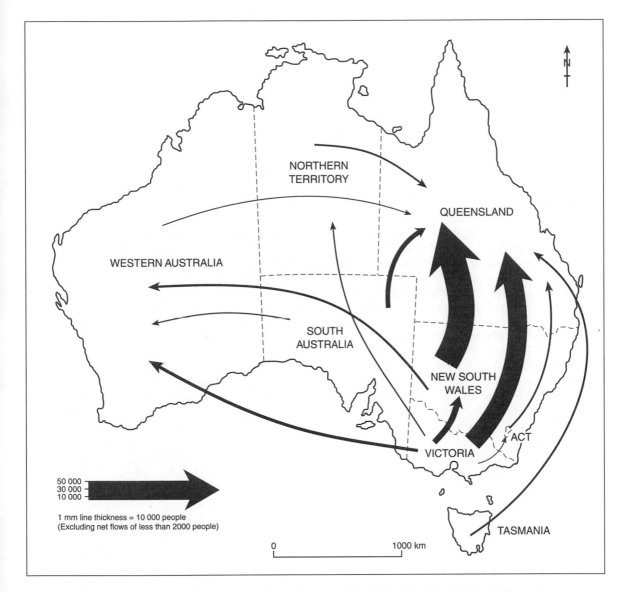

FIGURE 9.8 Main net interstate migration flows in Australia 1991–6 (ABS, 1998a: 5–6)

have gained significantly from this migration are Queensland and Western Australia. Queensland, which gained population from every state, recorded a net gain from internal migration of 145 500 people between 1991 and 1996. In Queensland, net interstate migration has exceeded natural increase in every year since 1988 (ABS, 1998a: 6). The main net migration flows between states are illustrated in Figure 9.8.

The fastest growing parts of Australia are coastal areas located close to a capital city. The reasons for the internal migration patterns are complex and involve economic and social changes. The economic changes include the decline in the importance of manufacturing (which has been very significant in Melbourne) and the rapid growth of jobs in the service industry which are more mobile. The lifestyle attraction of Queensland, with its favourable

---

Box 9.2 The Gold Coast

The Gold Coast is a popular holiday area on the southern coast of Queensland. Built on a narrow coastal strip, it became Australia's seventh largest city in the 1990s (ABS, 1998d). It is Australia's best expression of 'tourism urbanisation', a city built specifically to enable large numbers of people to visit for a short period for the consumption of fun (Mullins, 1993). The area is known for its high-rise development and its tourist attractions such as Jupiter's Casino and theme parks (Dreamworld, Seaworld and Movieworld). Sun, surf and sand abound in promotional material advertising the Gold Coast. The area is also home to a long-running tradition known as 'Schoolies Week': an annual pilgrimage made to the region by students who have just completed high school (Jones, 1986).

During the 1950s and 1960s the area grew in popularity as a holiday destination, and high-rise apartments replaced older weatherboard cottages. By the 1990s, in the central part of the Gold Coast (Surfer's Paradise) more than 80 per cent of the housing stock consisted of apartment dwellings. The Gold Coast also has some high-security US-style gated communities catering to the luxury market (see Chapter 11). The gated community concept was introduced to Australia in 1987 with the Sanctuary Cove development on the Gold Coast.

The Gold Coast represents a type of urbanisation that is unique to the late twentieth century. Cities always have some kind of economic base, be it manufacturing, commerce, and so on. The Gold Coast's economic base is tourism. More than a third of the Gold Coast's workforce is employed in sectors that rely on the tourist trade, such as recreation and real estate. In some parts of the Gold Coast, more than half of the population are tourists, many from overseas, and this figure is even higher in the peak holiday periods.

The Gold Coast is one of Australia's fastest growing cities. In the 1940s, the population of the Gold Coast was a mere 13 000. In 1986, it had risen to 249 145; and in 1996 it was 334 826; an average annual growth rate of 3.5 per cent. In comparison, the annual-growth rate for Queensland (a high growth state) for the same period was 1.5 per cent.

A high degree of transience and a volatile economy are common in cities with a consumption-based economy. In general, these cities have more low-income and casual workers, more newcomers, more adults, fewer children and more elderly people. The Gold Coast has lower average incomes and higher rates of unemployment than other Australian cities (Mullins, 1993). As holiday locations are considered desirable places in which to retire, cities such as the Gold Coast have an unusually high population of people over the age of 65. More than 14 per cent of the Gold Coast's population are over age 65. This figure is unlikely to decline as Australia's population in general continues to 'grey'.

---

climate and coastal environments, has helped make the Gold Coast (see Box 9.2) and Sunshine Coast (south and north of Brisbane respectively) very high growth rate areas. Much of the growth in these locations is related to tourism or retirement (Forster, 1995: 39–40).

In contrast to the rapid growth of coastal areas in Queensland, Australia's two largest cities are growing at much lower rates than Brisbane, Darwin and Perth. Indeed Sydney and Melbourne's population growth is highly dependent on international immigration (Hugo, 1996). Since the 1970s, both cities have experienced considerable negative net internal migration. The lowest population growth recorded in any state is for Tasmania, whose population is expected to decline during the first half of the twenty-first century.

In New Zealand the term 'the drift north' has referred to net migration flows from the South to the North Island. This flow has been typical of population movements within New Zealand for most of the twentieth century. However, in the 1986–91 period the drift north was reversed, with the South Island gaining around 1300 people from the North (Bedford et al., 1997). This 'drift south' intensified in the early-1990s, with a net gain to the South Island of almost 5000 people (Bedford et al., 1997) (see Figure 9.9). However, while the term 'drift

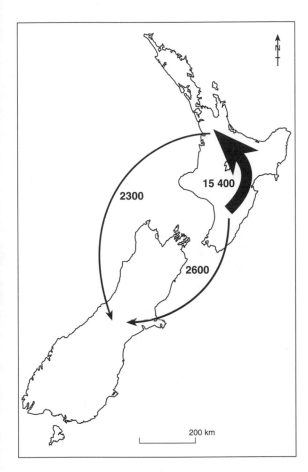

FIGURE 9.9 Internal migration in New Zealand 1991–6 (Cook, 1997: 21)

north' no longer applies to inter-island mobility, it still applies to population movements within the North Island.

Bedford *et al.* (1997) have identified some possible causal factors for the reversal of inter-island migration patterns from the late-1980s. First, the recession of the late 1980s was felt more deeply in Auckland firms, and thus the economic pull to the North Island would have been weakened. Also, even when the Auckland economy recovered in the 1990s, its pull was countered by socio-cultural factors, including retirement-type migration as well as possible 'white flight' from the disputations about Maori rights in the North Island (Bedford *et al.*, 1997). The 1990s' migrants from the North

Island to the South Island were attracted by 'talk of cheap housing, low crime rates, fresh air, absence of racial tension, surprisingly good schools and a relaxed pace of life' (McLoughlin, 1989; Bedford *et al.*, 1997: 22).

There has been some discussion in the 1990s of the phenomenon of 'counter-urbanisation'. In terms of internal migration, in both Australia and New Zealand there is an identifiable trend in the 1990s for a slight urban to rural population movement. However, while the data may indicate urban to rural population movements, many urban residents are moving to areas on the outskirts of existing urban areas. This is due to the growing attraction of a semi-rural lifestyle, where people live on larger blocks than can be found within existing city boundaries (Murphy and Burnley, 1996). Such residents are still functionally a part of the city population, even though they may be classified as 'rural' (SNZ, 1997b: 64). Some of the counter-urbanisation may be due to economic hardship, rather than choice or lifestyle reasons. In the 1990s, some New Zealand residents, especially tenants, have been forced to move to rural areas as a result of draconian changes to housing policies, putting many families in a situation where they were unable to afford to pay the increased rents (see Chapter 11). Those who have moved back to rural areas include many Maori, discouraged by extremely poor job prospects in urban areas (Waldegrave and Stuart, 1997).

There has also been a flow of elderly away from the main urban areas for a variety of reasons. This is partly due to the search for a more attractive environment for retirement and perhaps partly due to the lower living cost of smaller urban centres.

ETHNICITY AND IMMIGRANT GROUPS

Australia and New Zealand both have a high percentage of their population born overseas.

TABLE 9.7 Main ethnic groups in New Zealand, 1996

	Population
European	2 879 085
Maori	523 371
Pacific Island	202 233
Asian	173 502
Total*	3 618 300

Source: Cook (1997: 31).

*Note: Some people identified with more than one ethnic group.

With 23 per cent of its population born overseas in 1996, Australia has one of the world's largest immigrant populations. New Zealand's immigrant population of 16.5 per cent is still larger than many nations including the United Kingdom, Italy, Canada and the United States. New Zealand's immigrant population would have been even larger without periods of low and even negative net migration since the mid-1960s.

The ethnic composition of Australia and New Zealand is the result of a combination of immigration trends as well as the presence of indigenous populations. (Trends in ethnic diversity are discussed in detail in Chapter 10. Thus only a brief overview of existing ethnic groupings is provided here.) In New Zealand it is possible to examine the ethnic groups with which people identify, using questions in the census. In the 1996 census, respondents were asked to 'tick as many circles as you need to show which ethnic group(s) you belong to'. Using this information, the main ethnic groups can be identified as listed in Table 9.7. This shows that while the European ethnic group was still dominant, Maori represented 14.5 per cent of the population, a figure which has increased from the 1960s and is expected to continue to increase slowly for the next 40 years (SNZ, 1994). People of British descent dominate the European ethnic group. New Zealand has relatively few migrants from non-

British European countries, in contrast to the situation in Australia. Within the Pacific Islander ethnic group there were many smaller ethnic groups, each with distinctive age structures and settlement histories. The largest Pacific Island ethnic groups in the 1990s were: Samoan, Cook Islander, Tongan, Niuean, Fijian (excluding Fijian-Indian) and Tokelauan (SNZ, 1997b).

The Asian ethnic group in New Zealand grew rapidly from the late 1980s. The main ethnic groups within this Asian category were: Chinese, Indian, Korean, Filipino, Japanese, Sri Lankan and Cambodian. As a broad group, Asians are recent migrants to New Zealand. Although the first immigrants from Asia arrived in New Zealand over 100 years ago, over half of the Asians born overseas had been living in New Zealand for less than five years (SNZ, 1997b).

Australia's population evolved from an almost totally Aboriginal population in the eighteenth century to a predominantly Anglo-Celtic population in 1900. By the 1990s Australia had a multicultural mixture comprising approximately 74 per cent Anglo-Celtic, 19 per cent other European, 4.5 per cent Asian and only 2 per cent indigenous Australians (DIMA, 1998a).

Although the Australian census has no questions on ethnic group (unlike the New Zealand census), the ethnic make-up of Australia's population can be examined through statistics on first- and second-generation Australians (see Figure 9.10). Although this provides only a broad indicator of ethnic groupings, it does allow the identification of recently arrived groups. In the 1996 census, not only were 23 per cent of Australians born overseas, but of the remainder of the population a further 27 per cent were second-generation Australians. In other words, they had at least one overseas-born parent. In Figure 9.10 the recently arrived groups (e.g. Vietnamese) can be identified by the small number of second-generation Australians. In the long-established groups (e.g. Italians) second-generation groups account for more than half the

Box 9.3 Chinese in New Zealand

The Chinese were the first significant non-European and non-Maori ethnic group to develop in New Zealand. As early as 1874, they represented 1.4 per cent of the total population (SNZ, 1997a: 124). Chinese had settled in New Zealand since at least the 1840s, originally as isolated individuals. The first group of Chinese to come to New Zealand arrived in 1866, having been invited by the Dunedin Chamber of Commerce to help rework the goldfields after the first boom in gold mining had receded (Brooking and Rabel, 1995). However, even before they arrived in Dunedin, early settlers were protesting against the overrunning of their colony with 'ignorant' and 'treacherous' Chinese hordes (Ip, 1995). Racist attitudes towards the Chinese continued after their arrival. These attitudes were also reflected in official government policies that have since been referred to as part of New Zealand's 'White New Zealand' policy. Specific legislation was introduced to restrict Chinese immigration and even to exclude Chinese from the old age pension scheme (Brooking and Rabel, 1995).

The Chinese community survived overt racism and male-dominated sex ratios. By the end of the Second World War there were more than 3000 Chinese-born residents in New Zealand, though they still felt the need to remain a cohesive group to survive prejudice (Brooking and Rabel, 1995). It was not until 1952 that a small number of Chinese were naturalised.

From the 1950s through to the 1980s, the Chinese maintained a low profile. Some became successful professionals in the 1960s, but few had prominent leadership positions. They were also not noticeable in terms of negative statistics: they recorded very low levels of crime and prison sentences, as well as a low level of use of the health system. 'The Chinese ethnic community existed in such a low-key way that the country's homogeneous mainstream society hardly noticed it' (Ip, 1995: 196).

This position was about to change dramatically. The 1987 Immigration Act, which was passed during a period of massive economic restructuring, introduced a non-discriminatory approach to immigration. This opened New Zealand once again to an influx of 'non-traditional' migrants, and the Chinese began to migrate once again in large numbers.

The Chinese immigrants of the 1980s and 1990s were very different from the first Chinese immigrants. Rather than migrating directly from China, many of them were ethnic Chinese from places such as Hong Kong who were more likely to be successful business people. They were also more likely to be families intending to settle in New Zealand, rather than men seeing New Zealand as a short-term destination (as in the goldrushes). By 1996, there were over 80 000 people in the Chinese ethnic group in New Zealand (or about 2.2 per cent of the population) (SNZ, 1997b: 40).

New Zealand governments in the 1980s and 1990s welcomed the Chinese and other Asians because of their potential contribution to the economy. But, as in the nineteenth century, the New Zealand public did not universally welcome them (Bedford, 1996). The new Chinese migrants were treated with suspicion by some Maori groups, and 'middle' New Zealanders were envious of the wealth that many new Chinese migrants brought with them. In the schools, the high achievement rates of the Chinese students sometimes led to jealousy. Also there was a widespread belief that many Chinese were simply using New Zealand as a base for their families, while they carried on their businesses overseas. The populist press called some Chinese families 'astronauts' because of their frequent commuting, and also criticised Chinese for maintaining their own community ties, rather than making any attempt to 'assimilate' into New Zealand society. Many Chinese had difficulty setting up businesses in New Zealand. Sometimes this could have been due to discriminatory attitudes from other New Zealanders, but it could also have been due to unfamiliar laws and business conventions (Ip, 1995).

Chinese have often been the subject of populist denunciation as well as blatantly racist government policies. This has had much to do with the views of New Zealanders about their own cultural and/or racial superiority (Spoonley, 1994). The problems associated with recent Chinese immigration can be seen as part of the lack of a comprehensive policy on multiculturalism. Because of this, the Chinese (and many other immigrant ethnic groups) feel as though they do not have a place within the national identity of New Zealand. This issue is discussed further in Chapter 10.

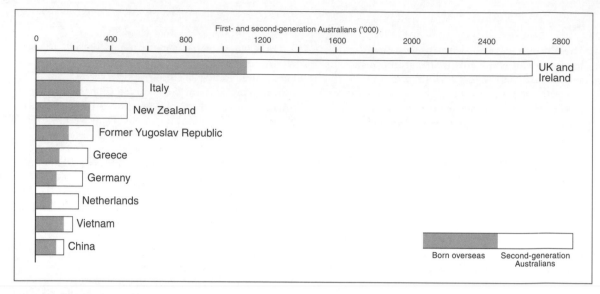

FIGURE 9.10 First- and second-generation Australians, 1996 (ABS, 1998d: 162)

total group. The figure shows the continued dominance of the United Kingdom and Ireland, as well as the growing importance of Asian countries and New Zealand.

One important statistic not shown on this figure is the percentage of the Australian population identifying themselves as of Aboriginal or Torres Strait Islander origin. These groups are Australia's indigenous population. In 1996, there were 386 000 people of indigenous origin, representing a mere 2.1 per cent of Australia's total population. Yet this was a substantial increase from the figure of 160 958 in 1976, although this is due partly to the increased willingness to be identified as one of these groups (ABS, 1994; ABS, 1997a). The percentage of indigenous people in the population varies substantially from state to state, ranging from a low of 0.5 per cent in Victoria, to 29 per cent in the Northern Territory in 1996 (ABS, 1998a: 15).

From the description of ethnic diversity presented above, some differences between Australia and New Zealand become evident. The most striking contrast is in the size of the indigenous population (14.5 per cent in New Zealand compared to about 2 per cent in

Australia). Another difference is the lack of a large non-British European contingent in New Zealand. This difference is due to New Zealand's ability to use Maori and Pacific Islander labour for its industrial development after the Second World War, while Australia encouraged immigration from Europe. These two differences have important implications for the development of policies for dealing with cultural diversity (biculturalism versus multiculturalism – see Chapter 10).

One similarity between the two countries is the recent rapid growth in Asian migration (which started earlier in Australia). This is fuelling some anti-immigration and anti-Asian sentiment among sections of the public. Related to this is the emergence in the 1990s of the One Nation political party in Australia, which stirred irrational fears of an 'Asian invasion' (see Chapter 10).

New immigrants in both Australia and New Zealand are concentrated in particular regions and cities. In New Zealand, more than half of all new immigrants were living in Auckland at the 1996 census (SNZ, 1997b: 31). In Australia, more than two-thirds of immigrants migrate to New South Wales and Victoria. New South

Wales, and Sydney in particular, has the highest number of new migrant settlers (White and Williams, 1996). Immigrants contribute to the colourful multicultural character of many Australian cities (see Chapter 12).

Religion

The source of immigrants to Australia and New Zealand is reflected in religious affiliations. In New Zealand in 1991 the main religious affiliations were Anglican (originally Church of England) (22 per cent), Presbyterian (16 per cent) and Catholic (15 per cent), suggesting English, Scottish and Irish heritage respectively. In Australia, in 1996, the main religions were Catholic (27 per cent), Anglican (22 per cent) and Uniting (7.5 per cent). The higher proportion of Catholics in Australia reflects both ongoing Irish immigration, as well as post-war immigration from European countries such as Italy. However, in both countries, the percentages in most of the Christian religions (particularly the ones listed above) have been in decline for the last 10 years or more. In contrast, there is a continuing trend for both Australians and New Zealanders to report themselves as having 'No religion'. The source of the most recent immigrants to both countries is reflected in the religions showing the most rapid growth in the 1990s: Hinduism and Buddhism. For example, in Australia, while the total number of Anglicans fell by almost 3 per cent between 1991 and 1996, the number reporting adherence to Buddhism increased by 43 per cent, and Hinduism by 54 per cent (ABS, 1998d: 165; SNZ, 1997a: 145).

The increasing variety of religious affiliations is an indicator of the increasingly multicultural nature of Australasian society. In the next Chapter, attention is focused on immigration and ethnicity, and the policies of multiculturalism in Australia and biculturalism in New Zealand.

FURTHER READING

Excellent summaries of various population characteristics are provided in the *Australian Social Trends* reports provided by the Australian Bureau of Statistics. These are now published annually, and replace the earlier *Social Indicators* report. Though there is now no similar report produced by Statistics New Zealand, the publication entitled *People and Places* does provide a very useful description of many population characteristics. Useful analyses of the New Zealand situation can also be found in various Chapters of Spoonley (1994).

10

MONOCULTURAL, BICULTURAL OR MULTICULTURAL?

Australia and New Zealand share a legacy of white colonialism. However, the two countries have had very different experiences in terms of the early relationships between European settlers and indigenous peoples, as well as in immigration policies and practices in the twentieth century. These different experiences are important in understanding responses to ethnic and cultural difference.

In New Zealand, the British made a treaty with the indigenous peoples in 1840, known as the Treaty of Waitangi. This treaty was to become the basis for the Waitangi Tribunal (see Chapter 5), which has been a fundamental component of biculturalism in New Zealand in the 1980s and 1990s. Because of the treaty, Maori people were recognised as *tangata whenua* (the people of the land).

Australia's history of European–indigenous relations was markedly different. In Australia there was no treaty. The first Europeans saw Australian aboriginals as the lowest order of races (Pawson, 1992). The British regarded Australia as *terra nullius*, a land owned by no one (Clarke, 1992: 13). The very category of 'Aborigine' assisted the process of colonisation, through the categorisation of Aboriginal people as the 'primitive other' (Stokes, 1997). Thus European settlers in Australia felt under little pressure to develop a bicultural policy.

Policies on immigration differed markedly between the two countries. The contribution of immigration to population growth was much higher in Australia (see Chapter 9) especially after the Second World War. More importantly, Australia encouraged very large numbers of non-British immigrants from various European countries in a bid to meet labour shortages in the 1950s and 1960s. During this time, immigration in New Zealand was much more dominated by British immigrants, and the small numbers of non-British Europeans were not conducive to a strong multicultural policy. Also, while Asian immigration to New Zealand remained very low until the late 1980s, in Australia migrants from Asia became an important component of the total intake in the 1970s.

Large-scale immigration has been a significant feature of most highly developed countries since 1945. Each of these countries developed specific responses to the issue of ethnic diversity. These responses can be summarised into three generalised models: differential exclusion, assimilation and pluralism (Castles, 1997). Differential exclusion exists where immigrants are allowed to participate in some facets of society (e.g. the labour market), but are denied access to other facets (e.g. welfare or citizenship). Assimilation is the policy of forcing new immigrants to adapt to the new culture by giving up their own language and culture, thus maintaining a monoculture. In a pluralist model, immigrant minorities retain their own culture, though they are usually expected to conform to key values of the host society. The pluralist model is well demonstrated in multicultural Australia.

The terms monocultural, bicultural and multicultural can each be used in two different ways. In a descriptive sense, they can refer to the diversity of ethnic, linguistic or religious groups in a population. In this sense, Japan is one of the most monocultural nations in the world, with only about 1 per cent of its total population being immigrants. New Zealand could be described as largely bicultural in the sense that its two main ethnic groups (European and Maori) comprise the majority of the country's population (about 80 per cent and 14.5 per cent respectively) (Cook, 1997: 31). (Even this descriptive use of bicultural is problematic: it does not recognise the considerable diversity within each of these two ethnic groups, nor does it acknowledge the existence of other ethnic groups.) Australia, along with many other nations, is multicultural in terms of having a population of diverse ethnic groups as a result of sustained high levels of immigration from countries throughout the world. However, Australian multiculturalism is based on a large number of very small ethnic groups (compared to the large Maori ethnic group in New Zealand) set into a predominantly Anglo-Celtic culture (Jupp, 1997).

Each of these three terms can also be used in an evaluative way to refer to an attitude or policy towards ethnic and cultural diversity within a society. In this sense, monoculturalism dominated New Zealand and Australia in the nineteenth century and for over half of the twentieth century. In a monocultural society, power is reserved for a culturally uniform elite, and groups who are labelled as 'other' are expected to conform to the values and practices of this dominant group. In the early history of both countries, some groups (e.g. Chinese) were denied access to citizenship, and were deterred or excluded from immigration under the model of differential exclusion (Graetz and McAllister, 1994; Pawson, 1992). In Australia until the 1960s, Aborigines were excluded from many areas of society, even from hotel bars in some country towns (Clarke, 1992: 302).

The definition of biculturalism in an evaluative sense as it applies in New Zealand is not universally agreed on. A limited definition is that it is 'a public policy giving official recognition to two peoples, Maori and European, and their cultures within the public institutions of a multi-cultural society' (Mulgan, 1989). However, some prominent Maori argue that biculturalism implies more than the simple recognition of two cultures and the respect of Maori values, language and culture: it should also indicate the sharing of power, resources and responsibility (Cheyne *et al.*, 1997: 122). Biculturalism has been strongly criticised by immigrant ethnic groups in New Zealand as being exclusionary. There is no place for non-Maori or non-British groups such as Chinese, Indians or Koreans in bicultural New Zealand.

In Australia, the response to ethnic diversity progressed from assimilation to multiculturalism. When the term first came to Australia from Canada, multiculturalism implied the retention of ethnic identity for minority groups (Jupp, 1997). Multiculturalism in Australia is an ideology that 'accepts diversity and supports policies of maintaining ethnic identities, values and lifestyles within an overarching framework of common laws and shared institutions' (Thakur, 1995). It now has the support of both major political parties in Australia, yet it has also been strongly criticised by both the Right and the Left of Australian politics (Castles *et al.*, 1992). Similarly, in New Zealand even a limited adoption of multiculturalism has been widely criticised, by both Pakeha and Maori groups.

There are many aspects of culture that a multiculturalist policy could address (e.g. class, gender, sexual preference and occupation) (Jupp, 1991). However, the debate on multiculturalism has usually focused on ethnic variety to the exclusion of these other factors.

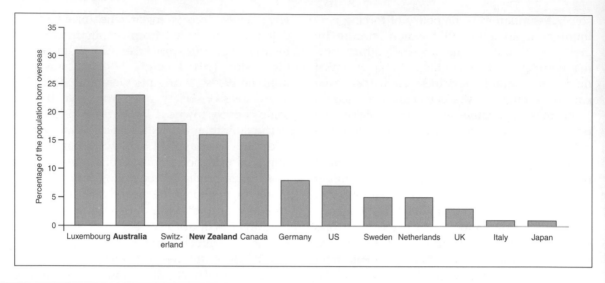

Figure 10.1 Immigrant populations in selected countries. 1993 data for all countries except Australia and New Zealand (1996), Canada (1991) and USA (1990) (ABS, 1997b: 15; Cook, 1997: 27)

Ethnicity itself is a complex term, which sometimes includes the concept of 'race', though the authors reject the notion that there is such a category. Ethnicity is used here to refer to a cultural identity based on shared history and cultural practices, often linked to a particular country of origin.

IMMIGRATION POLICIES AND ETHNICITY IN AUSTRALIA

In terms of ethnic mix, Australia remained an essentially monocultural nation until shortly after the Second World War, when large numbers of migrants arrived from many European countries. Between October 1945 (when Australia's post-war immigration programme began) and June 1997, 5.6 million people migrated to Australia. Now, with more than four million people born overseas (representing 23 per cent of Australia's 18.3 million people), Australia has the largest migrant proportion of all OECD countries except Luxembourg (ABS, 1997b) (see Figure 10.1).

At Federation in 1901, Australia's population was overwhelmingly British in origin. Though some concentrations of German settlers had developed, and large numbers of Chinese immigrants had arrived during the goldrushes in the 1850s, Australia's identity was firmly linked with 'the mother country'. One of the first acts of the new federal government was to affirm its commitment to a 'White Australia' policy (Castles, 1992). This was given legislative form through the Immigration Restriction Act of 1901, which put an end to the previous 112 years of uncontrolled immigration. The Act was supported by a language test to be 'used solely to exclude non-white immigrants' (Graetz and McAllister, 1994: 82). Largely because of this legislation, the ethnic composition of Australia changed little until the late 1940s.

After the Second World War the Australian government encouraged a massive influx of migrants to Australia. There were two main reasons for this government action. First, there was a fear (apparently made real by the

Japanese attempts to invade Australia during the war) that without a larger population Australia's huge continent would be indefensible. Second, there was a need for labour to meet the demands of growing manufacturing industries in Australia. The first Minister for Immigration, Arthur Calwell, justified the high levels of immigration to the Australian people through his now famous slogan 'populate or perish' (Collins, 1988: 21).

The racism of the White Australia policy, formally enacted at the time of Federation, was still an important element of the immigration policy immediately after the war (Lack and Templeton, 1995). The committee established to design guidelines for an immigration programme established that the British would be the priority, followed by Nordics and then, at the bottom, southern Europeans. Black or Asian immigration was considered totally unacceptable. The aim was to encourage 'assimilable types' who would rapidly become indistinguishable from other Australians (Castles *et al.*, 1992: 9).

It was apparent by 1947 that it would not be possible to have a predominantly British immigrant population. Not only were there problems attracting sufficient British migrants to Australia, but there were also pressures to accept refugees or 'displaced persons' from Europe. The International Refugee Organisation (IRO) was established by the United Nations to facilitate the refugee resettlement. Between 1947 and 1952 170 000 displaced persons came to Australia from the Baltic States, Ukraine, Russia, Poland, Hungary, Czechoslovakia, Yugoslavia, Romania, Bulgaria and Albania. The search for new migrants widened, and assisted passage agreements were signed with Malta and the Netherlands in 1948 and with other European countries, including Greece, Italy and West Germany, in 1962 (York, 1996: 4).

One important feature of migration in the 1950s and 1960s was the high number of immigrants from southern European countries, particularly Italy and Greece, and the concentration of these migrants in distinctive ethnic areas in Australia's major cities (see Chapter 12). This was most evident in Melbourne and Sydney. In 1971, people born in Italy or Greece comprised 7.3 per cent of Melbourne's total population and 4.0 per cent of Sydney's (Forster, 1995: 19).

The massive intake of European migrants during the early 1950s was seen as according with the 'White Australia' sentiment. Asian migrants were not encouraged. It was not until later in that decade that the Menzies Liberal government allowed Asians to apply for citizenship. In 1958, the infamous dictation test, which allowed immigration officers to exclude anyone on the basis of failing a test in any European language, was removed.

In 1973, the Whitlam Labor government removed racial restrictions in immigration policy. This opened the way for a new wave of migrants from Asia, with the selection criteria being based not on race, but on skills. The greatest increase in Asian migration, however, occurred as a result of the influx of Vietnamese refugees when the United States pulled out of the Vietnam War (Castles, 1992) (see Chapter 17).

The influx of migrants from Asia continued throughout the 1970s and 1980s. By 1980, 22 per cent of new immigrants were Asians, and in 1982–3, Asian migrants became the largest single component of all entries (36 per cent of net immigration), ahead of Britain and New Zealand (Castles, 1992).

Changes to immigration policy in the late 1990s led to a cutback in immigrant numbers, as well as more emphasis on 'delivering people with needed skills and experience' (DIMA, 1998b). Some observers claimed that this was to satisfy populist voters who were concerned about unemployment, and that Australia's immigration programme is aimed at a globally mobile elite of skilled people. It has also been claimed that the skilled migration program excludes about 99 per cent of the world's population.

The ethnic make-up of Australia's population has changed extensively over the time of European settlement, especially in the last 50

Figure 10.2 Cabramatta's 'Freedom Gate'. Cabramatta is a suburb of Sydney with a high proportion of south-east Asian born residents. The ethnic difference of Cabramatta, once a source of stigma, became the basis for the area's promotion as a tourist destination. It is now well known as an 'oriental' district (see also Box 12.4) (photo: P. Tranter)

years. This change is reflected in a country with over 100 languages, ethnic media, new foods and diverse religious and cultural activities.

Since the Second World War, the combined effect of Australia's immigration policies and international political and economic conditions has produced a vast change in the percentage of the population born in various overseas countries. As Table 10.1 indicates, the percentage born overseas has increased from a mere 9.8 per cent in 1947 to 23 per cent in 1996 (almost the highest of any country in the world). Of the overseas-born population, the percentage from non-English-speaking countries has increased from 19.2 per cent to over 60 per cent in 1996. The Asian-born population has grown rapidly, especially since the mid-1980s. In 1996, five of the top 12 birth-place groups were from Asian countries. Among these the largest increase was for Vietnam, increasing from a mere 2500 in 1976 to almost 150 000 in 1996. The changes in ethnic composition are even more marked if we

examine the countries of birth of settlers (Table 10.2). This shows the dramatic changes, especially since 1971–5 in the proportions coming from Asian countries. Over the 1996–7 year, the most rapidly growing birth-place groups were those from Singapore, Indonesia and China (ABS, 1998b). The Asian presence has been made more visible in the landscape through the active promotion and development of 'Chinatowns' and other ethnic enclaves (see Figure 10.2). It appears likely that Australia's ethnic composition will continue to be transformed over the coming decades.

Government responses to ethnic diversity in Australia

In the period following the Second World War, various policies were developed to deal with the ethnic diversity created by the immigration policies in Australia. These policies can be

TABLE 10.1 Overseas-born population : Top 12 birth-place groups (1947, 1976, 1996)

Countries	1947 %	Countries	1976 %	Countries	1996 %
UK and Ireland	72.7	UK and Ireland	41.1	UK and Ireland	28.7
New Zealand	5.9	Italy	10.3	New Zealand	7.1
Italy	4.5	Greece	5.7	Italy	6.1
Germany	2.0	Yugoslavia	5.3	Former Yugoslav Republics	4.4
Greece	1.7	Germany	4.0	Vietnam	3.6
India	1.1	Netherlands	3.4	Greece	3.4
Poland	0.9	New Zealand	3.2	Germany	2.8
China	0.9	Poland	2.1	China	2.5
Yugoslavia	0.8	Malta	2.1	Hong Kong/Macau	2.3
United States	0.8	USSR	1.9	Netherlands	2.3
South Africa	0.8	India	1.4	Malaysia	2.3
USSR	0.7	Lebanon	1.2	Philippines	2.2
Other	7.2	Other	18.3	Other	32.3
Australia	90.2	Australia	79.9	Australia	77.0
Overseas	9.8	Overseas	20.1	Overseas	23.0
MESC	80.8	MESC	46.5	MESC	39.5
NESC	19.2	NESC	53.5	NESC	60.5

MESC = Main English-speaking countries; NESC = Non-English-speaking countries
Source: ABS (1992a: 28; 1997b: 12).

summarised into three broad groups: assimilation (1901–1960s), integration (mid-1960s to 1972) and multiculturalism (since 1972). The assimilationist thinking was that minorities should quickly adopt all the cultural trappings of the existing Australian population. This policy was applied not only to new immigrants, but also to the indigenous people. Assimilation was a way of reassuring existing Australians that their way of life would not be destroyed by the influx of migrants with differing cultural practices. It was a way in which

TABLE 10.2 Main countries of birth of settlers

Countries	1947–51 % total intake	Countries	1971–5 % total intake	Countries	1991–5 % total intake
UK and Ireland	48.5	UK and Ireland	41.4	UK and Ireland	14.0
Poland	12.3	Yugoslavia[b]	7.2	New Zealand	9.1
Italy	7.4	Greece	3.9	Hong Kong/Macau	8.0
Baltic States[a]	6.2	USA	3.7	Vietnam	7.5
Netherlands	4.7	Italy	3.4	Philippines	5.1
Yugoslavia	4.1	New Zealand	3.4	India	4.6

[a]included Estonia, Latvia and Lithuania
[b]Former Yugoslav Republics
Source: ABS (1992a: 27; 1997b: 13).

the massive inflow of migrants could be legitimated.

Assimilation as a way of dealing with new immigrants soon proved to be problematic. Newcomers were left to their own devices to adapt to Australian society, and were given little assistance in this difficult task (Collins, 1988). They were expected to learn English and quickly assimilate, but they were not given any support to do this. The stress this placed on Non-English-speaking background (NESB) migrants was reflected in high rates of suicide and emotional breakdown. It also led to a high percentage of NESB migrants returning to their homeland: 13 per cent of Italian migrants had returned to Italy permanently by the late 1960s (York, 1996).

The assimilationist view was clearly directed at achieving a 'monoculture' and was antipathetic towards any type of pluralism. Yet such a view was also totally unrealistic. It was extremely difficult to force people to abandon their languages and customs. Thus, in the 1960s, assimilation started to give way to the slightly more enlightened idea of 'integration'.

Integration still aimed for a monoculture, but one in which it was acceptable to incorporate some aspects of the new ethnic groups' cultures. For example, new styles of cooking or even coffee could become part of the Australian society. Also associated with the idea of integration was the notion that NESB migrants should be provided with specific support services, such as migrant English courses. Integration was thus little more than a longer-term and better-supported assimilation process. Also, the adoption of the policy of integration was not based entirely on care and concern for the immigrants themselves. Another reason was that it was becoming difficult to keep migrants in Australia to provide the much-needed source of labour for new manufacturing industries.

The concept of multiculturalism first appeared in the early 1970s with the election of the Whitlam Labor government, and was incorporated into political structures more formally by the Fraser Liberal government later in the 1970s. An important political figure in the adoption of multiculturalism was Al Grassby, the colourful Italian-born Minister for Immigration under Whitlam. Grassby argued that non-English-speaking migrants were simply non-people in a monocultural society. Under the policy of multiculturalism, the distinctive cultures of the new ethnic groups were to be celebrated and encouraged, rather than submerged (Lack and Templeton, 1995: 143). Cultural pluralism replaced monoculturalism, or to put it in more colloquial terminology, the 'melting pot' was replaced by the 'salad bowl' (Collins, 1988).

When multiculturalism was first introduced into official policies in Australia, it was an 'ethnic group model' of multiculturalism. The central principle was the key role of ethnic groups, which were seen as having a relatively fixed cultural identity. This model was criticised because it gave the state the opportunity to fund groups and leaders that the government wanted to work with, while ignoring diversity within groups (Castles, 1997).

The Labor government of the 1980s and early 1990s introduced a different model of multiculturalism that moved away from the 'ethnic group model' to the 'citizenship model'. In the *National Agenda for a Multicultural Australia* (OMA, 1989) multiculturalism was defined not in terms of minority rights, 'but in terms of the rights of all citizens in a democratic state' (Castles, 1997). It was defined as a system that involved both rights and obligations. It included the acceptance of basic principles of tolerance and equality, as well as English as the national language. The new model also recognised the cultural rights of all individuals, as well as the fact that some groups were disadvantaged by lack of language proficiency, and discrimination by race, ethnicity and gender. The state accepted responsibility to combat such disadvantage. Thus multiculturalism was no longer seen as simply recognising and celebrating ethnic and cultural difference. It was also seen as a policy for linking cultural rights with policies of social justice (Castles, 1997).

In the early 1990s, the Labor Prime Minister Paul Keating embraced multiculturalism as the underpinning of Australian society. However, in March 1996, John Howard's Liberal–National federal government was elected, and subsequently multiculturalism has not enjoyed the same level of support. Some believe that the Prime Minister's rhetoric of free speech facilitated a resurgence of racism in Australia (Murphy and Watson, 1997). Despite this, there is still a commitment at the level of federal politics to multiculturalism.

Critiques of multiculturalism in Australia

Ethnic diversity has often been seen as a threat to national unity in nations throughout the world (Jones, 1996b). Australian governments have been acutely aware of this, but federal politicians from both sides of politics have actively supported immigration and various policies on the treatment of immigrants. Successive governments promoted immigration as a positive factor for nation building, not as a 'problem' (Fincher, 1997; Jordens, 1995). Despite the bipartisan support for immigration and for tolerance of difference in race or ethnicity, there has been strong and vocal criticism of Australia's immigration policies and policies relating to the treatment of ethnic minorities, in particular to the policy of multiculturalism.

Criticism of multiculturalism has come from sections of the general community as well as from the right and the left of the political spectrum (Castles *et al.*, 1992: 44). The extent to which the general community supports the concept and practice of multiculturalism is open to debate. Official government documents still present the view that the community does support such a policy. For example, a report of the Multicultural Advisory Council in 1995, *Next steps: multicultural Australia: towards and beyond 2000*, suggested that though there is 'some uncertainty and ambivalence about the concept, there appears reasonably widespread support for the under-

lying principles of multicultural policy' (Birrell, 1996). However, Birrell believes this to be misleading. Indeed, when opinion poll data are examined, even the ones used by the government in formulating its policy documents, it appears that the majority of both Australian-born and overseas-born respondents (62 per cent and 57 per cent respectively) still favoured an assimilationist option over a multicultural one in 1994. Birrell argues that for most migrants the main aim is to be treated as fellow Australians (or locals): 'any reference to ethnicity cuts across this objective' (Birrell, 1996). Other evidence suggests that very few Australians identify with a particular ethnic group (Betts, 1991).

Those on the right of the political spectrum in Australia suggest that multiculturalism is actually a radical conspiracy, which promotes ethnic groups at the expense of the Australian-born. The right argues that multiculturalism has provided ethnic groups with the power to influence government and to direct money away from other Australians. The historian Professor Geoffrey Blainey, for example, argued that multiculturalism would threaten social cohesion and denigrate Australia's British heritage (Blainey, 1984). He suggested that Australia's 'emphasis on granting special rights to all kind of minorities is threatening to cut this nation into many tribes' (Collins, 1988: 238). However, Jones (1996b) countered this argument, explaining that Australia will maintain its national unity because of the unity of the political elites, who have usually refused to appeal to ethnic and racial prejudice as a vote winning strategy. When politicians do appeal to such prejudice, the political elites have usually strongly defended their multicultural and anti-racist policies.

One of the most divisive forces in federal politics in the 1990s is the One Nation Party led by former Liberal Party candidate Pauline Hanson, elected to the Australian parliament as an Independent. Hanson maintains that Aboriginal Australians receive favourable treatment by the government. She also strongly disagrees with multiculturalism and current

Figure 10.3 Caroline Street Redfern. This area of Sydney provides one of the few noticeable urban concentrations of indigenous Australians. The disadvantage of the area is evident, even in the standard of maintenance of the nineteenth-century terrace housing (photo: P. Tranter)

immigration policies, and believes (despite evidence to the contrary) (Jones, 1996b) that her views are widely shared by mainstream Australians. She would like Australia to return to a monocultural state. In her first speech to the Australian parliament, Hanson explained:

> *I and most Australians want our immigration policy radically reviewed and that of multiculturalism abolished. I believe that we are in danger of being swamped by Asians ... They have their own culture and religion, form ghettos and do not assimilate. A truly multicultural country can never be strong or united (Australian House of Representatives, Hansard, 10 September 1996) (Jones, 1997: 287).*

Such views, which undermine the support for tolerance and an egalitarian society, are more likely to gain acceptance in times of poor economic performance. Australia has suffered considerably from the removal of trade protection, with huge job losses in manufacturing industries, widening gaps between rich and poor and high youth unemployment rates. In these difficult times, indigenous Australians and new immigrants (especially Asians) have been made scapegoats. This is despite the overwhelming evidence that indigenous Australians are incredibly disadvantaged on any indicator of well-being, and that many Asian migrant groups also have high levels of disadvantage (see Chapter 11). This disadvantage is evident in a low-income area of Redfern in Sydney (see Figure 10.3), where there are concentrations of indigenous Australians as well as Asian migrants.

Multiculturalism became more controversial when Asian migration increased. The first significant inflow of Asian migrants, from Vietnam, was at first welcomed as an unfortunate group who had survived unthinkable deprivation and trauma in the Vietnam War. However, this attitude was soon replaced, largely due to talkback radio which developed such ideas as 'an invasion by stealth from the north' (Waldren and Carruthers, 1998). The Vietnamese became an easily identifiable target for racist attacks.

As well as the criticisms of multiculturalism coming from the right (e.g. Blainey) there has also been criticism from the left. Jamrozick et al. (1995) and Jakubowicz (1984), for example, have suggested that multiculturalism is a neo-conservative policy aimed at containing the working class and ensuring that it is not disruptive to the capitalist system (Jakubowicz, 1984; Jamrozick *et al.*, 1995). They argue that multiculturalism's function is to keep migrants in their place by funding conservative sections of the ethnic communities, and that multiculturalism failed to achieve the equity for migrants that it promised. Indeed, despite two decades of multiculturalism, migrant disadvantage persists, especially for NESB migrants. From a feminist perspective, multiculturalism is seen as a way of maintaining sexist social and cultural practices. By retaining ethnic heritage, multiculturalism also reinforces patriarchy (Jupp, 1997).

Multiculturalism has also been criticised for paying too little attention to the needs of indigenous Australians. Though Aborigines were mentioned in the *National Agenda*, most of the policy initiatives relate only to immigrants. Not surprisingly, there is a lack of support for multiculturalism from indigenous Australians who do not want to see themselves as just another ethnic group (Castles, 1997; Pearson and Ongley, 1996).

We can ask whether, on balance, multiculturalism has been successful. It has been able to enrich Australian life without leading to any significant dis-benefits for Australian society. For example, it has contributed to the avoidance of the development of ethnic ghettos. It has helped avoid much of the violence associ-

TABLE 10.3 The ethnic allegiances of Australian soccer clubs

Clubs	Ethnicity	Location
Apia Leichhardt	Italian	Sydney
Marconi Fairfield	Italian	Sydney
Parramatta Melita	Maltese	Sydney
Sydney Croatia	Croatian	Sydney
Sydney Olympic	Greek	Sydney
Heidelberg Alexandria	Greek	Melbourne
Melbourne Croatia	Croatian	Melbourne
Preston Macedonia	Macedonian	Melbourne
South Melbourne Hellas	Greek	Melbourne
Brisbane United	–	Brisbane
Adelaide City Juventas	Italian	Adelaide
West Adelaide Hellas	Greek	Adelaide
Newcastle Breakers	–	Newcastle
Wollongong City	–	Wollongong

ated with racial issues that has been found in other countries including Britain, Germany and the United States (Smolicz, 1995). Given the reality of the cultural diversity in Australia, as well as the increasingly diverse ethnic base, multiculturalism seems to be a more effective response than assuming Australia can return to an increasingly distant British past.

However, although Australia can be held up as a model for the rest of the world in terms of the way that it has been able to incorporate millions of migrants from diverse cultures into a democratic society (Jupp, 1991), it is less satisfactory as a model for the equitable treatment of its indigenous peoples, as will be discussed in Chapter 11.

Australia's ethnic mix will continue to change as the percentage of the population of Anglo-Celtic origin declines. In such a climate, it is important that the bipartisan political support for multiculturalism is continued, even if popular support for the policy lags behind. Continued official support for multiculturalism cannot always be guaranteed, however, as the case study of ethnicity in Australian soccer in Box 10.1 illustrates.

Box 10.1 Multiculturalism and Australian soccer

The status of soccer in Australia relative to other countries throughout the world is indicated by the way in which the generic term 'football' does not apply to soccer in Australia, but instead is used to describe sports such as Rugby League, Rugby Union and Australian Rules. Soccer was established in Australia by 1880. However, Australians developed a strong attachment to their local varieties of 'football' and soccer attracted the label of a 'foreign' sport. 'To play soccer was proof of one's division from mainstream society' (Mosely, 1995). When large numbers of non-British European migrants began arriving in Australia in the 1950s, they brought with them their love of soccer. The new migrants quickly outnumbered the Anglo-Australians in the soccer establishment, and the game became an ethnic preserve.

In the national competition, teams formed links with particular national origins. In 1992, the teams which comprised the National Soccer League, had the names and ethnic allegiances shown in Table 10.3 (Hughson, 1992: 13).

Officials at the most senior level of the sport, who wanted Australian soccer to become accepted by a wider group of Australians, construed the ethnic nature of soccer as a problem. Thus there have been ongoing attempts to 'de-ethnicise' the game of soccer. At first it was believed that, given enough time, migrants would assimilate into mainstream society, ethnic loyalties would wane, and it would become easier to change the image of soccer to an Australian game rather than an ethnic one (Mosely, 1995). However, the ethnic nature of Australian soccer has proved to be more enduring than at first expected.

In 1992, the Australian Soccer Federation decreed that teams competing in the National Soccer League would not be permitted to have 'foreign or ethnic names' (Hughson, 1997). This was a significant phase towards 'Australianising' national soccer. It was followed by the removal of any ethnically based club logos. Yet the battle to 'de-ethnicise' soccer continues. The president of Soccer Australia argued on an ABC television programme that: 'for soccer to be a commercial success, we must abandon the multiculture for monoculture' (*Multicultural Marketing News*, 1996).

The national media, representing the dominant discourse, have not fostered the popularity of national soccer in Australia. Instead, they have further marginalised soccer by focusing on its ethnicity, and particularly on ethnic violence. Rather than simply reflecting a general lack of interest in soccer, commercial television's treatment of soccer creates an image of soccer that makes it hard for Australians without an ethnic background to relate to it.

Australian soccer provides an anomaly in Australian multicultural society. In the 1950s and 1960s, when assimilation was the unofficial policy on ethnic minorities, soccer was clearly a multicultural sport. Yet in the late 1990s, when official policies on multiculturalism may not have widespread support from the Australian population (Birrell, 1996), Australian soccer is being forced to return to an assimilationist position (Hughson, 1997).

IMMIGRATION POLICIES AND ETHNICITY IN NEW ZEALAND

New Zealand has gained almost half a million people through immigration during the twentieth century. However, there have been major fluctuations in the volumes and in the direction of migration, particularly over the last three decades. Net migration has ranged from a net loss of over 26 000 in 1979 to a net gain of 36 000 in 1996 (Cook, 1997). New Zealand immigration policies have influenced these fluctuations, and they have also had marked effects on the source of immigrants, especially since the late 1980s.

It is widely accepted that New Zealand's immigration before 1945 was based on a 'White New Zealand policy', similar to the 'White Australia policy'. Specific legislation was passed to exclude or limit Asian migration (Pawson, 1992: 25). In some respects, the New Zealand policy went further than the Australian policy, in that there was a deliberate bias against Catholic Irish. Because of this exclusive policy, until the Second World War 96 per cent of non-Maori New Zealanders were of British origin. By the end of the war in 1945, most of the non-British migrant groups had been assimilated, and New Zealand life was 'overwhelmingly monocultural' apart from occasional reference to Maori culture (Brooking and Rabel, 1995).

From the late 1940s to the 1960s, a number of schemes were introduced to assist migrants to come to New Zealand. These schemes continued the 'White New Zealand' theme, and focused on immigrants from Britain, though Northern Europe was also targeted. From 1945 to 1965 New Zealand had net migration gains of 200 000 people.

The ethnic mix of migrants to New Zealand in this period contrasts markedly with Australia, where migrants came from a broader range of areas, including areas in southern and eastern Europe. Although New Zealand also experienced labour shortages after the Second World War, these were largely met by rural to urban migration of Maori workers, as well as labour migration from nearby Pacific Islands (mostly from Tonga, Samoa, Fiji, Niue and the Cook and Tokelau Islands).

In 1961, Britain still dominated European immigration to New Zealand. The non-British European groups were so small that they were unable effectively to develop any distinctive cultures, and thus assimilation was the only option.

In the early 1970s there was a surge of net migration to New Zealand. This wave was also from the 'traditional source' countries of the United Kingdom, Australia, North America and the Pacific Islands. Even so, the sheer scale of the immigration (70 000 net gain from 1973 to 1975) at a time of rapidly rising unemployment, led to hostility towards the new immigrants. Much of this hostility was directed against the Pacific Islanders (Brooking and Rabel, 1995).

Many of the Pacific Islanders in the 1960s were on temporary permits, having been brought into New Zealand to do jobs that New Zealanders were unwilling or unavailable to

do. Many overstayed their permits, sending vital remittances back to their island homes. Despite the fact that Pacific Islanders experienced far more than their share of poverty and unemployment, populist opinions held these people responsible for everything from unemployment to the decline of law and order. Pacific Islanders were regarded in a similar way to the West Indians in England, the North African migrants in France and the Turks in Germany (Cheyne *et al.*, 1997: 114).

Such opinions were reflected in official government action. Pacific Islanders were targeted in the 'overstayers' campaign of the mid-1970s, when police and immigration officers conducted 'dawn raids' and random street checks (Spoonley, 1994: 88). Subsequent relaxation of the permit scheme meant that relatives of the Pacific Islanders could migrate to New Zealand, and concentrations of Islanders developed in parts of Auckland (see Chapter 12).

During the election campaign of 1975, populist politicians played the anti-immigration card, and the National government of Robert Muldoon came to power with a perceived mandate to reduce the migration flows. Immigration policy was tightened considerably. Net migration losses were recorded for the next decade, which was also a time of severe economic recession in New Zealand.

Until 1987, New Zealand strictly controlled the numbers of migrants as well as the countries they came from. Most of those who were allowed in, apart from a small number of refugees, came from Britain under the 'traditional source' policy, which could be interpreted as a British-only or white-only policy. Even the refugees were expected to assimilate quickly with New Zealand society. However, in 1987 and again in 1991, important changes were made to New Zealand's immigration policies. The changes had dramatic effects. The 1987 Immigration Act, based on the 1986 Immigration Policy Review, led to significant changes in the source countries of migrants, with a substantial increase in migrants from various countries in Asia (Farmer, 1996; Ip,

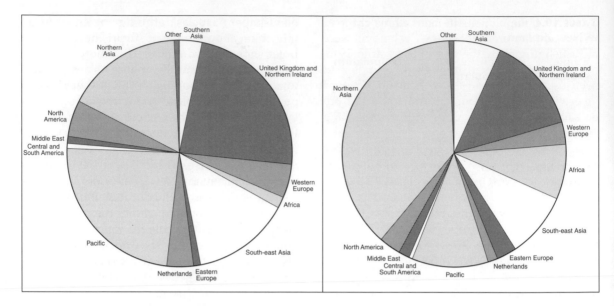

FIGURE 10.4 Immigration to New Zealand by Region, 1982–91 (left) and 1991–4 (right) (percentage of total migrant approvals) (Trapeznik, 1995: 94–5)

1995). The 1987 Act specifically removed discrimination based on race and national or ethnic origin. The review on which it was based also signalled the end to assimilation as a strategy for responding to new immigrant ethnic groups, and the beginnings of a multi-cultural ethos that saw 'positive value in diversity and the retention of ethnic minorities of their cultural heritage' (Bedford in Le Heron and Pawson, 1996: 355). However, despite these changes, there was still a slight net migration loss until the 1991–5 period, when net migration surged.

The emphasis on non-traditional source countries continued after more changes to immigration policy in 1991 (Trapeznik, 1995). The 1991 policy was known as a 'targeted' immigration policy, which sought to attract larger numbers in the 'general investment' and 'business investment' categories of migrants from a wider catchment of countries (Farmer, 1997). The 1991–5 period was the first period of significant net immigration to New Zealand since the early 1970s: 163 005 immigrant appli-

cants were approved, and over half of these were for Asian migrants (Cook, 1997).

In the 1980s, the 'traditional source' countries (Britain and the Pacific Islands) dominated immigration to New Zealand. However, from 1991, migrants from these countries were outnumbered by those from Asian countries, as Figure 10.4 shows (Trapeznik, 1995: 94–5). North Asia became the main source region, with 38.2 per cent of total migrant approvals for the 1991–4 period.

The significant changes to New Zealand immigration policies in the late 1980s and the early 1990s should be seen within the context of the changes in linkages in the global economy (Farmer, 1996; Kelsey, 1995). New Zealand, like many other countries including Australia, has sought to attract foreign investment and become competitive in expanding markets in Asia. As a way to facilitate this, the New Zealand government has encouraged the migration of both people and capital from Asia, and has promoted a 'limited and liberal form of multiculturalism' (Spoonley and Berg,

TABLE 10.4 Populations in main ethnic groups in New Zealand

Ethnic group	Census population		Population change
	1991 000	1996 000	1991–6 %
European	2783	2879	3.5
Maori	435	523	20.4
Pacific Island	167	202	21.0
Asian	100	174	73.9
Total	3374	3619	7.2

NB. The figures are based on census data in which it is possible to indicate belonging to more than one ethnic group. Thus the figures do not sum to the totals. *Source:* Cook (1997).

1997). Perhaps, as Mitchell (1993) suggests, multiculturalism can be seen as simply a part of a broader strategy to facilitate global capitalism.

The changes in the relative sizes of ethnic groups in the 1990s in the New Zealand population are indicated in Table 10.4. Though New Zealand is still dominated by European ethnicity, the large proportional growth in the Asian ethnic group has been the focus of populist debate.

Asian migrants were blamed for many of New Zealand's problems in the 1990s (Spoonley and Berg, 1997). It was claimed that they were 'heating up' the economy, driving up interest rates and housing prices and putting extra pressure on schools (Le Heron and Pawson, 1996: 350–60). Some New Zealanders believed that many of the new immigrants were just taking advantage of New Zealand and still had loyalties to their original countries. Indeed, some families were seen as 'astronaut' or 'bi-local' families, who would return to their own countries for their business activities (Friesen in Le Heron and Pawson, 1996). In Auckland, the suburb of Howick (which had a large Asian population) was given the stigmatised sobriquet of 'Chowick'.

Newspaper articles arguing against Asian migration abounded. Not surprisingly, opinion polls showed that almost half the population believed there were too many Asians in New Zealand. This anti-Asian sentiment was fuelled by the rhetoric of the controversial politician Winston Peters, who chose Howick as the location of a speech decrying an invasion of migrants.

Friesen (in Le Heron and Pawson, 1996) points out that while many New Zealanders stereotype Asians as rich and well educated, this is a gross over-generalisation that ignores the considerable heterogeneity within the Asian ethnic group. Some Asians, notably the Indians, have been concentrated in low status occupations, including dairy (local shop) owning, which is normally an economically marginal business activity.

Bedford (1996) questioned the reality of the supposed 'Asian invasion', showing that such a term is grossly misleading. Using data from 1995 (when migration levels were higher than any preceding year) he explains how a total net migration of about 23 000 Asians (from all parts of Asia) could hardly be described as an invasion in a country of 3.6 million. He also questioned why a total net gain from white nationalities of 24 100 in the same year was not seen as a 'white invasion' (Bedford, 1996).

Ironically, many of the targeted immigrants were also disillusioned. Many of them experienced difficulty finding jobs, often because their qualifications were not recognised by the New Zealand authorities. Many also found it very difficult to set up businesses.

In 1995, in a climate of a public backlash against Asians in New Zealand, the National Party government announced major changes to immigration criteria, including the hard-to-pass English language test for all migrants over 16 from a non-English-speaking background. The policy change had a substantial effect. The net migration gain for the year to August 1997 was 14 490, just half the net gain of 28 390 for the year to August 1996 (McLoughlin, 1997). Asian migration was reduced to less than half, and the main sources of new migrants became Britain (again) and South Africa. One result of these changes has been a huge downturn in the level of investment being brought into the country by migrants.

The emergence of biculturalism in New Zealand

A multiculturalist ideology, at least in the simple descriptive sense of encouraging a more ethnically diverse population, was behind New Zealand's post-1987 immigration policy (Pearson and Ongley, 1996). However, a concern for the interests of immigrant groups was replaced by a focus on Maori interests and the Treaty of Waitangi. By 1985, official practice began to promote biculturalism and to move away from the models of multicultural-ism that evolved in Australia and Canada.

Unlike the situation in Australia, when governments carefully justified increased levels of immigration, the New Zealand governments did not prepare the public (Pakeha or Maori) to receive and tolerate the new migrants. There was little active promo-tion of a viable multicultural policy (Farmer, 1997) and little attempt to incorporate multi-cultural ideas into the school curriculum (Greif, 1995). Compared to the administrative support and the various entitlements for ethnic minorities in Australia, new immigrants to New Zealand were very poorly resourced (Pearson and Ongley, 1996).

In New Zealand the concept of multicultur-alism was criticised by those supporting Maori interests. Under a multicultural policy Maori would be regarded as only one of a multitude of ethnic minorities, with no recognition of their special status. One prominent Maori, Ranginui Walker, has suggested that while the changes to immigration policy in 1991 were primarily aimed at stimulating economic growth through an encouragement of Asian migration, there was also an underlying 'agenda of countering the Maori claim to first-nation status as *tangata whenua* [people of the land]' (Walker, 1995). Under a multicultural policy, Maori argued that new immigrants

were not obliged to respect the Treaty of Waitangi (Cheyne *et al.*, 1997: 117).

The reasons for the development of a bicul-tural ethos, rather than a multicultural one, in New Zealand include:

- the large Maori population;
- the legal implications of the 1840 Treaty of Waitangi;
- the growing activism of Maori leaders;
- the introduction of the Waitangi Tribunal; and
- the desire for a unique New Zealand national identity (Pearson and Ongley, 1996).

Unlike Australia, where indigenous people are a tiny minority in most states, by the 1970s Maori represented a large and visible minority group in New Zealand. Though the process of Western colonialism had suppressed their culture, many Maori had maintained their tribal affiliations, even among the highly urbanised population.

The Treaty of Waitangi, which was signed by a number of Maori chiefs and representa-tives of the British Crown, was revered by Maori. Although it was largely ignored by the early colonists, it was never forgotten by Maori. There were some differences between the two versions of the treaty (English and Maori). The English version secured British sovereignty over New Zealand, while the Maori version guaranteed Maori *te tino rangati-ratanga* (the chiefly power over the land) while providing only governorship to the Crown (Pawson, 1992). Despite these differences, the treaty provided an important foundation for Maori claims and activism.

Maori culture looked in danger of being lost to Pakeha culture in the 1960s when large numbers of Maori moved into urban areas. By the 1970s, both Maori and Pacific Islanders were heavily concentrated in urban areas, particularly in Auckland, and these groups were disproportionately concentrated in unskilled or semi-skilled occupations (Spoonley, 1994: 87). Thus when the economic downturn hit New Zealand in the early 1970s, the Maori and Pacific Islanders were the most severely affected in terms of unemployment and various social problems. Ironically these groups were also targeted as scapegoats for the economic problems of New Zealand.

Partly as a response to the adverse economic and social circumstances of Maori, including their construction as part of 'the problem', Maori nationalism and political activism rapidly developed in the late 1960s and 1970s. The closer contact between Maori and Pakeha in urban areas also made Maori deprivation more obvious to Pakeha. The concept of *tangata whenua* was able to connect groups of Maori who had diverse tribal backgrounds, and thus united action from Maori became possible (Spoonley *et al.*, 1994: 89). This new focus celebrated Maori culture and stressed the need to preserve Maori language and customs, including the importance of *iwi* (a confederation of tribal groups). There was a renewed pride in being Maori and having Maori culture (see Box 10.2).

The Waitangi Tribunal was established by the Treaty of Waitangi Act in 1975, the same year that a Maori land rights march was held down the length of the North Island to parliament. The tribunal was given the task of interpreting the treaty of Waitangi, and hearing claims from Maori regarding perceived breaches of the treaty. In 1985 the Treaty of Waitangi Act was amended to enable the tribunal to hear Maori claims about treaty violations back to 1840. This opened the way for a multitude of claims, with 650 being held by 1997 (Pawson, 1992; 1999). The Waitangi Tribunal provided a potent symbol of the potential for Maori to provide real input into decision-making in New Zealand: 'No longer is it possible for the Pakeha majority to assume that Maori are or should be invisible within the broader society, "integrated" or "assimilated" in ways that were political policy until two or three decades ago' (Bedford in Le Heron and Pawson, 1996: 359).

Pearson and Ongley (1996) believe that another important reason for the movement away from multiculturalism towards biculturalism relates to a desire of both Maori and some Pakeha to create a New Zealand identity that is distinct, not only from Britain, but also from the multicultural identity existing in Australia. Thus the idea of a co-operative 'majority/minority partnership' was supported by many (though also highly contested by some Pakeha) as a way of presenting New Zealand as a model of cultural harmony (Pearson and Ongley, 1996).

The New Zealand government has recognised Maori concerns and claims for retaining language and culture. This biculturalism is reflected, for example, in its official bilingualism: all passports are now printed in both English and Maori; all government departments have a Maori name; a Maori welcome (*powhiri*) is included as part of official ceremonies; and regional councils are required to consult the local *iwi* on a range of issues (Pawson in Le Heron and Pawson, 1996). It is also exemplified by the increasing use of the name Aotearoa as an alternative to New Zealand.

At the same time that government policies were being radically altered to promote a bicultural ethos that benefited Maori, other government policies involving economic restructuring were having profoundly negative impacts on Maori (much more than on non-Maori). Thus the rhetoric of co-operative biculturalism is weakened by the reality of continuing Maori disadvantage (Pawson, 1992; Smith, 1992).

Maori groups themselves see biculturalism in two ways. It can be an empowering force, allowing them to maintain their distinctive culture and overcome the dominance of Pakeha values (Pearson and Ongley, 1996). Alternatively, it can be seen as a way of 'managing' political pressure by co-opting Maori leaders, or simply appearing to consult with Maori without being committed to the outcome of this consultation (Walker, 1995).

Another issue for Maori is the way in which biculturalism constructs Maori as a 'unity' or the 'other' in a way that overlooks the range of groups within Maori society. Before the arrival

Box 10.2 Popular music and Maori culture

Since the 1980s, Maori have shown a renewed pride in their cultural traditions and they have also worked on regenerating *te reo Maori* (their Maori language) (Mitchell, 1996). Popular music has played an important role in this process. It has been important in providing Maori with a means of expression. For example, Moana Maniapoto-Jackson (leader of the group Moana and the Moahunters) has argued that her music is making a statement about Maori simply by using the language. She believes that 'if you do anything that's got a strong Maori flavour to it, that in itself is a strong political statement' (Cruickshank, 1998). Moana's 1998 album, *Rua*, incorporates traditional Maori aspects (forms of *waiata, te reo Maori, haka* slaps and stomps) as well as more contemporary music styles such as rap and dance beats and melodies.

The music of Moana and the Moahunters also has lyrics with a political message, thanks to tracks like *Treaty*, inspired by arguments heard at *hui* (Maori meeting) throughout New Zealand (Cruickshank, 1998). Moana is well known for her stance on Maori issues (Jackson, 1992), and she uses the power of music to motivate people. As well as promoting awareness of Maori history and language, she also uses her music as a way of promoting an anti-smoking message to her people (who have over twice the rate of smoking of Pakeha).

At the same time that influential Maori were encouraging increased adoption of Maori culture, Maori youth were faced with appalling education and job prospects, as well as having limited understanding of their own culture. In response to this, many Maori and Pacific Islander youth turned to break dancing, rap music and hip hop culture as a way of associating with an international black identity. New Zealand youth not only appropriate musical styles from oppressed black communities in the USA, but they have also appropriated other cultural activities, including a form of graffiti known as 'tagging' (Lindsey and Kearns, 1994). Maniapoto-Jackson argued that part of the reason for the adoption of American black culture was that local

TV and radio stations featured very little Maori music or language, so youth adopted an American black identity as the 'next closest thing' (Mitchell, 1996).

However, when Maori musicians adopted rap music, they also added their own style, instruments, lyrics and often language. Upper Hutt Posse, a 'hard-core' rap group inspired by Public Enemy from the US, released the first hip hop record in New Zealand in 1988 entitled *E Tu* ('Stand Proud'). This was a tribute to rebel Maori warrior chiefs of New Zealand's colonial history (Mitchell, 1996). Thus while they appropriated US black culture, they also added their unique Maori style and Maori themes.

Commercial radio usually presents a monocultural style of music, dominated by an Anglo-American influence, and features the same music that parents would have first heard in their own youth (Davis, 1997). The low proportion of Maori who can speak their language fluently (3 per cent in the 1990s) makes it difficult even for Maori radio stations to maintain listeners when playing Maori music, so most include considerable amounts of US and UK music (Mitchell, 1996). This monoculturalism of commercial radio makes it difficult for Maori or Pacific Islander bands to get wide coverage. It is rare for such bands to become a commercial success. One exception is OMC, an Auckland-based band whose track *How Bizarre*, has been a number one hit in New Zealand and Australia, and also became popular in the US. This music reflects life in Auckland's largely Polynesian area, South Auckland. Indeed, one critic commented that *How Bizarre* 'couldn't have come out of anywhere other than South Auckland' (Laurie, 1996).

Popular music may have an important role in the future of biculturalism in New Zealand. It provides a means of expression for Maori, but probably more importantly it provides a way to communicate the importance of Maori culture to Maori youth, as well as to Pakeha.

of Europeans, Maori did not see themselves as a unified group (Wall, 1997). Instead, there was a multitude of different tribal groups, whose interests often conflicted with each other

(Bedford, 1996). Also, the term Pakeha has been strongly contested by many non-Maori (Spoonley and Berg, 1997). As Pawson (1999) explains: 'There are therefore, not just two

New Zealands, but multiple New Zealands, with different memories, identities and places'.

For many small ethnic groups, the development of biculturalism may represent little improvement on the monocultural model still favoured by some Pakeha. Biculturalism appears simply to increase the marginalisation of immigrant ethnic minorities. As a young Indian New Zealander explained on a television documentary, with hurt, frustration and anger; 'this is a bicultural country; I don't exist' (Greif, 1995: 16). Greif believes that there is a need for another term to describe the people who are not covered under a bicultural model. He suggests that the term Tauiwi (meaning Other Tribe) may be appropriate, though it does not yet have wide usage or recognition. New immigrant minorities are facing increasing alienation from both Maori and Pakeha. The promotion of biculturalism has also led to growing political activity among immigrant minorities who recognise their marginalisation under the current bicultural model. The New Zealand Chinese population, for example, has become more assertive, especially in Auckland where racial tensions are high (Pearson and Ongley, 1996).

CONCLUSION

The operation of multiculturalism in Australia and biculturalism in New Zealand has led to greatly contrasting situations for different minority groups in the two countries. In Australia indigenous people are now calling for particular recognition in a multicultural model that was designed largely with new immigrant groups in mind. Although indigenous peoples can be included within Australia's multiculturalist framework, they are simply one of a number of different ethnic groups. Most Australians regard multiculturalism as separate from issues relating to indigenous culture. In contrast, in New Zealand, immigrant ethnic groups are seeking recogni-

tion in a society where biculturalism still appears to have dominance over multiculturalism (Pearson and Ongley, 1996). Thus biculturalism ignores the reality that New Zealand in the 1990s is an increasingly multiethnic society, but one in which members of many ethnic groups still feel that they are nonpeople.

In both countries there has been considerable debate about the real effect of the two policies. Are they simply about recognising different cultures, or are they about a real sharing of economic and political power between different groups? In New Zealand, if biculturalism is really to work, it must be more than simply incorporating Maori customs into public ceremonies, or using Maori names for government departments. It must involve the incorporation of very different ways of thinking and very different cultural values. In Australia, there is considerable doubt as to whether multiculturalism is any more than the commodification of the cultures of different immigrant groups. There is little doubt that multiculturalism has enriched Australian society, even through the proliferation of different ethnic restaurants for example. However, when ethnic cultures are seen to conflict with the values of dominant discourse (as in the Australian soccer case study) the value of multiculturalism is easily overlooked.

One of the strongest indicators that neither biculturalism nor multiculturalism has involved any real redistribution of power in society is the fact that Maori in New Zealand and indigenous and various immigrant groups in Australia are still significantly disadvantaged. This issue will be discussed in the next Chapter. Also, some places have been more directly affected by immigration and by the growing activism of indigenous people. In New Zealand, Auckland has the highest concentrations of both new immigrants and Maori. In Australia, high concentrations of Asians and Aborigines are found in areas such as Cabramatta and Redfern in Sydney. These two places will be discussed in Chapter 12.

FURTHER READING

Castles *et al.* (1992) provide an entertaining and insightful analysis of multiculturalism in Australia. A detailed historical perspective on immigration, national identity and multiculturalism in Australia, accompanied by selected extracts from valuable documents, is provided by Lack and Templeton (1995). For a recent account of how Australia has dealt with ethnic diversity, see Castles (1997) which also discusses the relevance of multiculturalism for indigenous Australians. Recent issues of the journal *People and Place* (e.g. Birrell, 1996; Farmer, 1997) provide engaging debates about multiculturalism and immigration in Australia and New Zealand. Pearson and Ongley (1996) who discuss Australia, New Zealand and Canada, provide an interesting international perspective on multiculturalism and biculturalism. For an excellent explanation of the historical and cultural context for Maori–Pakeha relations in New Zealand, see Pawson (1992; 1999). Grief (1995) discusses the history of immigration in New Zealand, with interesting case studies of specific immigrant groups. Finally, Le Heron and Pawson's (1996) Chapter 11 ('Senses of Place') deals very effectively with the topics of international migration and Maori identities.

11

NEW SOCIETIES?

The 1970s marked the start of a period of major economic and social change throughout the world (Harvey, 1989). Australia and New Zealand were entangled in this massive change, as they both left behind the 'long boom' of the 1950s and 1960s and entered a period of economic restructuring in response to global economic forces. Since the 1970s the changes have been so pronounced that it is possible to think of the creation of 'new societies' in Australia and New Zealand.

The context for these changes in society can be found in the phenomenon of 'globalisation' and the particular response to this process adopted in both Australia and New Zealand: 'neo-liberalism' (also described as 'economic fundamentalism' and often referred to as 'economic rationalism' in Australia) (Badcock, 1997; Kelsey, 1995; Stilwell, 1997; Walmsley and Weinand, 1997a). The important features of globalisation include the internationalisation of labour, enormous growth in the impact of transnational corporations and global flows of capital, information and people (Burnley et al., 1997a: 5).

Neo-liberalism is an ideology of the 'new right' that emphasises individualism and free choice. It posits that the free market is the most effective instrument for maximising the efficiency of an economic system. In the free market, it is argued, individuals will be best able to pursue their particular self-interest.

Neo-liberalism is consequently highly critical of any state interference, as this will simply disrupt the operation of the free market (Cheyne et al., 1997: 80–90).

The influence of neo-liberalism in many capitalist countries such as Australia, Britain and (particularly) New Zealand over the 1980s and 1990s is difficult to overstate (Badcock, 1997). Yet the promotion of neo-liberalism in Australia was relatively slow compared to the wholesale adoption of this ideology in New Zealand (Kelsey, 1995): 'New Zealand embarked on its neo-liberal, regulatory experiment with a frightening theoretical certainty that assumed the path was right before any other nations or groups of nations had moved anywhere near as far' (Moran in Le Heron and Pawson, 1996: 387).

Governments and economists in both Australia and New Zealand argued that the only rational response to globalisation was to embrace the ideology of neo-liberalism. Stilwell has questioned this logic, suggesting that globalisation itself is an 'ideological construct, designed to legitimise, or at least make appear inevitable, particular patterns of structural economic change and socio-spatial change' (Stilwell, 1997: 8). Other researchers share this view, and suggest that governments deliberately overstated the inevitability of the globalisation process to justify their neo-liberal policies, including the substitution of market forces for state action. This has led to the deregulation of the economy, encouragement of individual autonomy and arguments that welfare should be the responsibility of families and communities rather than governments (Walmsley and Weinand, 1997a).

The economic restructuring that resulted from the rise of neo-liberalism can be linked to a number of changes or trends in Australian and New Zealand society, which will be discussed in this Chapter. First, there are changing employment patterns, which involve new types of employment (and unemployment) as well as changing roles for women. It is also possible to identify increases in levels of inequality in society, which have been exacerbated by changes in the state provision of welfare and housing. Minority groups (e.g. Maori, Aborigines and new immigrants) have been particularly disadvantaged by some of these changes. Finally, there is evidence of a trend towards more privatised societies in Australia and New Zealand, where people are retreating from public spaces to the supposed 'safety' of their own homes, cars or privatised shopping malls. The most extreme manifestation of this trend is the rise of 'gated communities' (see Box 11.4 below). This privatisation can be interpreted as part of an 'Americanisation' of Australian and New Zealand society.

NEW EMPLOYMENT PATTERNS

Since the 1970s, the workforces of Australia and New Zealand have changed substantially in a number of ways. Like many other industrialised countries, there has been a progressive shift towards a service economy, where employment has moved away from manufacturing towards service sector jobs in such areas as financial services, information, retailing and tourism. This reflects a change in global capitalism from a Fordist to a Post-Fordist mode of production. There has been a considerable rise in part-time and casual work, as

well as in unemployment rates. Some researchers have identified the emergence of a 'two-nations cleavage' between a core, well-paid, highly skilled and high-consuming 'service class', and a marginalised group whose traditional industrial base can no longer provide jobs (Spoonley, 1994; Thorns, 1994). There has also been a noticeable feminisation of the labour force.

Various factors have contributed to these changes. These include changes in technology, economic restructuring, an increased demand for services previously carried out within the family home (e.g. child care and meal provision), and a recognition that many service industries have the capacity to earn export income (e.g. tourism). New technology has had a profound impact on the workforce, especially with computerisation, mechanisation and automation, and the resulting replacement of labour and creation of new types of jobs (ABS, 1997b). Economic restructuring has involved a number of changes that have affected the nature and the level of employment. In both countries there was a gradual removal of protection for domestic industries, removal of restrictions on international capital flows, industrial relations reforms (which emphasised workforce flexibility), and a restructuring of government employment, often through the privatisation of government activities (McKinnon, 1997: 93; Murphy and Watson, 1994).

The majority of new jobs created since the 1970s has been in the service sector, which accounted for 63 per cent of all employment in New Zealand in 1991, and 72 per cent of all employment in Australia in 1996 (ABS, 1997b: 95; Larner in Le Heron and Pawson, 1996). The main growth of service sector jobs has been in areas such as property and business services, accommodation and restaurants, and personal services such as child-care workers, travel stewards and tourist guides. However, the jobs gained in the service industries have not been able to replace all the full-time jobs lost in manufacturing (see Chapter 15).

Jobs in the service sector range from highly skilled positions such as computer software design to labour-intensive jobs such as cleaning. Because of the increased demand for the highly skilled jobs (especially in information technology, telecommunications and finance), pay rates have been high and increasing in this area. Conversely, for the unskilled labour-intensive jobs, pay rates have remained low, and jobs are routine and insecure. For many, poverty has resulted from inadequate wages. The largest number of jobs in the service industry are less skilled jobs, such as clerical or sales and service workers, and the number of people working in temporary, part-time or casual work is also much higher than in manufacturing (Birrell *et al.*, 1997; Larner in Le Heron and Pawson, 1996).

In both Australia and New Zealand there has been a growing disparity between a relatively small group of highly paid, highly skilled workers, and a growing number of people who find it difficult to get full-time or well-paid work. This is evidenced by the increased income gap between the top and bottom level of income earners. Larner (in Le Heron and Pawson, 1996) identifies a distinctive geography to this pattern of polarisation. In New Zealand, while the highly paid jobs are concentrated in the central areas of Auckland and Wellington, unemployment is highly concentrated in outer suburban areas or rural areas. 'One consequence of this spatial division ... is that few of those articulated into the high-skill and high-paying sectors of the economy are confronted with the "tragedy of the market" on a regular basis' (Larner in Le Heron and Pawson, 1996: 96). Similar patterns of segregation are becoming evident in Australian cities (see Chapter 12). Another aspect of the changes in employment patterns is the differential effect on Maori and Pacific Islanders. These groups have been over-represented in manufacturing jobs since they migrated to the cities to provide unskilled labour in factories. Consequently, job losses for these groups have been huge. In 1991, unemployment rates for Maori were 23.3 per

TABLE 11.1 Employment rates of women aged 15–64 in Australia and New Zealand as a percentage of all women in that age group

Australia	%	New Zealand	%
1968	42	1966	40
1998	59	1996	69

Source: ABS (1998a: 111); Cook (1997: 37).

cent, compared to 8.4 per cent for non-Maori (Spoonley, 1994: 94).

Women's participation in the workforce

In both Australia and New Zealand there has been an unprecedented increase in the participation of women in the labour force from the 1950s and 1960s to the 1990s. The changes in the employment rates for women are illustrated in Table 11.1. By the late 1990s, women made up 45 per cent of all those employed in New Zealand (1996), and 43 per cent of all those employed in Australia (1998) (ABS, 1998a: 111; Cook, 1997: 37).

There are many interrelated reasons for the increase in women's participation in the labour force. First, there is the increase in service sector jobs relative to manufacturing jobs, which has reduced the proportion of jobs involving manual work, and increased the number of jobs that have characteristics associated with 'women's work' (requiring 'attributes such as caring, sociability and "people skills"') (Larner in Le Heron and Pawson, 1996a: 99). Second, the increase in part-time jobs has enabled more married women to join the workforce. In Australia, for example, the labour force participation rate for married women aged 15–64 almost doubled between 1968 and 1998 (increasing from 34 per cent to 63 per cent) (ABS, 1998a: 112). Third, social attitudes to the roles and the rights of women have changed since the 1950s and 1960s. During the 1950s very few women were employed, and of those that were, few were married. At that time, strong gender role differentiation existed, where women were expected to provide unpaid domestic labour rather than be part of the paid workforce. Significantly, however, women still do more unpaid work in the 1990s than men, even if they also work: 'the increasing participation of women in paid work is not being offset by a decrease in their levels of unpaid work' (Larner in Le Heron and Pawson, 1996a: 100).

There are still strong concentrations of women in certain occupational groupings. In New Zealand, the vast majority of women are employed in service industries (79 per cent) and two-thirds of all female employment is in the two largest groups in the service sector (community, social and personal, and wholesale and retail trade). This horizontal segregation is accompanied by vertical segregation, where women have less senior positions and earn less money on average. Women in Australia are subject to similar levels of horizontal and vertical segregation. In 1998, more than half of women workers were employed in clerical, sales and service areas, and significantly outnumbered men in these areas (ABS, 1998a, 114).

In New Zealand, ethnicity has been found to interact with gender in terms of employment opportunities. While Pakeha women are gaining ground in male-dominated occupations (e.g. professional and managerial) Maori and Pacific Island women have high and increasing levels of unemployment. This is just one example of increasing polarisation in society.

DIVIDED SOCIETIES?

The social policies of Western governments have generally been unable to substantially reduce the inequalities that exist in society (Graycar and Jamrozik, 1993: 311). In fact, inequalities in most Western industrialised countries have actually increased since the 1960s. Australia and New Zealand are no exceptions.

There exists a perplexing range of conflicting findings about poverty and inequality in Australia (Whiteford, 1997) and New Zealand. Though the consensus may be that inequality has been rising significantly since the 1970s, this is not an uncontested view (Walmsley and Weinand, 1997b). There is considerable complexity in the data on inequality, and differences between countries in the way in which the data are analysed can make valid comparisons very difficult. Thus, we should heed the warning made by Whiteford (1997) and be content with broad conclusions about poverty and inequity, rather than making 'implausibly precise claims'.

Notwithstanding these caveats, it is clear that both Australia and New Zealand are becoming increasingly unequal societies, even if they may not yet have the levels of polarisation evident in countries such as India. 'Evidence exists now to suggest that inequality has widened more extensively in New Zealand than in any other OECD country' (Cheyne *et al.*, 1997: 184–5). Kelsey (1995: 271) sees this as the result of the zealous pursuit of free-market economic policies in New Zealand, as in Britain and the United States. In Australia, though the gap in incomes has widened, the living standard of the poor has tended to be maintained in real terms, by better social security systems, especially under Australia's federal Labor governments (Argy, 1998: 185). In this sense, changes in inequality in Australia are broadly similar to the Canadian experience, and thus Australia lies in the middle ground between OECD countries showing dramatic rises in income gaps between rich and poor (Britain, United States and New Zealand) and those where inequalities have changed little in the 1980s and 1990s (Sweden, Norway and the Netherlands) (Badcock, 1997).

In Australia, 'the trend towards greater inequality has accelerated since the mid-1970s' (Graycar and Jamrozik, 1993: 42). Graycar and

TABLE 11.2 Income distribution in Australia, 1986–96

	1986	1990	1996
Share of gross income going to top quintile (of all income units)	45.3	46.2	47.9
Share of gross income going to bottom quintile (of all income units)	4.7	4.8	3.6
Gini coefficient	0.41	0.42	0.44

Source: ABS (1997b: 114).

Jamrozik argue that the growth in the inequality of income distribution has been most marked among families with dependent children. When they examined the distribution of income for married couples with dependent children in Australia, they found that the share of income going to the top decile increased from 22.1 per cent in 1973–4 to 23.6 per cent in 1985–86. Over the same period, the share going to the bottom decile fell considerably, from 3.9 per cent to 2.5 per cent (Graycar and Jamrozik, 1993: 43)

The trend toward inequality continued throughout the 1980s and 1990s in Australia. The share of income going to the top quintile increased from 1986 to 1996, while the share going to the lowest quintile decreased, and the Gini coefficients (an index for measuring inequality) continued to rise (see Table 11.2).

In New Zealand there is a widespread belief that there has been an 'explosion' of poverty, as an 'underclass' has grown rapidly during the 1980s and 1990s (Lees and Berg, 1995). This belief is supported by quantitative data on increasing income disparities between the rich and poor, as well as by qualitative data. Kelsey (1995: 273) cites reports from welfare agencies throughout New Zealand who report poor people increasingly going without food, clothing, heating, health care and education opportunities. Cheyne *et al.* (1997: 189) report a study that provides some alarming qualitative data on the effects of drastic welfare cuts on the spending patterns of welfare beneficiaries. The data indicate that after welfare cuts were introduced in 1991, 75 per cent of beneficiaries in the study reported cuts in their food and transport expenditure, and the same percentage

reported that they had stopped spending altogether on house maintenance and sport. Such a situation may lead to slum conditions emerging in areas with high reliance on welfare benefits.

The effect of increases in income disparities in Australia and New Zealand may be lessened slightly by the more equitable access to outdoor leisure resources. Residents of Sydney, Melbourne, Auckland or Wellington have ready access to a natural environment that provides many free leisure opportunities. Large areas are set aside for active sport and leisure. This contrasts with the situation in cities such as London or New York, where many leisure facilities are mainly available to those who can afford them, and this reinforces income disparities.

Apart from cuts to welfare spending (which were particularly savage in New Zealand) and cuts to spending on health, housing, public transport and education, there are a number of factors leading to increases in inequalities in Australia and New Zealand. First, the growth in service sector jobs brings with it increases in the income disparity of workers, as well as in the security of employment (Badcock, 1995; Birrell *et al.*, 1997). There are huge contrasts between the incomes of senior managers or professionals and unskilled workers in this sector. Second, there is an increase in single-parent or lone-elderly families or unemployment for lower-income earners, at the same time as an increase in dual-income families for high-income earners (Graycar and Jamrozik, 1993: 323). Third, those who are disadvantaged in terms of income are increasingly being spatially polarised as well (see Chapter

Figure 11.1 Public housing in Otara. This is an outer suburban area of Auckland, dominated by government housing that was originally built to provide housing for employees on new industrial estates (photo: P. Tranter)

12). The most disadvantaged groups are increasingly being forced to live in outer suburban or rural areas with poor access to services and employment (see Figure 11.1), thus compounding their disadvantage (Murphy and Watson, 1994). Fourth, some ethnic groups, particularly Maori and Pacific Islanders in New Zealand, are becoming increasingly marginalised in terms of employment. Finally, there is a phenomenon known as 'the rural crisis', where both farmers and residents of country towns have experienced increasing loss of income as the profitability of farming has decreased (Walmsley and Weinand, 1997b).

There is little disagreement that the neo-liberal policies implemented in Australia, and more forcefully in New Zealand, have increased polarisation in society. Differences emerge, however, in the evaluation of this polarisation. Critics of neo-liberalism argue that inequality is an inevitable outcome of neo-liberal policies, because those who are already deprived of skills and resources, or who are discriminated against, are unable to make the 'best' market choices. The neo-liberals agree, but they also see inequality as both 'inevitable and desirable' (Cheyne et al., 1997: 83). They argue that inequality can provide an incentive for those who are less well off to better themselves.

Kelsey (1995) has been particularly critical of the impact of neo-liberalism on inequalities in New Zealand, and the changes in values that have arisen in governments. In discussing the effects of New Zealand's radical restructuring since the mid-1980s, she explains:

The result of a decade of radical structural adjustment was a deeply divided society. The traditionally poor had been joined by growing numbers of newly poor ... The burden fell most heavily on those who already had least. This was neither coincidence nor bad luck. It was the calculated outcome of a theory which many New Zealanders viewed as morally and ethically bankrupt (Kelsey, 1995: 271).

The increasing polarisation of society, that has been enhanced by neo-liberal policies in Australia and New Zealand, has led some teenagers to deliberately reject society as a response to the perception that society has rejected them. One result of this is the presence of 'street kids' (see Box 11.1).

THE CHANGING ROLE OF THE WELFARE STATE

After the Second World War, in most democratic industrialised nations, the state began to play a more active role in the protection of the well-

---

Box 11.1 Street kids

One of the distinguishing characteristics of any social group is the way in which the members of that group use and demarcate space. Homeless teenagers – 'street kids' – in Australian and New Zealand cities have distinct landuse patterns that emphasise their 'otherness' from mainstream society. Research in Auckland (Lindsey in Le Heron and Pawson, 1996) and Newcastle (NSW) (Winchester and Costello, 1995) reveals that street kids are an identifiable part of the cultural landscape in Australasian cities, even though youth homelessness is less obvious than in many American cities.

Street kids live within a distinct culture. In addition to their lack of a 'home', street kids adopt many attitudes and behaviours that make them part of that culture. A lack of respect for any kind of authority is a distinguishing cultural feature. Authority is seen to be represented by police officers, shop owners (from whom the street kids steal), welfare workers and parental figures. This disrespect for any kind of authority is reflected in the fact that groups of street kids do not usually have a nominated leader. However, there is a hierarchy of status that is established by street knowledge, fighting ability and the length of time a person has lived on the streets. Gender does not necessarily play a role in determining this hierarchy. Girls are as likely to have the respect of their peers as boys. This has been attributed to the fact that specific gender roles are among the middle-class norms that are rejected by street kid culture. Unlike women in mainstream society, female street kids are not tied to domestic responsibilities: the skills that are necessary for survival as street kids are not bound by gender. It is interesting to note that this phenomenon is not repeated in homeless communities in the USA (Winchester and Costello, 1995).

Group solidarity is seen to be very important by street kids. Information about services and possible dangers is shared, both verbally and by means of graffiti left in commonly used squats. By looking out for each other, street kids protect themselves from outsiders and thus reinforce their own cultural boundaries. The distinction between insiders and outsiders is common to all cultures, but it is particularly important to transient communities, such as that of the street kids.

As street kids lack a permanent residence, squatting in disused buildings is common, and trespassing on to private property is not seen as a serious offence. Street kids also rely on services such as refuges and shelters to provide meals and occasional temporary accommodation. Disused buildings are viewed by street kids as part of their 'turf', as are parks and other spaces within the inner city. As with less tangible aspects of street-kid culture, the establishing and maintenance of territory boundaries allows street kids to identify their place within the wider community. The space occupied by street kids is not exclusively theirs – for example, homeless adults use refuges and other groups of teenagers and young adults 'hang out' in little-used inner-city areas. However, the patterns of spatial use by street kids mark them as a group separate from the others. Squats used by street kids may be abandoned rapidly but reclaimed at a later date. Transience makes street kids less of a target for police and other users of space who may wish to harm street kids.

In Western societies there is increasing polarisation between economic groups. Street kids are a discrete group amongst the urban poor. By defying mainstream society and creating their own cultural norms, street kids assert their right to use space and reclaim their power to do so from authorities who would remove that power from them.

being of citizens. This included increased responsibility for full employment, economic equality and the care of those suffering economic hardship. The welfare state was accepted as an integral part of economic development. While the nature and extent of the welfare state varied from nation to nation, in all democratic societies it had some effect in increasing equality and decreasing social unrest (Saunders, 1994: 1). In both Australia and New Zealand, the welfare state was always seen more as a residual or reluctant one, compared to its more extensive counterpart in post-war Britain (Murphy and Watson, 1994; Shirley, 1994). This 'reluctant' welfare state was meant to provide support only after the market or the family had failed. Welfare was provided through ensuring high employment levels and adequate working wages.

The economic crises of the 1970s provided pressure from the 'new right' to reduce the welfare state. This was related to rising unemployment and reduced tax revenue as well as increasing proportions of the elderly. This pressure was felt strongly in Australasia. Some economists argued that 'it does not follow that cutting government expenditure in general, and welfare spending in particular, is essential for economic recovery' (Saunders, 1994: 14). Yet the response of governments in Australia and New Zealand has been to implement rigorous cuts to welfare spending in the 1980s and 1990s. These cuts were applied to welfare states that were already markedly less generous than those in some European countries, especially Scandinavia, which demonstrate that there are alternatives to the new right policies implemented in Australia and New Zealand.

In the 1980s and 1990s, the application of neo-liberal policies in Australia and New Zealand led to the privatisation of welfare service provision, partial or full cost recovery practices, or complete withdrawal of some welfare programmes. According to neo-liberals such as David Green (1996), welfare should not be the responsibility of the state, but should be provided by voluntary groups, private charity or by the family. In many cases, while respon-

sibility for services was transferred to voluntary agencies, there was little transfer of any resources. As an Australian geographer has suggested, in such circumstances 'preaching the virtues of self-reliance and voluntarism as an alternative to "welfare dependency" is a studied insult' (Badcock, 1997: 254). Increasingly, as the welfare state has been cut back, welfare benefits were aimed at the very poor, rather than either the general population or the slightly disadvantaged.

It appears that in Australia and particularly in New Zealand, 'reluctant' welfare states are becoming both more reluctant and more punitive (Shirley, 1994). This is illustrated in the following discussion of housing policies, as well as in Box 11.2, which focuses on health care services in rural New Zealand.

Public housing reforms

Housing has been recognised as being a particularly important component of welfare: 'in many ways, access to the housing market is the litmus test of inequality in contemporary capitalist societies' (Kearns and Smith, 1994: 126). In the 1990s, there were three main housing tenures in Australia and New Zealand: home ownership (by far the dominant tenure), private rental and public rental (from state housing authorities). Home ownership provided significant financial advantages. Private tenants were the most disadvantaged, paying high rents and having limited security of tenure. Public tenants were generally better off than private tenants. However, during the 1990s, changes to housing policy, first in New Zealand and then in Australia, have considerably weakened the status of public housing. Public housing in both countries has progressively changed from housing for workers from nuclear families, to 'welfare housing', as those in serious housing need were increasingly targeted by public housing authorities (Forster, 1995; Murphy, 1997).

In New Zealand, access to adequate and affordable housing was recognised as a right of citizens since the late nineteenth century (Kearns *et al.*, 1991). By 1949 the state had become the major provider of rental housing for the poor, and it had become well established that the government would ensure adequate housing for all New Zealanders. By the 1980s, when New Zealand's economy was experiencing high inflation, the state's resources were shifted toward the more 'productive' sectors of the economy, away from 'less productive' sectors such as public housing for the poor (Kearns *et al.*, 1991: 370).

One of the most striking examples of the effects of changes in the welfare state is the change to public housing in New Zealand in the early 1990s. After 1991, state housing in New Zealand was shifted to a market-based system where public tenants were forced to pay market rents (to provide more equity between private renters and public renters), and in which the only form of housing assistance available to low-income earners was an accommodation supplement. This supplement was available to private and public renters as well as to home buyers. The supplement, however, was insufficient for most households, and led to low-income households paying an increasing percentage of their income on accommodation costs, even those in public housing (Murphy, 1997). At the same time as the supplement was introduced, the government significantly reduced the income received by those on benefits. It can be argued that this demonstrates the way in which the application of neo-liberal policies has increased housing poverty, exacerbated spatial polarisation and ignored the particular problems faced by some groups in society.

Welfare groups saw the changes in state housing as being responsible for substantial increases in reliance on 'foodbanks', which provide food parcels for people suffering poverty. 'In 1995 more than 120 foodbanks operated in the Auckland area alone, with one of these supplying 7000 parcels annually compared with 600 five years earlier' (Cheyne *et al.*, 1997, 190). In the 1990s, an increasing number of those relying on foodbanks were employed rather than receiving benefits, but

on such low wages that housing costs were leaving them in poverty.

Instead of improving the living standards of public housing tenants, it is likely that the accommodation supplement has simply been 'incorporated into the rent setting strategies of private landlords and has become a subsidy for landlords operating in the lower end of the market' (Murphy, 1997: 13). Also, by ignoring the social reality of discrimination, the accommodation supplement makes no allowance for the particular difficulties faced in the housing market by some groups (e.g. Maori and Pacific Island households, single-parent households, and people with disabilities) (Murphy, 1997).

The changes to housing assistance in New Zealand have also had discernible impacts on the social polarisation of areas within New Zealand cities. To understand these impacts, it is necessary to note the particular residential geography created by state housing policies in New Zealand cities, dominated by a focus on large-scale outer suburban public housing areas, often linked with the location of manufacturing. This geographical pattern was found in all of New Zealand's main cities, to greater or lesser degrees (Morrison, 1995).

Because all tenants had to pay market rents, which reflect the value of a property, housing in these outer suburban state housing areas was usually cheaper than more central locations. Consequently, those who could least afford accommodation were concentrated in these areas. Some tenants have moved out of urban areas altogether, to get cheaper housing in rural areas (Waldegrave and Stuart, 1997). There is also increased pressure on Housing New Zealand Ltd to sell better located housing and purchase stock in low rent areas (Morrison, 1995). Thus the outer areas are increasingly being occupied by the most disadvantaged groups, including single parents, unemployed and Maori and Pacific Island groups. The net result of the changes to housing policies in New Zealand is to 'exaggerate a residentially segregated social geography that has developed around the major pockets of state housing' (Morrison, 1995: 54).

Box 11.2 Health care in rural New Zealand: restructuring, inequity and community activism

Attempts to introduce market forces into services provided by the state are often contested. The restructuring of health care in New Zealand has been strongly opposed by some sections of the community. As Kearns and Joseph (1997: 19) explain: 'the collective provision of health care has invariably been regarded as a cornerstone of the welfare state'. In particular, hospitals provide 'very tangible expressions of community welfare' (Le Heron and Pawson, 1996: 224). Consequently, restructuring leading to the closure of hospitals often resulted in organised opposition.

The restructuring of health care services in New Zealand in the early 1990s exacerbated pre-existing inequities in the provision of health care. In New Zealand, these inequities have long been strongly associated with ethnicity: Maori people have the worst health status as well as the poorest access to medical care. Further compounding this disadvantage, many Maori are now involved in return migration to rural areas, which have lower levels of access to medical care.

In 1993, the National government in New Zealand introduced controversial reforms to the provision and funding of health care, based on the ideology of increasing competition and thus supposedly increasing the 'efficiency' of health care provision. Newly created Regional Health Authorities were given the task of purchasing a range of health services from a mix of public and private providers, including the publicly owned Crown Health Enterprises. The theory behind the changes was that competition for contracts among the various providers would encourage efficiencies in the delivery of health services. Whether such a theory will produce the expected practical outcomes has been questioned (Fougere, 1994). Also, even if providers do become more 'efficient', this may simply mean that some areas experience a reduction in medical services. One reason for this is that the attempt to 'rationalise' medical services has led to the closure of many small rural hospitals, severely limiting access, especially in areas where public transport services have also declined (Kearns and Joseph, 1997).

Although health services in the majority of rural areas in New Zealand have declined as a result of restructuring, there are some exceptions to this generalisation. Kearns and Joseph (1997) provide an encouraging case study of community activism amongst the largely Maori population of Hokianga, a rapidly growing but economically deprived district in Northland. Hokianga was the first SMA (special medical area) in New Zealand. SMAs were developed to improve health care in areas that were recognised as disadvantaged. Fears of a withdrawal of health services from Hokianga, as part of the National government's restructuring initiatives, prompted a carefully constructed opposition from both Maori and Pakeha residents, including a threatened sacred walk to the parliament in Wellington. The government was influenced by the protests, and the local community negotiated a position where it now runs its own health service. Kearns and Joseph suggest that the activism concerning health service provision at Hokianga was not simply a protest about the provision of local services. They believe that some of the activists were also objecting to the whole ideology of the health reforms.

The history of public housing in Australia shows some similarities to that in New Zealand. In the early part of the twentieth century, governments in Australia concentrated on expanding home ownership. The first state housing authorities in Australia were formed in the late 1930s and early 1940s. In 1945, the first Commonwealth State Housing Agreement (CSHA) was established, providing a commitment from the Federal government to provide financial support for public housing, and to raise the quality of housing, particularly for low-income households. This agreement was very significant in the history of housing in Australia, as it legitimated public housing as a viable tenure. However, by the 1970s, budgetary pressures encouraged the Commonwealth to reduce its support for public housing. Late in the twentieth century, public housing in Australia was considerably weakened, largely by changes in the 1995 CSHA.

TABLE 11.3 Disparities between Maori, Pacific Islanders (PI) and Pakeha (%)

	Maori	PI	Pakeha
Children aged less than five living in one-parent households	43	31	14
Unemployed seeking full-time work as a percentage of full-time labour force, aged 15–19	50	50	27
Adults aged 20–39 owning their residence	50	35	77

- Less than 2 per cent of Maori hold university degrees (7 per cent for non-Maori).
- 51 per cent of households in serious housing need are Maori.
- Sudden infant death syndrome (cot death) for Maori infants is twice that for non-Maori. This is linked to a higher prevalence of cigarette smoking for Maori adults (44 per cent compared with 26 per cent for Pakeha).
- Although a smaller number of Maori drink alcohol regularly, those that do drink do so more heavily than non-Maori. This difference is reflected in mortality statistics, which show that alcohol-related deaths among Maori men were more than double the number for non-Maori.

Source: Davey (1993); Davis *et al.* (1997); Spoonley (1994); SNZ (1994).

Changes to public housing in Australia in the late 1990s followed the lead of New Zealand in terms of a shift away from direct intervention in housing production towards subsidies paid directly to tenants, and hence more reliance on the private sector (Hayward, 1996; Yates, 1997). Tenants are being referred to as clients or customers and public rents are market-based, with the aim of encouraging wealthier tenants to leave public housing. Security of tenure in public housing is also being weakened. As Hayward (1996: 32) explains: 'these reforms are heavily imbued with economic rationalist assumptions and logic'. Also, because the level of spending on public housing in Australia is very low by world standards, and assuming that the current economic situation does not improve, waiting lists for public housing are likely to grow dramatically. The changes to Australian public housing in the 1990s have weakened a form of housing tenure that was already considerably inferior to home ownership (Hayward, 1996).

Recent changes in housing policies have had a particularly marked effect on some ethnic groups in Australia and New Zealand. The relative disadvantage of some of these groups is discussed in box 11.2 opposite.

DISADVANTAGED MINORITIES

One test of the success of policies on multiculturalism or biculturalism (see Chapter 10) is the extent to which members of certain ethnic groups can maintain their culture without paying the price of social and economic disadvantage. We can ask: have policies on migrant and indigenous groups been able to provide 'new societies' of social justice in Australia and New Zealand, with a fair distribution of economic resources and equal access to essential services such as housing, health care and education? In this section, this question will be answered in regard to indigenous Australians, migrant groups in Australia, and Maori and Pacific Islanders in New Zealand.

Indigenous Australians

When people from overseas countries observe Australia's standing on human rights, as they did during the Olympic 2000 bid, they do not find any extreme examples of state-initiated violence such as at Tiananmen Square or Sharpeville. However, Australia has a long history of discrimination. There are those who argue that Australia is still a racist society (Tatz, 1995). This racism is indicated by extreme levels of disadvantage for indigenous Australians with regard to economic and social conditions, as well as mortality (see Chapter 9). On almost any economic or social indicator, Aborigines and Torres Strait Islanders represent an extremely disadvantaged group in society.

Indigenous Australians have chronically high unemployment, with an official unemployment rate (38 per cent in 1994) about four times the Australian average (Castles, 1997). Also, many indigenous people are classified as not in the labour force, because they have given up looking for jobs. Annual incomes are also much lower, with only 7 per cent of indigenous Australians earning more than $25 000 in 1991, compared with 23 per cent for non-indigenous Australians (ABS/AIHW, 1997: 103).

There are huge discrepancies between Australian indigenous people and non-indigenous people in terms of educational opportunities. Indigenous Australians are much less likely to have post-secondary school qualifications. In 1993, when 7.7 per cent of non-indigenous Australians had a degree, the corresponding figure for indigenous Australians was a mere 0.8 per cent. Indigenous Australians are also more likely to experience housing disadvantage, with only 25 per cent owning or purchasing their home, compared with 71 per cent for Australians overall. Household crowding is also more common, and inadequate housing and need for housing assistance are much higher. Almost 40 per cent of indigenous households have insufficient income to meet their basic needs or afford adequate housing. In many areas of indigenous housing, water is deemed to be unfit for human consumption (ABS/AIHW, 1997).

Smoking is twice as prevalent among indigenous people, and of those who drink alcohol, indigenous Australians are much more likely to drink at unsafe levels (79 per cent of current drinkers compared to 12 per cent for non-indigenous). Other factors contributing to

disadvantage include poor access to health facilities, higher rates of substance abuse and higher levels of exposure to violence, including child abuse. Aboriginals are much more likely to be in prison than the total population. In Victoria, Aboriginals are almost 19 times as likely to be in prison. As the Governor-General of Australia explained in a 1996 speech, the statistics on indigenous health and welfare 'tell us a story of present human sickness, suffering, dying and death which can be attributed to the past dispossession, oppression and injustice' (ABS/AIHW, 1997: 5).

Maori and Pacific Islanders

A similar story of disadvantage emerges for New Zealand Maori and Pacific Island people, though the contrasts with Pakeha are not quite as extreme as the contrasts identified above between indigenous and non-indigenous Australians. Despite the Maori cultural renaissance and the activities of the Waitangi Tribunal (Chapters 5 and 10), there remain major disparities between the economic and social wellbeing of Maori and Pakeha. According to some researchers, the situation is worsening despite, or perhaps because of, the neo-liberal policies of recent New Zealand governments (Smith, 1992; Spoonley, 1994). Maori unemployment rates are more than twice as high; they earn only half of the average income of Pakeha; they comprise half of the prison population; nearly half of all Maori women are dependent on social security; one in 10 Maori go into further education, compared with one in four Pakeha; and 45 per cent of Maori heads of household own their own homes as opposed to 75 per cent of Pakeha. Some of the disparities between Maori, Pacific Islanders and Pakeha are listed in Table 11.3.

As Davis *et al.* (1997) report, the differences between Maori and non-Maori in terms of health and mortality cannot be explained solely in terms of differences in socio-economic status. Ethnicity appears to have some explanatory power of its own: 'if working in a lower-status job, being poorer and less educated already significantly increases the risk of death, the social consequences of being Maori adds further to that risk' (Fougere, 1994: 147).

Davey's (1994) conclusion, having examined disparities between Maori and Pakeha, was that 'a century and a half of colonisation and fifty years of welfare state policies have not produced equality of outcome for Maori. This suggests that unrealistic and standardised approaches to social policy have failed' (Davey, 1994: 104). However, she did also pick out some 'success stories':

* *Te kohanga reo* (Maori language nests) began in the early 1980s as a community initiative. This is now the most popular form of early childhood education for Maori.
* In 1992, 10 per cent of primary schools offered full (Maori) immersion or bilingual education.
* Over half of primary school children are taught basic aspects of Maori culture, songs, greetings and simple words.
* Special housing schemes (e.g. Papakainga Housing Scheme) and special financial arrangements (e.g. Homestart) provided housing on Maori-owned land from 1985.

Migrant groups in Australia

Typically, in immigrant countries, migrants have been exposed to many forms of discrimination, including slavery, forced labour, racial discrimination, and inferior socio-economic and legal status. With more migrants in the population than virtually all other countries, it is of interest to examine Australia's record on migrant disadvantage. Australia has a long history of racism towards non-Europeans. The White Australia policy was not abolished until quite recently (1973). But even non-English-speaking (NESB) Europeans were discriminated against in terms of job opportunities. After the Second World War, government policies on European immigration kept NESB migrants in a secondary status working in

TABLE 11.4 Indicators of disadvantage amongst NESB migrants in Australia

* In 1991, less than 7 per cent of legislators and government officials in local, state or federal government were of first or second-generation non-English-speaking background, when NESB people comprised 25 per cent of the total population.
* NESB people were poorly represented in the public service, and particularly in the senior executive service.
* NESB people represented only 6 per cent of directors and executives in major companies.
* NESB unemployment rates were, in general, higher than average. In 1994, the NESB rate was 14 per cent, compared to 9 per cent for the total population.
* Some groups had very high unemployment rates (19 per cent for Lebanese and 32 per cent for Vietnamese). 'These are not recently arrived groups; their unemployment rates have been high for many years, and government labour market measures seem to have done little to improve the situation' (Castles, 1997: 17–18).

remote areas or in heavy industry or infra-structure projects (Castles, 1992; Fincher, 1997). While the NESB migrants were mainly employed in unskilled and unappealing jobs in manufacturing, migrants from English-speaking backgrounds were employed in a much wider range of occupations.

Castles (1997) has provided some indicators of the extent to which NESB migrants are disadvantaged in Australia. These indicators include participation in government and senior management and unemployment (Table 11.4).

The reasons for the differences in unemployment levels between NESB migrants and the total population may not be due to high levels of discrimination in the labour market. Instead, English language proficiency is a crucial determinant of an immigrant's chances of finding a job (Burnley *et al.*, 1997a: 127; Burnley, 1989).

There are also important gender factors to be considered in the analysis of disadvantage of

Box 11.3 Children's lost freedom

When parents in Australia and New Zealand are asked to reflect on their own middle childhood experiences, they usually remember having more freedom to travel around their own neighbourhood or city without adult supervision than their own children have today (Cunningham et al., 1994; Tranter, 1996; Tranter and Whitelegg, 1994). They are likely to claim that they had high levels of freedom to travel to school without an adult, visit friends alone, cycle around their own neighbourhood, catch buses (or trains), go to the shops, or even to go out after dark. They were given these 'licences' of 'independent mobility' at a younger age than today's children (see Figure 11.2). Data from longitudinal studies in England support such anecdotal evidence that there has been a substantial decline in children's travel freedoms over the last generation (Hillman et al., 1990).

The reasons for these reductions are complex, but the increase in the levels of motorised traffic is likely to be at least an important part of the explanation. Over the last generation, not only has there been a considerable growth in the levels of traffic in Australasian cities, but the average speed of this traffic has also increased (Moriarty and Beed, 1992: 13). As traffic levels increase, more and more people (adults as well as children) cease to use the streets as pedestrians. This is partly a response to traffic danger, but also a response to the loss of local shops and services, and hence greater reliance on the motor vehicle for access to shops, schools and even playgrounds. Eventually, residential streets are perceived as being deserted, lonely and hence dangerous places for children, in terms of the fear of assault and molestation. There are few adults around on the streets to provide surveillance and support for children. In particular, there are few adults who know their neighbours' children and can look out for them.

Available data indicate that children's independent mobility in Australian and New Zealand cities is much lower than in cities in Germany, and slightly more restricted than children in England (Hillman et al., 1990; Tranter, 1996). Children in Australian and New Zealand cities are more car-dependent than children in Germany or England. Figure 11.3 provides an indicator of the level of difference between children's freedoms in Australian and New Zealand and German and English cities. (The differences between Australian and New Zealand cities are relatively minor.)

There are also marked international contrasts in terms of the gender differences in children's independent mobility (Tranter, 1995). In England, Australia and New Zealand, boys are given significantly more freedom than girls for every indicator of children's independent mobility. However, in Germany, there appears to be very little difference between boys and girls. The gender differences in Australasian and English cities are also evident in the parents' stated reasons for restricting their children's freedom to travel to school or to places other than school alone. For the parents of boys the most common reason for restricting their travel was concern about traffic danger. However, for the girls, the most common concern was fear of assault and molestation.

Perhaps the reduction in children's freedom is part of a more general trend towards the retreat from public spaces in society, to the supposed 'safety' of the private house, the enclosed shopping mall or even the walled estate: a trend towards a fortress society.

migrants in Australia (Castles, 1992; Fincher, 1997). One extremely disadvantaged section of the labour force consists of 'outworkers', many of whom are migrant women. Outworkers generally work in their homes and are not usually covered by award conditions as they do not belong to any union. 'They are perhaps the most exploited of all sections of the Australian work-force' (Castles, 1992: 65).

Outworkers are concentrated in industries such as textiles, footwear and electronics. The high proportion of migrant women working as outworkers in such industries is part of the increasing marginalisation and exploitation which has arisen from global economic restructuring (Castles, 1992).

The reasons for the high levels of disadvantage for indigenous Australians, Maori and

FIGURE 11.2 Children being escorted to primary school in Christchurch, New Zealand. Children are now less likely than in previous generations to be allowed to walk or cycle to school without adult supervision (photo: P. Tranter)

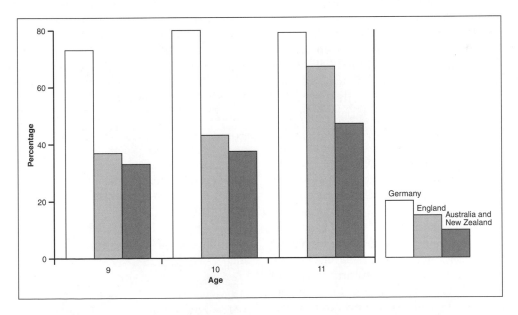

FIGURE 11.3 Percentage of children allowed to travel to places other than school alone for cities in Germany, England and Australia and New Zealand. (source for German and English data: Hillman *et al.*, 1990: 131; for Australian and New Zealand data: Tranter, 1996)

Pacific Islanders, and NESB migrants are complex. English language proficiency is part of the explanation, as is the history of involvement in particular industries that were most affected by restructuring. However, discrimination against ethnic groups is, unfortunately, still likely to be part of the explanation.

Ethnic groups are not the only groups who are discriminated against in Australia and New Zealand. Children are also disadvantaged, at least in terms of their freedom to use public space independently (Valentine, 1997) (See Box 11.3).

Privatised societies: fortress societies?

A slow, but perceptible trend is emerging in Australian and New Zealand society in terms of the way citizens use the spaces in their cities and suburbs. In the 1960s, if a person walked along a suburban street in any New Zealand or Australian city, he/she could expect to see children playing on or near the street and adults talking to each other or watering their front lawns or gardens, as well as seeing other pedestrians (or cyclists) in the street. In the 1990s, fewer people are to be seen in the suburban street. They are more likely to be behind the closed doors (and fences) of their homes. Streets have changed, both in terms of the lack of people in them (apart from the occupants of cars), and in the design of the houses facing the streets. In some New Zealand cities, an emerging trend has been the construction of high front fences that divide the public space of the street from the private space of the houses and yards. The fences have the effect of preventing any passive surveillance of the street, and hence add to the feeling that the streets are lonely, deserted and dangerous places. Even in suburbs without fences, such as in Canberra suburbs (where front fences are not allowed), new streetscapes have emerged in which the frontages of houses are dominated by double garage doors: 'the most noticeable element of street life is the opening and closing of remote-controlled garage doors' (Legat, 1995a: 58). There is often little interaction between the house and the street.

The reasons for these changes are complex. They include changes in the proportions of women in the workforce, changes in technology that allow more communication and entertainment from within the home itself, and changes in the perception of the dangers in society, both fear of crime and traffic danger.

Over the last 20 or more years, there has been an increasing fear of crime in Australia and New Zealand, fuelled by tabloid press and television coverage of 'gruesome stories of strangers assaulting, raping, robbing and murdering their victims in public places' (Morgan, 1993: 386). The reality, however, is poorly represented by such media coverage. While there may have been a real increase in some types of crime (e.g. burglary), the dangers to people on residential streets have probably changed little over the last 20 years (Pratt, 1994). Despite the media hyperbole, neither Australia nor New Zealand are particularly dangerous societies by international standards. In terms of murder/manslaughter rates, Australian and New Zealand rates are about four times lower than the rate for the USA (see Table 11.5). Gang violence in Australia and New Zealand is minimal compared to gang-related activity in US cities.

TABLE 11.5 International comparisons of murder/manslaughter rates (1995) (selected countries)

	Rate per 100 000
United States	8.2
Canada	2.0
Australia	2.0
New Zealand	1.8
Germany	1.5
England and Wales	1.4

Source: ABS (1997b: 172).

Another important observation, rarely acknowledged in the media, is that the majority of all violent crime takes place in private homes, often the victim's own home (Pratt, 1994). This applies to crimes against both adults and children. Thus when politicians promise to make the streets safe for citizens, they are ignoring the problems associated with the private lives of citizens.

Despite evidence that crimes in public spaces are not (on a per capita basis at least) increasing dramatically, people in Australia and New Zealand are following an American trend towards restricting their daily activity to privately controlled spaces. They are retreating behind the closed walls of their own homes, privatised shopping malls and walled estates. People are protecting themselves from the supposed danger outside these privatised spaces (Hillier and McManus, 1993).

In some Australian suburbs there is a growing use of walls to keep out unwanted visitors or intruders, as well as an increasing use of video surveillance equipment to provide extra security for people in public and private spaces (Crang, 1996; Hillier, 1996). Also, an entire security industry has developed to provide not only products (alarms, deadlocks and bars) but also services for those who can pay (security patrols and personal escorts).

Private security firms and the developers of walled estates have cleverly marketed the need for security by focusing on the fear of crime. Security is being commodified as something that can be bought and sold. Enclosed shopping malls and walled estates offer exclusivity and a supposed respite from an outside world increasingly portrayed as being violent, dangerous and disadvantaged.

Privately owned, enclosed suburban shopping malls are becoming an increasingly common retail form in Australia and New Zealand. In Australia, the first drive-in shopping centre was opened at Chermside in Brisbane in 1957, followed soon after by Top Ryde centre in Sydney in the same year. Melbourne's first centre was opened in the early 1960s (Spearritt, 1995). The first in New Zealand was opened in Auckland in 1962, followed by another in Christchurch in 1965 (Bowler in Le Heron and Pawson, 1996). Since then, enclosed suburban malls have proliferated in Australia and New Zealand cities.

Despite their apparent use as public space, these enclosed shopping malls are in fact privately owned spaces. Indeed, a legal decision was made in the USA to allow mall owners to define an Oregon mall as private space (Sandercock, 1997). Part of the mall culture concerns the separation of the middle class from what have been constructed as the city's dangerous public streets. Malls have become popular in response to the perception of 'Main Street's inability to cope with the problems of modern life' (Duffy, 1994: 30). The mall management restricts the actions of shoppers, as well as the shop tenants. People who behave in unacceptable ways, or whose behaviour is perceived to be intimidating to other shoppers, may be asked to move on. Groups of teenagers are often targeted in this way.

Suburban shopping malls are only one response to the perceived decline in safety of public space. Gated communities represent another response that is growing in popularity in Australia (see Box 11.4).

This Chapter has explored some broad changes in Australian and New Zealand society over the last few decades, many of which are similar to those that have occurred in the UK and USA. Some of the changes identified in Australian and New Zealand society have a particular expression within cities, where spatial patterns can enhance broad trends. The following Chapter focuses on such patterns in the urban environments of Australia and New Zealand.

FURTHER READING

Badcock (1997) provides a very useful summary of recent research on the increasing polarisation of Australian society. Kelsey

Box 11.4 Gated communities

Gated communities (or walled estates or enclave estates) are still relatively rare in Australia and New Zealand, compared to the USA where millions of people, mainly white and high income, live in these communities. However, some already exist, and there is an increasing trend in many Australian cities towards such communities. In some of these estates property values have appreciated more quickly than in surrounding areas. Security has been an important selling feature and can be extreme by the standards of Australian housing estates. Some estates feature high fences, infrared alarm systems and entrance gates where identity is checked at a guardhouse.

The first gated community in Sydney, The Manor, was built in the early 1990s. This has a two-metre-high chain mesh fence and massive iron gates at the entrance. Since The Manor was built, about a dozen more gated communities have been built in Sydney, with more planned. Other Australian cities now have similar communities catering to affluent groups, though many of their residents do not see themselves as contributing to division in society. One resident of The Manor explained: 'There's nothing elitist about us. There might be millionaires here, but they're very down-to-earth millionaires' (Hills, 1998).

There are already planned communities in Auckland (e.g. Broadway Park) that have elements of a siege mentality, where the wealthy congregate in their own 'ghettos' and where casual visitors are discouraged by a single entry point with gateway architecture (Hansen, 1995). However, as yet, New Zealanders have largely avoided the move to high-security walled estates.

The promoters of such estates argue that this approach is the way of the future, a way of protecting people and providing security from increasing fear of drug-related crime. Many local government councils also see these estates favourably, because they can provide a means to privatise council responsibilities such as garbage collection, as well as providing higher rates.

Despite the popularity of the gated communities with some Australians, they are also subject to considerable criticism (Hillier and McManus, 1993). They are seen, for example, as contributing to the polarisation of society, where the rich live behind walls and guards, and the rest of the community

lives in an increasingly deprived society. Also, residents of such communities often have to forfeit some of their freedom of choice so that they do not disturb the homogeneity of the environment. This problem has been highlighted in the 1995 video-movie *The Colony*, directed by Rob Hedden, which depicts the extreme social control exercised by the owners of a fictitious Los Angeles gated community.

Another problem with the proliferation of these communities is that the dangers (or perceived dangers) for those outside the walls are increased, so that the public spaces (streets, parks and public squares) become even less attractive for people. This helps to undermine hope that the city can be reclaimed by people. It destroys the public realm of cities: it threatens the whole purpose of cities as places for exchange. The following two quotes illustrate the contrast between humane livable cities, and the cities that may result from a rapid growth of walled estates:

> *Cities were invented to facilitate the exchange of information, friendship, material goods, culture, knowledge, insight, skills, and also the exchange of emotional psychological and spiritual support (Engwicht, 1992: 17).*
> *Anyone who has tried to take a stroll at dusk through a strange neighbourhood patrolled by armed security guards and signposted with death threats quickly realises how merely notional, if not utterly obsolete, is the old idea of the 'freedom of the city' (Davis, 1992: 250).*

Fortunately for those who oppose such developments, other models of residential development are also being implemented. One positive example of this is Stanhope Gardens, a new estate in Sydney's western suburbs. This estate is based on a holistic approach to urban design, rather than a retreat from society behind a set of walls. The designers aimed to create a 'permeable space which maximises passive surveillance on the street and open spaces' (McKenzie, 1998). Traffic is slowed, and children are encouraged to play in the street spaces, which also increases the passive surveillance of the street and houses.

However, if the trend towards walled estates in Australia is to be reversed, more than model residential design alternatives will be needed. There is also a need for social changes, aimed at creating positive community interaction in an environment where 'social capital' (Cox, 1995) is valued as much as financial capital, and where governments and society show more concern for disadvantaged groups, rather than increasingly marginalising them (Hillier and McManus, 1993).

(1995) presents an excellent critique of the neo-liberal policies that have been implemented in New Zealand. (Note that this book has also been published under the title: *The New Zealand experiment – a world model for structural adjustment*). Green (1996) provides arguments in favour of neo-liberalism in New Zealand.

Good overviews of social change in New Zealand are provided by Spoonley *et al.* (1994) and Cheyne *et al.* (1997) and for Australia by Graycar and Jamrozik (1993). The disadvantage of indigenous Australians is highlighted in ABS/AIHW (1997). For ethnic disadvantage in New Zealand see Davey (1993, 1994).

12

URBANISATION AND URBAN FORM

Overseas visitors are often surprised by the considerable contrasts between cities in their own country and cities in Australia or New Zealand. Visitors from Europe and Asia are impressed by the low density of Australian and New Zealand cities, with their prevalence of separate single-storey houses and their generous supply of open space. For visitors from the US, other contrasts are more conspicuous. Compared to US cities, there is less disparity between rich and poor suburbs, with few areas of slum housing, a lack of ghettos and 'no go' areas for middle-class residents, and a scarcity of walled estates or gated communities.

What is not apparent to a casual observer is the history behind the urbanisation process and the changes occurring within cities in Australia and New Zealand in the 1990s. This Chapter explores the distinctive urbanisation processes in Australasia, and the current urban forms typical of their cities. Various attempts to contain and channel urban growth are discussed, including the highly contested strategy of urban consolidation. The current trend for large-scale urban redevelopment projects is described, using as a case study Auckland's controversial 'Britomart' development on Waitemata Harbour. Then attention is focused on urban social areas. The increasing spatial polarisation of cities in the two countries is examined. It is characterised by a growing trend for the rich to concentrate in particular areas (including inner suburbs) and the poor to be increasingly marginalised in outer suburbs. Part of this trend involves the gentrification of inner suburbs. Ethnic concentrations within cities are also examined. The debate over whether or not ethnic ghettos exist in Australia is illustrated with a case study of Cabramatta, a suburb in Sydney with a high proportion of Vietnamese immigrants.

THE URBANISATION PROCESS

Urban development in Australia proceeded in a very different way from that experienced in New Zealand. In the Australian colonies,

TABLE 12.1 Populations of major cities in Australia and New Zealand (1986 and 1996)

	1986 000	1996 000	% growth 1986–96
Sydney	3472	3879	11.7
Melbourne	2967	3283	10.6
Brisbane	1217	1521	24.9
Adelaide	1004	1079	7.5
Perth	1050	1295	23.3
Auckland	822	998	21.4
Wellington	326	335	2.8
Christchurch	300	331	10.3
Hamilton	140	159	13.5
Dunedin	108	112	3.7

Sources: ABS (1998d); SNZ (1997a).

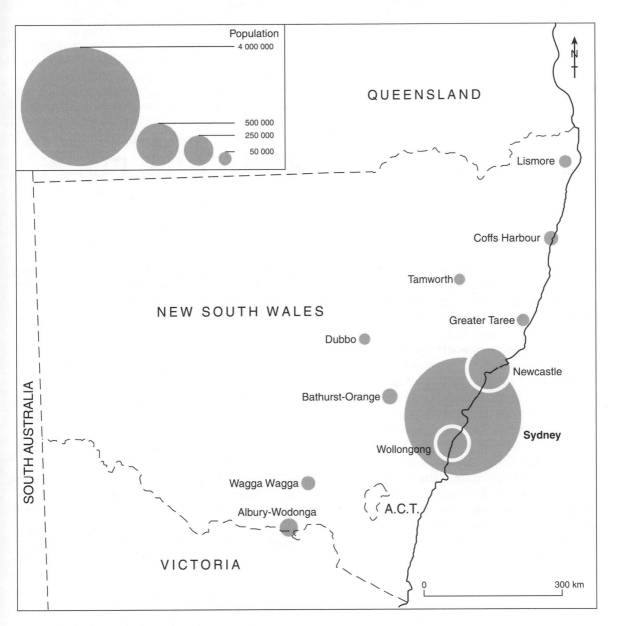

FIGURE 12.1 Populations of major urban areas in New South Wales (ABS, 1996 census)

which later became states, the urban hierarchy was marked by a strong primacy, where the largest city has many times more people than the second largest. Figure 12.1 illustrates this metropolitan primacy for the state of New South Wales. At the national scale, Australian cities have a more even size distribution, and

this is becoming more so, as smaller state capitals such as Brisbane and Perth are growing faster than Sydney and Melbourne (see Table 12.1). In New Zealand there is a much more even spread of city sizes than within any Australian state. The reasons for this relate to Australia's evolution from its colonial past, as well as to the topographical differences between Australia and New Zealand (see Chapter 2). In Australia, each of the major colonies began as a port town. These towns were destined to be dominant. They were the centres of administration, all trade passed through them, and when the railways were built in the mid-nineteenth century, the networks focused on these centres (Forster, 1995: 7–8) (see Chapter 6). In terms of topography, Australia had a relative lack of barriers (apart from long distances) to easy communication between the dominant cities and the rest of the colony.

In contrast, New Zealand's complex, mountainous landscape provided many barriers to communication between districts. Consequently, New Zealand developed a number of towns that dominated relatively self-contained regions. For many years there were four cities of approximately equal importance in New Zealand: Dunedin, Christchurch, Wellington and Auckland. Even now, despite Auckland's clear dominance in the urban hierarchy, there are four other cities in New Zealand with sizeable populations (with Hamilton being a post-Second World War addition to the list of major cities).

Auckland now shows the highest growth rate for any New Zealand city (see Table 12.1). Not only does it have the largest city population, but it also has considerable economic, political and cultural influence (Perkins *et al.*, 1993). It has many of the country's corporate head offices and a significant number of political representatives, and is the base for much of the country's cultural media (especially television). Given these factors, as well as its attraction for overseas immigrants, Auckland is likely to increase its dominance, creating considerable pressure on its environment.

In the 'long boom' of the 1950s and 1960s, cities in Australia and New Zealand showed remarkable growth. In the period from 1947 to 1971, the populations of Australia's five major cities grew by between 65 per cent (Sydney) and 158 per cent (Perth). These figures translate to average annual growth rates ranging from 2.7 to 6.8 per cent (Forster, 1995: 15). As Forster explains for Australian cities, 'the sprawling, decentralised, automobile dependent, ethnically diverse cities most of us live in today are mainly a legacy of the 1950s and 1960s'. The urbanisation process that occurred in Australia and New Zealand during this period was typical of many industrialised nations after the Second World War. The boom was boosted by a massive inflow of foreign capital, and involved a move to a Fordist society of mass-production and mass consumption.

During the post-war period the impact on urbanisation of overseas immigration differed markedly between New Zealand and Australia. In New Zealand, external migration had relatively little impact on the population of cities, though the internal (rural–urban) Maori migration did have a considerable impact (see Chapter 10). In Australia, while internal migration contributed little to city growth, overseas immigration had a profound effect. In Australia between 1947 and 1971, immigration accounted for well over half of the population growth of Sydney and Melbourne (55 per cent and 59 per cent respectively) compared to 42 per cent of national population growth during this time. In this period the total populations of Melbourne and Sydney grew much faster than had been predicted, due to immigration and high fertility levels. This unpredicted population growth created major urban planning problems, including shortages of housing, schools and hospitals, as well as rapid urban sprawl (Burnley *et al.*, 1997a: 14–16).

One urban feature evident in some Australian states, but not in New Zealand, is the emergence of large-scale conurbations of cities. Three such conurbations are emerging in Australia: Newcastle–Sydney–Wollongong; the Sunshine Coast–Brisbane–Gold Coast; and Melbourne–Geelong. There are now people who regularly commute from Newcastle and Wollongong to Sydney, from the Gold Coast to Brisbane and from Geelong to Melbourne. Even by world megacity standards, the spatial extent of these urban agglomerations is impressive. From the southern edge of Wollongong to

FIGURE 12.2 Suburban Wellington. Like other cities in Australia and New Zealand, many Wellington suburbs are dominated by separate single-storey houses and a generous supply of open space. The two-storey dwellings in the foreground are multi-unit public housing (photo: P. Tranter)

the northern edge of Newcastle is over 200 km. In terms of population, however, even the largest of these (based on Sydney) is still under five million, a figure which excludes this conurbation from the list of the world's megacities (Fuchs *et al.*, 1994).

Distinctive urban forms

The most noticeable feature of the physical form of Australasian cities is their very low densities compared to cities in most other countries. While major Australian cities average 14 persons per hectare (pers/ha) (the same as in the US), European and Asian cities average 54 and 157 pers/ha respectively (Kenworthy, 1995). Within Australia, urban densities range from 18 pers/ha in Sydney to about 10 or 11 in Perth, Canberra and Brisbane (Newman and Kenworthy, 1991). Densities in New Zealand cities are also very low, but are slightly higher in Auckland, with densities ranging from 22 persons per hectare in Auckland City to 14 in Manukau City Council (Auckland City Council, 1997).

Most Australian and New Zealand cities were developed at a time when suburban living in detached housing was practicable. There was very little high-density pre-industrial housing (except in Sydney). High wages, high employment, cheap land and good public transport encouraged suburbanisation in low-density suburbs in the late nineteenth and early twentieth centuries.

The suburbanisation process was encouraged by the rapid growth in car ownership after the Second World War. In 1945, there was approximately one car for every ten persons in Australia. By 1970 there were almost five. Public transport, which had allowed the suburbs to develop before widespread car ownership, had led to star-shaped cities, with development spreading out along the tracks of radial tram and train routes. With high levels of car ownership, the areas between these public transport routes were developed, and the sprawl of detached houses on large blocks engulfed large areas of the city (see Figure 12.2).

The rapid development of residential suburbs was accompanied by a suburbanisation of manufacturing, retailing and other services. Governments in Australia and New Zealand assisted the process of suburbanisation, not least through involvement in the provision of housing (see Chapter 11). Outer suburban public housing estates were provided in virtually all Australasian cities, often for employees on new industrial estates. Some examples include: Green Valley and Mount Druitt in Sydney; Broadmeadows in Melbourne; Inala in Brisbane; Elizabeth in Adelaide; Otara in Auckland; and Porirua in Wellington.

The progressive suburbanisation of residences, employment and retailing was associated with an increasing reliance on private transport. The use of public transport (and walking and cycling) declined fairly consistently from the 1950s in most Australasian cities. Growth in car ownership and use was supported by the demand-modelling strategies employed by government transport departments. These strategies assumed that increased car demand would continue, and that the 'best' option would be to provide road space to cater for that demand (Alexander, 1986). Narrow cost-benefit analyses were used to justify road building, which in turn created an induced demand for yet more traffic.

While private motor vehicles were being encouraged, public transport services in most areas of the cities were allowed to deteriorate, which in turn led to lower public transport patronage. The extent of the decline in public transport passenger numbers is well illustrated in Auckland, where the percentage of commuters who use public transport declined from 22 per cent in 1971 to 7 per cent in 1991 (Heal, 1997). These figures compare with percentages of over 50 in many Asian cities, and over 30 in many European cities (Newman and Kenworthy, 1991).

Private motor vehicles are exceedingly well provided for in Australian cities. In terms of the length of road per person, Australian cities (with 8.7 metres/person) have more roads than cities in the US (with 6.6 metres/person). Car ownership levels and car parking spaces are also very high, though slightly lower than in the USA (Kenworthy, 1995). The high reliance on private motor vehicles is associated with high levels of energy use in Australasian cities (Kissling and Douglass, 1993).

Transport planners have used the low-density urban form of the cities to justify cuts to public transport services and increases in provision for the car. However, public transport has been found to work well in some low-density cities (such as Toronto in Canada, for example). As Mees (1995) explains in relation to one Australian city: 'Urban form is not Melbourne's major transport problem; rather it is an excuse used to perpetuate the real malaise, which is a refusal by transport planners to treat seriously transport modes other than the car.'

There are exceptions to the general decline of public transport services. For example, Perth, an extremely low-density and car-dominated city, has successfully upgraded its rail system, providing a significant boost in public transport patronage. The new Northern Suburbs line in Perth attracted 40 per cent higher patronage than the bus service it replaced (Newman, 1995).

Policies to contain and channel urban growth

Planners in Australia and New Zealand have attempted to reduce the negative impacts of cities, including urban sprawl, traffic congestion, air pollution and long journeys to work. Three main strategies have been implemented: decentralisation, urban consolidation and multi-centralisation (Forster, 1995: 67). Decentralisation aims to reduce the rate of growth of large cities by moving population to smaller cities or to new cities. Urban consolidation involves increasing the density of urban areas, through urban infill, redevelopment of low-density areas, or building medium- and high-density housing in new areas. Multi-centralisation encourages the development of significant sub-centres of employment and retailing within existing cities. None of these strategies have been particularly successful in achieving the planners' policy goals. Nor have they been uncontested.

Despite a long history in Australia, decentralisation policies have been largely ineffective (Forster, 1995). Under the new cities programme of the Whitlam government in the 1970s, an attempt was made to develop Albury–Wodonga and Bathurst–Orange as regional growth centres (see Chapter 15). However, the slight population growth in such cities had negligible impact on the growth of Sydney and Melbourne. In total, these two cities grew by over 1.8 million between 1971 and 1996, equivalent to 19 times the current population of Albury–Wodonga.

In Australia and New Zealand, governments have enthusiastically embraced the concept of urban consolidation as a way of making cities more ecologically sustainable. The argument for urban consolidation is that higher densities will reduce urban sprawl as well as energy use and pollution, especially for transport, because public transport, walking and cycling would be encouraged (Kenworthy, 1995; Newman and Kenworthy, 1989; Newman and Kenworthy, 1992). However, critics of urban consolidation (Maher, 1997; McLoughlin, 1991; Stretton, 1994; Troy, 1996) claim that these arguments are unsound, and that housing density would need to increase dramatically to have any significant impact on suburban expansion.

As well as criticism from researchers, urban consolidation has been strongly resisted by local communities (Hansen, 1994). The great majority of citizens still aspire to owning their own conventional suburban home in the suburbs, sometimes described in Australia as 'the Great Australian Dream' (Archer, 1996).

One component of urban consolidation that has met less community opposition involves increasing the number of people living in the central business districts (CBDs) of cities. In Auckland in the late 1990s, more than 4000 people live in the CBD. This has created a new vibrancy for the inner city, with street cafés,

nightclubs and casinos. In Auckland, the 1987 property crash provided the opportunity for developers to convert office space into residential blocks in the early 1990s, or even to build new high-rise residential apartments. Local authorities are keen to encourage this as a way to rejuvenate demand for inner city services (Morrison and McMurray, 1997). In Auckland, the city council relaxed planning controls, allowing residential development throughout the central city area, even offering rent relief to developers for converting office buildings to residential use (Lees and Berg, 1995). Similar trends have been occurring in Australian cities, especially Sydney. High numbers of residents living in the city centre provide advantages, not only for sustaining a broad range of retailing and entertainment and improving the sense of safety, but also for reducing transport pressures.

Some success has been achieved with policies of multicentralisation in some Australian cities. Melbourne's 1980s metropolitan strategy plan identified 14 suburban 'district centres' on public transport routes, as locations to channel new retailing and employment (Robinson, 1994c). However, only one of these, Box Hill, has been successful (see Chapter 15). Government employment was not decentralised to support these areas, and large retail stores preferred car-oriented centres on their own sites. Sydney's multicentralisation had marginally more success, largely due to the relocation of government offices. However, in only one Australian city has multicentralisation come anywhere near the planners' original conception, and in this city, Canberra, there have been considerable problems in the development of new town centres (See Box 12.1).

City promotion and urban redevelopment schemes

One of the consequences of the neo-liberal response to globalisation in Australia and New Zealand is that cities are open to the processes of competition, and governments are less committed to ensuring a fair distribution of

Box 12.1 Canberra: an urban planning experiment

Canberra's reputation as a well-planned city dates from Walter Burley Griffin's creative vision of a city where the built environment would complement the natural environment (Morison, 1995). Canberra has been developing as a multicentred city since the late 1960s. Its growth since then has been based on the so-called 'Y-Plan', which provided for a series of relatively self-contained new towns (Fischer, 1984). Each town was to have significant levels of employment and retailing in its town centre at an early stage of development. Towns would be separated from each other by Canberra's hills and ridges. The towns were not meant to be completely self-contained, so a network of peripheral parkways (freeways) was developed, along with a public transport system running through the centre of the towns (see Figure 12.3).

Canberra had a number of characteristics that should have made it possible to implement the Y-Plan. It had what was probably the most powerful planning organisation in the Western world, the National Capital Development Commission (NCDC). It was based on a green-field site, so that no previous development interfered with the plan. The topography of Canberra, with its ridges and valleys, provided natural separations between the towns. The federal government owned all the land, and was the main employer. Finally, it was the national capital, and because of this, it was important that it provided a model of urban planning for the rest of the world.

Despite these advantages, the Y-Plan was not implemented as planned. The main problem was that the town centres developed far too slowly. When other cities in Australia were decentralising their employment, Canberra was concentrating its employment in its city centre, at the expense of the new town centres. The development of major retailing and employment in the town centres

lagged further and further behind residential development with each new town (Morison, 1987). The federal government became reluctant to follow the Y-Plan in terms of locating its office buildings in the new towns (especially in the southern-most town of Tuggeranong). The result was that outer suburbs in Canberra became dormitory suburbs, where many employees must commute all the way into the central city area, despite the objective of the Y-Plan to provide significant local employment. This situation is placing considerable stress on Canberra's transport system and is creating substantial costs for the local government.

The reasons for the deviation from the plan are complex, but relate to the vested interests of particular power holders in the city. First, there was significant financial advantage for developers in redeveloping sites in the city centre for office space, rather than developing new sites in the outer areas. Second, inner city retailers, who saw new office development as a way to stimulate retail trade, supported the developers. Third, the heads of public service departments may have resisted the siting of their offices in the outer (and less prestigious) new towns (Morison, 1995). Fourth, in an attempt to save money in the short term, the Department of Finance chose to rent some office space in central Canberra, rather than building more of its own office buildings in the new town centres. Finally, the technical advice from within the NCDC, which explained the large extra transport costs involved in allowing an over-development of Canberra's inner area employment, was ignored (Morison, 1987).

Canberra does provide an impressive urban landscape that Australians can be proud of. However, while Canberra appears to be a multicentred city, the effectiveness of the newer centres has been compromised by a failure satisfactorily to implement the Y-Plan.

resources and employment between cities. Governments are also less capable of providing employment, as many of their former activities have been privatised. Consequently, each city must create its own employment opportunities and promote itself in an international

marketplace. Different cities have devised various strategies to identify themselves in a positive way. Some, such as Dunedin and Christchurch, have attempted to create a distinctive image through the promotion of their historical and lifestyle characteristics

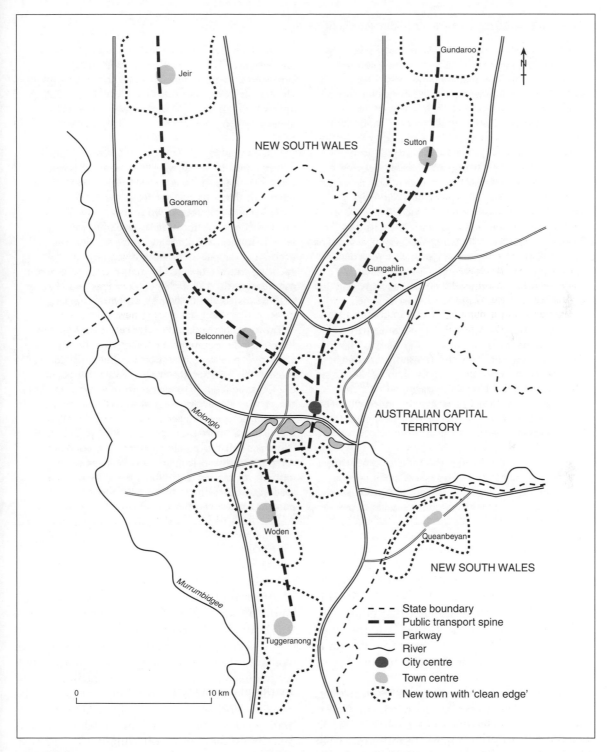

Figure 12.3 The Y-Plan for Canberra, adapted from Sparke (1988: 155) as published in *Architecture Australia* in 1968

Box 12.2 Auckland's Britomart development on Waitemata Harbour

One example of how redevelopment sparks controversy and heated debate is the proposed Britomart redevelopment in Auckland. The Britomart development plan was first announced in December 1994 (Hyde, 1996). It is part of the Waterfront 2000 project designed to modernise and revitalise the inner area of the city (Auckland City Council, 1998). It is hoped that the area being redeveloped in the Britomart site will become a hub for Auckland's public transport. The Britomart Transport Centre will consist of a five-level underground transport centre, including 2900 car parking spaces and bus and train interchanges. The Transport Centre will also allow ferry users to link up with other transport systems. Above ground, the Britomart development will have office and residential high- and medium-rise buildings. It is estimated that the project will cost around NZ$350 million.

Part of the rationale behind the Britomart redevelopment was to answer the demand of Auckland residents that the foreshore and wharf area be made more accessible to the public (Legat, 1995b). Auckland City Council hoped that this redevelopment would revitalise the inner city area, and create a vibrant focus for transport, business and recreation within Auckland (see also Chapter 15).

However, the plans for the redevelopment have not been uncontested. There have been several major points of dissent (Gunder, 1996). Originally, the Transport Centre was to have been above ground. However, the council argued that an underground Transport Centre would allow the land above to be used in a more cost-effective manner, as office and residential buildings – built by private developers – would allow it to be self-funding (Hyde, 1996). This is unlikely to be the case.

Opponents to Britomart also argue that the development is not serving the needs of public transport users in Auckland (Hucker, 1995). They point out that it is too far from the CBD to make it convenient for people who work in the CBD to use the Britomart location. In addition, there are fears that the large amount of car parking will encourage people to drive into the city, thus reducing the usefulness of a public transport hub. It has been posited that there are better and cheaper alternatives to the Britomart scheme (Gunder, 1998).

Throughout the planning process of the Britomart development, opponents of the scheme claim that public consultation has been inadequate. Hyde (1996) argues that not allowing the public a chance adequately to express their views regarding the proposed development was part of a process of 'manufacturing consent' – that is, artificially creating the impression that a lack of public response equated public support. The controversy surrounding the Britomart development is one example of the ways in which the uses of urban space can be contested, and how the state can exert control over the public.

(Pawson in Le Heron and Pawson, 1996). This strategy rests on the argument that cities with the most appeal to investors become the cities with the cleanest air and water, lowest crime, and best recreational and environmental assets (Stilwell, 1997).

New Zealand's two largest cities, however, have followed international trends in waterfront redevelopment programmes as a way to attract investors and hence employment. In Wellington, the city council has been actively involved in a redevelopment that emphasises the appeal of the waterfront (Page in Le Heron and Pawson, 1996). In Wellington's Lambton Harbour development project, the waterfront has been 're-invented' as a landscape of consumption. The new development will provide restaurants, leisure facilities and residential complexes in an attempt to enhance tourism and make the city attractive to international investment. But the development has not proceeded with full community support. The development programme was slowed considerably in the late 1990s by public protest from Wellington residents who were unhappy with the public space and harbour access designs for development. They were also unhappy with the loss of waterfront land to a proposal for a casino–hotel complex. The whole development project placed considerable financial costs on the city. Similar problems may face Auckland's Britomart development (see Box 12.2).

Australian cities have witnessed a number of redevelopment projects that imitate similar developments in North American cities. The best known of these include Darling Harbour in Sydney and Southgate in Melbourne. While such projects are invariably seen as positive by the state governments involved, they also bring with them significant opportunity costs. While investment is poured into these developments, there is a lack of funding for important public investments, including hospitals, schools, police stations and community centres. Thus one effect of these new developments is to increase polarisation between the high-profile, innercity areas and the increasingly disadvantaged outer suburbs. Even the year 2000 Olympics may add to social polarisation, through an increase in the general cost of living in Sydney, and further pressure on the housing costs of the poor in central locations: 'although Sydney may strengthen its global city status through staging the Olympics, a prudent question may be at what social costs?' (Baum, 1997: 32).

Redevelopment in inner city locations often means the removal of old buildings and the remodelling of an area that already has a distinct identity within the urban fabric. This is the situation in the Britomart redevelopment near Auckland's CBD (see Box 12.2). Urban redevelopment schemes can be an important factor in the explanation of urban social areas in Australia and New Zealand. Identifiable patterns of socio-economic status, as well as ethnicity and family status, have long been an important aspect of their cities. The next sections explore some salient features of the social geography of poverty and ethnicity in these cities.

SOCIO-ECONOMIC STATUS – DISADVANTAGED OUTER SUBURBS?

Although many Australians and New Zealanders still cling to the myth that their societies are truly egalitarian and relatively classless (Stilwell, 1997), identifiable social areas have developed within their cities since the early stages of their settlement (Damer, 1995). In many New Zealand cities (especially Wellington, Auckland and Dunedin), segregation by class was linked with topographical characteristics. The more affluent residents were concentrated on the hills and ridges, and the homes of the 'workers' were 'crowded into valleys and gullies, such as Freeman's Bay and Parnell in Auckland, or on the "flats" such as Te Aro in Wellington' (Damer, 1995: 32). The spatial polarisation of cities in both Australia and New Zealand became more evident during the so-called Fordist era of mass production, when large-scale manufacturing was associated with a clear pattern of high- and low-income areas.

Cities in Australia and New Zealand are still much less segregated than cities in the US. Part of the reason for this may be due to the smaller sizes of the former. Although the reasons why larger cities have more inequalities are unclear, there is a clear empirical connection (Murphy and Watson, 1994).

An important contrast exists between the patterns of socio-economic status found in Australian and New Zealand cities and the patterns found in cities in the US, even though cities in each of these countries have been subject to the same global forces of economic change. While in the US cities the poor and ethnic minorities are heavily concentrated in inner city areas, this pattern is no longer evident in Australian or New Zealand cities (Murphy and Watson, 1994). In fact, affluence has been concentrating in cores of wealth in inner city areas since the 1970s, in a reversal of the US experience (Badcock, 1997). However, while more low-income people in absolute terms are found in the outer areas of

Australasian cities, there are still some inner areas with high proportions of low-income households. Those living in poverty in the inner city are more likely to be single households living in private rental accommodation, while low-income families with children tend to be found more in the suburbs. Thus, poverty tends to be spatially sorted into outer and inner areas on the basis of household characteristics (Gleeson et al., 1998). Also, while there are (as yet) few obvious signs of desperate social disadvantage (such as beggars and people sleeping in cardboard boxes on the streets), Australasian cities do have concentrations of particular disadvantaged groups in the inner city (e.g. the homeless and people with chronic mental disability). Such concentrations are evident even in small cities such as Dunedin in New Zealand (Gleeson et al., 1998).

It is possible to identify marked increases in inequality within Australian and New Zealand cities since the 1970s (Badcock, 1997: 245). For example, Hunter and Gregory's study of small census areas in Australian cities of over 100 000 people indicates that, between 1976 and 1991, average household income in the lowest 5 per cent of areas ranked by socio-economic status fell by 23 per cent. During the same period, the top 5 per cent of areas recorded increases of 23 per cent (Hunter and Gregory, 1997). Generally, the larger the city, the greater the level of polarisation. Wealthiest and poorest areas in cities are much more polarised in the 1990s than they were in the 1970s (Badcock, 1997: 245).

Urban inequality in New Zealand cities increased from 1981 to 1991, in an even more dramatic way than occurred in Australia. Morrison (1997) found that the gap between unemployment rates in different groups of areas in Wellington had increased dramatically, as demonstrated in Table 12.2. Unemployment was particularly high in some of the outer suburban areas of early state housing. Morrison also noted the importance of what is called 'non-employment' (the withdrawal of individuals from the active labour market). When this measure was examined, it became

TABLE 12.2 The gap between highest and lowest unemployment rates in area units within Wellington, 1981–91

Area unit	1981	1986	1991
Highest unemployment rate	11.2	16.2	24.9
Lowest unemployment rate	0.6	2.1	2.3
Difference	10.6	14.1	22.6

Source: Morrison (1997).

clear that the higher the unemployment rate, the lower the propensity of males to declare themselves unemployed. Thus, the spatial differences in unemployment understate the differences between areas in terms of the under-utilisation of labour (Morrison, 1997).

There are a number of reasons why disadvantage seems to be increasing more rapidly in the outer suburbs of Australasian cities. First, there are increasing disparities in access to jobs. Between 1981 and 1990, for example, the resident labour force in Sydney's western suburbs grew by nearly one-third, but the number of jobs increased by less than one-fifth (Burnley et al., 1997b). Similar differences in 'occupational opportunity' have been noted for New Zealand cities (Morrison in Le Heron and Pawson, 1996). A second, related issue is that outer suburban areas typically have poorer public transport services, and this reduces access to both jobs and services (Murphy and Watson, 1994).

A third factor is the general lack of adequate provision of services such as hospitals, universities and child care centres in many outer suburban areas. Hodge (1996) relates this inequity in service provision in western Sydney to the construction of this area as Sydney's 'other'. Universities, for example, were seen by some decision-makers in the 1960s to the 1980s as not being necessary in disadvantaged outer suburban areas, because of a supposed 'inherently low ability for higher education amongst western Sydney residents'

(Hodge, 1996: 40). Hodge notes that this assumption ignored the likelihood that the disadvantages experienced by the 'others' in western Sydney were brought about by structural inequities, rather than any innate disadvantage. These structural inequities include the historic under-provision of services. A fourth issue, which is particularly important for Sydney, is the way in which new migrants tend to locate in outer suburban areas. Because recent non-English-speaking (NESB) migrants are more likely to be unemployed, or to work in lower-paid, entry-level service jobs, there may be a 'link between increasing polarisation and immigration' (Baum, 1997: 28).

One feature of the segregation of the rich and poor in Australasian cities is its increasing scale. The regions of contrasting wealth have become much larger. Thus large areas of western Sydney or south Auckland are becoming disadvantaged, with high levels of poverty and disturbingly high unemployment rates. Other areas, such as Sydney's north shore, are becoming the preserves of the wealthy. It should be noted that such preserves are not homogeneously Anglo in population composition. There are significant proportions of overseas-born, particularly those from northeast Asian countries (ABS, 1998c). Figure 12.4 shows the segregation of high-income earners in Sydney, where the inner and northern suburbs comprise large areas with high proportions of high-income earners, while the western and southern suburbs have relatively few high-income earners (apart from some isolated pockets).

One effect of this large-scale residential segregation is the accentuation of polarisation through the redistribution of resources. Stretton (1989) cogently explains the way in which resources are quietly shifted from the poor to the rich when residential segregation increases. This occurs with both privately provided services (e.g. doctors) which are less likely to be provided in poorer areas, as well as public services, such as hospitals and schools. Areas with mainly poorer people are less able to attract positive resources such as

hospitals, or prevent the location of undesirable facilities (e.g. waste dumps) in their areas. Local councils in high-income areas are also able to raise more rate revenue, and are thus better able to provide services such as parks, swimming pools and libraries (Stretton, 1989). Increasingly in Australian and New Zealand cities, where people live really matters in terms of standard of living, opportunities for education and employment and access to recreational and cultural resources.

Gentrification: changing patterns of socio-economic status

While the outer suburbs of Australian and New Zealand cities have become increasingly disadvantaged, the corollary of this is the increasing concentration of wealth in inner areas, often through a process known as gentrification:

> *Gentrification commonly involves the invasion by middle-class or higher income groups of previously working-class neighbourhoods or multi-occupied 'twilight areas' and the replacement or displacement of many of the original occupants. It involves the physical renovation or rehabilitation of what was frequently a highly deteriorated housing stock and its upgrading to meet the requirements of its new owners. In the process housing in the affected areas . . . undergoes a significant price appreciation. Such a process . . . commonly involves a degree of tenure transformation from renting to owning (Hamnett, 1991: 175).*

This definition applies well to the process of gentrification in Australian suburbs such as Glebe and Paddington in Sydney and Carlton in Melbourne, as well as to suburbs such as Ponsonby and Kingsland in Auckland. To understand why gentrification has taken place in such areas, it is necessary to consider at least three themes: changing values and lifestyles of

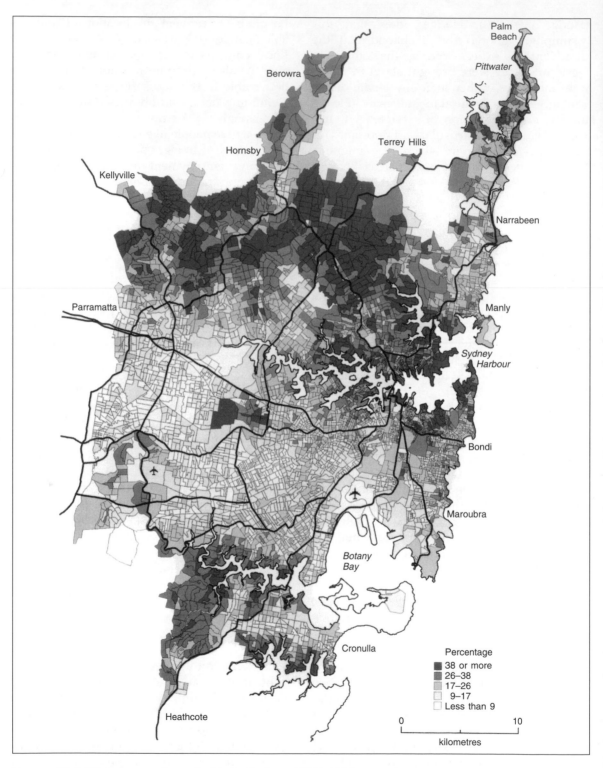

FIGURE 12.4 High-income households in Sydney (ABS, 1996 census)

FIGURE 12.5 A section of the Glebe Estates, with single-storey nineteenth century terrace housing, rehabilitated by the New South Wales government in the 1980s. This now provides public housing for low-income households in an inner city location (photo: P. Tranter)

the gentrifiers, global flows of capital, and the role of banks and building societies in providing loans for inner city housing rehabilitation (Horvath and Engels, 1985).

Gentrification would not have occurred without a major change in housing preferences. In the 1970s, middle-class residents started to appreciate the lifestyle opportunities of inner city suburbs. They began to value the heritage buildings, the cosmopolitan mix of different groups of people, the better access to employment and entertainment, as well as the real-estate investment opportunities of the inner city. However, this in itself was not enough to lead to widespread gentrification. Finance had to be available to invest in inner city housing, and this became more widely available in the 1970s. An inflow of foreign capital into Australasia prompted banks and building societies to relax lending restrictions on inner city housing.

In Glebe, an inner Sydney suburb, the gentrification process was assisted by an innovative public housing scheme. In an area known as the Glebe Estates, the Church of England had owned a large tract of terrace housing that had been rented out to low-income earners. By the 1970s this housing was so run down that it was generally regarded as a slum. The federal government bought the Glebe Estates, and handed the area over to the New South Wales state government for rehabilitation as a model public-housing estate. Rather than clearing the area, the housing was carefully renovated, providing some of the most sought-after public housing in Australia. This revitalisation of the Glebe Estates gave confidence to private investors to gentrify adjacent private housing through the 'neighbourhood effect' (Horvath and Engels, 1985). Glebe now provides a rare combination of privately owned gentrified housing, alongside public housing in renovated nineteenth century terraces (see Figure 12.5).

Several suburbs in Auckland have also undergone gentrification. The best known of these is Ponsonby, located immediately south-west of the city's CBD, on the other side of the motorway. Ponsonby's houses date back to the

Box 12.3 Gay communities – gay spaces in cities

The study of gay and lesbian geography is a relatively new sub-field of research. The development of homosexual geography had a similar point of entry to feminist geography. It is the insertion of an 'other' into the schema of social, economic and political geographies that already exist. Much work has been done that examines the specific ways in which gay men and lesbians use and shape space (Adler and Brenner, 1992; Bell, 1991; Knopp, 1990; Lauria, 1985; Valentine, 1993). One of the best examples of this is the growth of gay and lesbian parades and associated events. The Sydney Gay and Lesbian Mardi Gras and Auckland's Hero parade are two such events.

The Sydney Mardi Gras started in 1978. The first Mardi Gras was held on 24 June: International Gay Solidarity Day. The single-day event was attended by 1000 to 2000 people and ended in 53 arrests. The march was held on Oxford Street, in Sydney, and has continued to be an annual event. The first 'official' Mardi Gras Festival was held in 1985, with over 20 events. Since then the focus of Mardi Gras has become more symbolic and celebratory, although politics has not entirely been removed. The 1998 parade had more than 200 entries. More than 500 000 people lined Oxford Street to watch the parade, with another one million or so people watching the televised event. The month-long festival that precedes the parade comprised over 100 events, including film festivals and dance parties.

The Hero parade in Auckland began in 1990 as a fund-raising event for HIV/AIDS services. It was modelled on the format of the (by then) well-established Sydney Mardi Gras. Initially, it was predominantly an event for gay men, but lesbians have become increasingly involved. Like Sydney's Mardi Gras, Hero is a month-long succession of events, culminating in a parade and party.

The use of city space by homosexual people for events that are specifically 'gay' in focus has not been uncontested. In Auckland, a city councillor explained the refusal of a grant for the Hero parade: 'I don't want Auckland to become known as the city of the homosexual parade' (Crawshaw, 1998). More importantly, for a number of reasons, in 1996 the route for the Hero parade was moved from Queen Street to Ponsonby Road. It is significant that Ponsonby Road has a number of gay and lesbian businesses, and is popularly considered to be a 'gay' space. Queen Street, on the other hand, is a main street and is 'straight' space. The contrasting opinions about where it is appropriate for people to be openly 'gay' have been hotly debated in many situations (Johnston, 1997).

Despite the controversy, gay and lesbian pride parades have become a very important part of the urban scene. The company Sydney Gay and Lesbian Mardi Gras contributes significantly to the Australian economy. This event is a very large part of the Sydney cultural scene. International visitors time their trips so as to be in Sydney for the parade. Similarly, the Hero parade is growing in international importance, in spite of its short history (Crawshaw, 1998). Both events attract large crowds of heterosexual people as well as the gay and lesbian community. Sydney's Mardi Gras has helped to promote Australia as a tolerant place to visit, countering the publicity for the One Nation Party (see Chapter 10).

late nineteenth century. In the early twentieth century it was a high-status area, but became a working-class suburb after the Second World War. By the 1960s it was seen as an ethnic enclave, with a large Maori and Pacific Islander population. In the 1970s, the area became attractive to middle-class residents, who valued not only its central location and the character of its housing, but also its interesting multicultural mix and inner city lifestyle (Heal, 1994). However, 'as more and more white middle-class gentrifiers moved in they have pushed out many of the Maori and Pacific Islanders, transforming the ambience' (Lees and Berg, 1995: 35). Property prices have increased significantly. Houses bought for NZ$30 000 in the 1970s may be worth over NZ$300 000 in the 1990s. Some of these homes were bought by Maori and Pacific Islanders when prices were low. For these households, especially those with large families 'the lure of the money, and the chance to buy a bigger, cheaper property in the suburbs, is irresistible' (Heal, 1994: 86). The gay community (see Box 12.3) has also had an impact on gentrification in Ponsonby, as it has had in many inner city areas in Europe, North America and Australia.

The distinctive multiethnic composition of Australasian society (Chapter 10) is reflected in particular ethnic concentrations within cities. The following sections examine the clustering of Maori and Pacific Islanders in Auckland, and new immigrant groups in Australian cities.

Maori and Pacific Islanders in Auckland

The very successful New Zealand film *Once Were Warriors* draws attention to the link between socio-economic status and ethnicity in New Zealand cities. The film starts with a spectacular scene of mountains in the South Island. But when the camera pulls back, the scene is simply a billboard beside a motorway and factory in Auckland. The film powerfully illustrates the deprivation suffered in an area of South Auckland where the largely Maori population has been disaffected by the economic restructuring process. The director of the film explains that he wanted to represent a particular urban environment: '... a hard, almost treeless urban experience which most New Zealanders would never have been aware of. But people do live in such an environment – even in New Zealand they live right next to motorways and under power pylons' (Lees and Berg, 1995).

When Maori first started to migrate to Auckland in large numbers, they were initially heavily concentrated in inner suburbs such as Freeman's Bay, just west of the city centre. However, since the 1950s, many Maori have moved out of these inner areas, especially to new areas of public housing in the southern suburbs of Auckland (McKinnon, 1997: 91) (see Chapter 15). The concentration of Maori in many outer suburbs in south Auckland has been increasing in the 1980s and 1990s. Figure 12.6 illustrates the distribution of Maori in

Auckland in 1996. European New Zealanders are moving out of many previously Pakeha working-class suburbs, while Maori and Pacific Island groups have been moving in. This has occurred in such south Auckland suburbs as Penrose, Glenn Innes and Mangere. This concentration is also linked to the location of public housing areas in outer areas of Auckland (Morrison, 1995). The central Auckland suburbs have witnessed a very different process, with the decline of Maori and Pacific Islander populations and an increase in Pakeha. There is also evidence that this increased ethnic residential segregation is accompanied by Pakeha parents pulling their children out of schools that have a predominantly Maori and Pacific Islander enrolment. Lees and Berg (1995) believe that such increases in ethnic segregation in suburbs and in schools require investigation before they become problematic, as they have in the United States.

Ethnic concentrations in Australia

Post-war immigration provided the basis for a complex multicultural mix in Australian cities. Initially, immigrants from specific countries tended to settle together in particular areas of cities, before they dispersed. British migrants and migrants from northern and eastern Europe tended to settle in the outer suburbs. Southern Europeans (Greeks, Italians, Yugoslavs and Maltese) initially concentrated in the inner suburbs, creating identifiable ethnic enclaves in suburbs such as Leichhardt in Sydney and Carlton in Melbourne. More recent migrants from south-east Asia and Latin America, notably the Vietnamese, have concentrated in particular middle and outer suburbs (e.g. Cabramatta in Sydney, Springvale and Dandenong in Melbourne, and Enfield in Adelaide).

Much of the public criticism of multiculturalism in Australia (see Chapter 10) has been fuelled by the existence of clusters of visibly different ethnic minority groups. These clusters have been disparagingly described by some

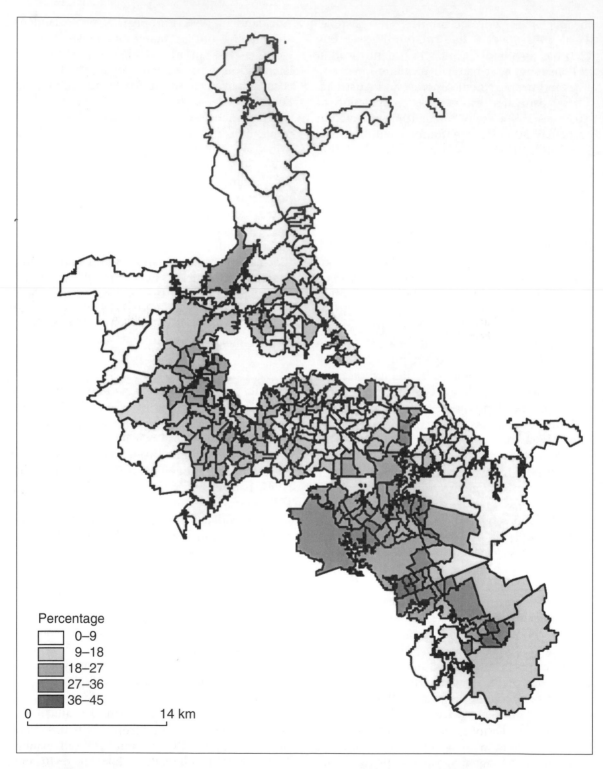

FIGURE 12.6 Distribution of Maori in Auckland (1996) (compiled by SNZ, July 1998)

Box 12.4 Cabramatta

Is Cabramatta really an ethnic ghetto, characterised by Vietnamese gangs, drug dealing and violence? Media coverage in the 1980s of this western Sydney suburb, in the Fairfield Local Government Area, certainly promoted this notion (Waldren and Carruthers, 1998). However, by the 1990s, this image was largely replaced with a carefully 'commodified image' of Cabramatta as a tourist destination – a 'South East Asia in Australia' (Powell, 1993). Also, Cabramatta's supposed Vietnamese 'ghetto' status can be easily refuted by 1996 census statistics showing that Vietnamese-born make up only 25 per cent of a population that has considerable diversity. This diversity stems from successive periods of settlement by different ethnic groups.

Until the 1950s the Fairfield area was predominantly farmland (Dunn, 1993). After the Second World War, many European migrants (mainly from southern Europe) settled in low-cost housing in the area, near immigrant and refugee facilities. The main shopping precinct, located in Cabramatta, reflected the tastes and habits of the southern European migrants.

During the 1970s and 1980s Indo-Chinese migrants and refugees settled in the Fairfield district, especially in Cabramatta, to be close to migrant and refugee services and facilities, like the previous wave of immigrants (Dunn, 1998). In the 1990s over 50 per cent of Fairfield's population were born overseas, in 133 different countries. Cabramatta is a particularly vivid example of the ethnic diversity of the area: it has become known as the 'multicultural capital of Australia' (Powell, 1993: 138).

In recent years, the tourist potential of Cabramatta's ethnic diversity has been strengthened by Fairfield City Council. Council members are representative of many different European and Asian cultures. The shopping precinct has been beautified and revitalised in a distinctly Asian manner. Street furniture and statuary that is evocative of the orient has been installed, with the clear aim of benefiting from the 'difference' of the area (see Figure 10.2). A similar redevelopment of Sydney's inner city Chinatown predated Cabramatta (Anderson, 1993). The decision by the council to 'Asianise' the shopping precinct in Cabramatta was a deliberate move to attract tourists to the area. Even buildings owned by non-Asians have adopted this theme (Dunn, 1993; 1998).

Ethnic clustering has been a subject of much debate within Australia (Burnley et al., 1997a: 36). Those who wish to limit immigration point to areas such as Cabramatta as examples of immigrants' non-conformity with Australian norms. However, others argue that suburbs with high concentrations of overseas-born residents are places where immigrants adjust to their new surroundings, and are 'jumping-off' points for immigrants to establish themselves within mainstream society (Jones, 1996a).

Services provided in areas of high ethnic concentrations allow migrants to adapt to their new environment with the minimum of difficulty. Migrants can find employment with own-language employers, business can be transacted in a familiar manner and more established residents can assist those who have newly arrived. This informal network reduces the reliance of migrants on state-supported services. Also, ethnic clustering makes it easier for governments and other bodies to provide services that are required by a specific group. These services may include English classes, bilingual bank tellers and immigration offices (Dunn, 1998).

Australians, including federal politicians, as ethnic ghettos.

One way of examining ethnic segregation is to use the index of dissimilarity (ID). This ranges from 100 (total segregation for a group) to zero (where a specific group would be distributed among areas in exactly the same way as Australian-born people). Scores over 50 are interpreted as showing a very significant level of segregation (Forster, 1995: 103). In Australia, while migrants from English-speaking and northern European countries have IDs of less than 30, migrants from the Middle East and Asia are highly segregated. In Sydney and Melbourne the Vietnamese had IDs of 67 and 55 respectively in 1991 (Forster, 1995: 106).

However, even the Sydney figure of 67 is not much higher than the figure of 57 for Greeks in Melbourne in the 1960s (Jones, 1996a). Thus, ethnic residential concentration in Australian cities is by no means a new phenomenon. Migrant groups have tended to disperse eventually. Even the Vietnamese may do this, as did the Greeks, Dutch and Italians (Jones, 1996a). Also, the IDs for blacks and Hispanic groups in US cities were much higher than for any groups in Australia. Every major city in the United States recorded figures of between 70 and 80 for these indices (Marcuse, 1996).

While the Vietnamese groups show noticeable concentrations at present, they certainly do not qualify as living in 'ghettos', as do some ethnic groups in US cities. Ghettos have been defined as districts which are 'almost exclusively the preserve of one ethnic or cultural group' (Johnston *et al.*, 1995: 231). This definition clearly does not apply to any suburb in Australia, although it is often invoked, by police and some policy-makers, to discredit the Aboriginal claims to the central Sydney space of Redfern (Anderson, 1999). Also, ghettos in the US are havens from discrimination and persecution. 'While Harlem in New York may be regarded as a ghetto in this sense, immigrant concentrations in Australian cities have not become places of last resort' (Burnley *et al.*, 1997a: 56).

Some researchers see distinct advantages accruing to ethnic groups in their concentration within enclaves, as the case study of Cabramatta illustrates (see Box 12.4).

This Chapter has demonstrated how urban areas in Australia and New Zealand are characterised by distinctive urban forms and social patterns. In the next Chapter, some of the broader economic contexts for the development of these particular characteristics are explored.

FURTHER READING

For a summary of urban themes in New Zealand, Le Heron and Pawson's (1996) *Changing places: New Zealand in the nineties* is a valuable source with some excellent case studies (Chapter 9 is particularly relevant). For a range of well-presented information on urbanisation in New Zealand, see McKinnon's (1997) *New Zealand historical atlas*. While Statistics New Zealand has little published information at the city level, the Australian Bureau of Statistics' *Social Atlas* series is a well-presented source of descriptive information on all Australian capital cities. A highly readable account of urbanisation and current urban policy issues in Australia is provided in Forster's (1995) book: *Australian cities: continuity and change*. A very useful analysis of ethnicity in urban areas is provided by Burnley, Murphy and Fagan (1997) in *Immigration and Australian cities*. Stretton's (1989) *Ideas for Australian cities* provides thought-provoking ideas on social issues within Australian cities.

13

GLOBALISATION AND THE AUSTRALASIAN ECONOMIES

One of the most far-reaching developments in both Australia and New Zealand during the last two decades has been the way in which a combination of domestic economic policy reforms and changes in the world economy have transformed manufacturing activity. Furthermore, the new sets of relationships created between capital, the nature of production and the workforce have had ramifications for all productive activity, contributing to expansions in the service sector and to clear geographical outcomes with respect to the location of industry and services. Various aspects of these changes are examined in this and the succeeding two Chapters. Initially, consideration is given to the reversal of policy whereby widespread protectionism has been replaced by deregulation and the pursuit of 'economic liberalism'. The interaction of the newly exposed manufacturing sector to outside competition and globalising tendencies in the world market are explored in this Chapter, with case studies drawn from the motor vehicle industry, the sugar industry in Australia, the Australian iron and steel industry, and the dramatic reforms in New Zealand known as 'Rogernomics'.

LEGACIES OF TARIFF PROTECTION

Tariff protection was employed in Australasia from early in the twentieth century as a means of achieving economic diversification to create employment and encourage population growth. Although Australian state and federal governments supported diversification into new primary industries, notably dairying, dried fruits, sugar and wine, export-oriented manufacturing remained small, as did structural change in the economy. Both countries were highly sensitive to changes in the level of world demand for their narrow range of traditional exports. Tariffs were increased in the early 1920s, with the aim of protecting industries established during the First World War and to help provide job opportunities for the growing population. The tariffs reduced the openness of the two economies and, along with wage fixing promoted by powerful trades unions, impaired the international competitiveness of their exports. It is generally agreed that long term reliance on these tariffs slowed the rate of response to changes in the world economy (Schedvin, 1990).

In Australia an increasing share of resources was focused on lower yielding activities, including wool production, closer settlement policies and manufacturing industry which was becoming more uncompetitive. Indeed, it was not until the mid-1960s that the value of wool exports was exceeded by those from manufacturing. Tariffs were aimed at maximising import replacement, but industry was slow to respond for several reasons: the inter-war economic slump, lack of a sizeable national market, slow growth in demand for staple exports, and high labour costs.

The protectionism operated by successive Australasian governments shielded not only domestic manufacturers, such as Broken Hill Proprietary (BHP), but also foreign firms operating within the protected Australasian environment. For five decades after the First World War the latter were largely British and American firms who dominated certain manufacturing sectors from their local production facilities. These had little incentive to be competitive internationally and hence productivity growth was not as high as in North America and some parts of Western Europe. Coupled with gradually declining terms-of-trade for the staple primary industries, this meant that per capita income in Australia gradually fell from the world leader status that it had enjoyed before 1914. This fall was gradual and Australia is still just amongst the top 20 leading countries in terms of income per capita, as is New Zealand, which occupied third place in the late-1940s.

In Australia, the nineteenth century pattern based on wool was repeated by substituting minerals for wool but with few domestic forward or backward linkages. The pattern of strong tariffs on the products of domestic industry was maintained post-1945, so that effective rates of protection remained amongst the highest of the world's advanced industrial nations. Import replacement largely failed to reduce the dependence upon traditional export commodities, and from the 1960s has gradually been curtailed. Mineral exports became much more significant in the Australian economy

from this time, but much of the processing was done abroad, especially by Japanese companies. However, post-1945, Australia's most rapid economic growth occurred in the 1960s (around 6 per cent per annum increase in gross fixed capital formation), coinciding with a peak in the proportion of the workforce employed in manufacturing. Subsequently it has been the service sector that has grown most rapidly in both countries. By the mid-1990s Australia had only 23.9 per cent employed in manufacturing and construction whilst 63.8 per cent were in the service sector (Table 13.1). Significantly, the biggest successes in manufacturing exports in recent years have been in high-technology goods which have low levels of protection and which have relied on comparative advantages created by virtue of the highly literate and skilled workforce. The highest levels of protection were accorded to the automobile, textiles, clothing and footwear industries, long regarded as the most inefficient sectors of domestic manufacturing.

Tariffs and the motor vehicle industry

The changing fortunes of the automobile industry reflect the about-turn in economic policy in the early 1980s when the long period of protectionism for the manufacturing sector was replaced by various moves towards a more open and less closely regulated economy. In fact, tariffs on vehicles and vehicle components have remained amongst the highest in operation, but the restructuring that has occurred in the industry has certain common features with those in other parts of manufacturing.

Australia

The motor vehicle industry first received government protection in Australia in 1907. Quantitative restrictions on imported motor bodies were introduced during the First World War, encouraging the establishment in 1917 of

TABLE 13.1 Employment and gross production within Australian businesses by sector (1995/6)

	Number (000)	%	Employment Full-time 000	86–96 % ±	Part-time 000	86–96 % ±	Gross production A$ million	%
Agriculture	110.4	12.0	319	−3.4	103	+18.7	13 558	3.6
Mining	2.8	0.3	87	−7.5	4	+185.7	17 967	4.8
Manufacturing	69.4	7.5	1012	−1.9	107	+23.2	64 623	17.2
Utilities	0.1	0	66	−51.6	3	+36.4	13 449	3.6
Construction	151.5	16.4	510	+17.2	93	+51.5	27 033	7.2
Service industries	588.6	63.8	4242	+18.4	1773	+65.6	239 130	63.6
Wholesaling	53.9	5.8	423	+10.9	71	+55.6	40 941	10.9
Retailing	138.7	15.0	706	+8.3	533	+74.2	30 008	8.0
Accommodation	28.5	3.1	212	+52.0	166	+66.4	7892	2.1
Transport/storage	45.5	4.9	341	−2.4	52	+43.6	23 724	6.3
Communications	8.0	0.9	144	+0.4	20	+40.6	13 467	3.6
Finance/insurance	21.8	2.3	263	−3.5	50	+41.3	17 034	4.5
Property and business services	135.3	14.7	621	+73.1	192	+98.3	33 698	9.0
Education	16.7	1.8	404	+20.2	186	+36.7	19 153	5.1
Health and community services	56.3	6.1	473	+21.5	293	+60.8	22 042	5.9
Culture/recreation	27.0	3.0	111	+26.3	78	+100.5	9106	2.4
Personal and other services	56.9	6.2	224	+32.7	81	+40.6	6839	1.8
Defence and government administration			320	+5.5	52	+132.0	15 226	4.0
Total	922.8	100.0	6236		2083		375 760	100.0

Source: Australia Yearbook (1997).

TABLE 13.2 Major motor vehicle manufacturers in Australia, 1995

Company	Number of production facilities	Number of employees	Total motor vehicle production[a]
Ford	2	6300	101 731
General Motors Holden	1	5772	107 326
Mitsubishi	3	5000	44 548
Toyota	1	3800	60 547
Totals	7	20 872	314 152

[a]Includes vehicles for export.
Source: Bamber and Lansbury (1997: 83).

Holden Motor Body Builders in Adelaide, which became the leading Australian-owned motor manufacturer in the 1920s. Tariffs were maintained after the war and contributed to Ford (of Canada) and General Motors (GM) (of the US) setting up plants in Australia. In 1931 GM took over Holden to form General Motors Holden (GMH), subsequently agreeing to manufacture a car with at least 95 per cent of its components provided by Australian companies, though with duty-free import of equipment and tools. Similar government support followed for Chrysler and Ford who were joined in the 1950s by Volkswagen and six British firms assembling cars in Australia under government protection. This protection raised the proportion of the local supply of components to the car industry from 45 per cent in the mid-1950s to 85 per cent by 1990 (Fagan and Webber, 1994; 111). Smaller companies, which included Nissan, Renault, Toyota and Volvo, were allowed to assemble vehicles with less local content.

Since the mid-1960s the rise of Japanese car manufacturing has transformed the industry in Australia. Initially this was seen by a rising tide of imported Japanese vehicles, so that by 1975 the Japanese controlled nearly one-quarter of the market and had put some of the smaller Australian-based firms out of business. These imports came to a small market in which it was difficult for local car manufacturers and assemblers of imported components to realise

any economies of scale. By 1985 only five vehicle manufacturers remained, collectively producing 13 models.

Despite the fact that the car-makers in Australia were foreign companies, the Australian government, in seeking to protect jobs, continued its support for the industry, notably in the 1985 Passenger Motor Vehicle Plan (known as the Button Plan after former Industry Minister, Senator John Button) and its extension in 1991, and through continuing tariffs and export facilitation schemes. The latter has enabled GMH to build an engine plant in Melbourne, earning 'credits' which they could use to import components and cars. However, quota restrictions on imported vehicles were abolished in 1988 and the nominal tariff rate on imported vehicles and components will have been cut to 15 per cent in 2000, from 57.5 per cent in 1985.

To encourage vehicle exports, exporters were allowed to import duty-free components to the same value as their exports. However, exports required longer production runs, a better product and cheaper supplies, which therefore necessitated more incorporation of the higher-quality foreign-produced components. As a result, approximately one-third of domestic component suppliers went out of business in the 1970s and 1980s.

There has been substantial rationalisation since the mid-1980s. Nissan ceased manufacturing in Australia in 1991 and various assem-

TABLE 13.3 Shares of the Australian motor vehicles market by local manufacturers, 1987–95 (%)

Company	1987	1989	1991	1993	1995
Ford	28.6	24.5	20.6	20.9	21.5
General Motors					
Holden	18.7	16.5	16.8	17.4	19.7
Mitsubishi	11.8	11.9	11.3	13.7	10.1
Nissan[a]	9.3	11.3	10.0	4.7	3.8
Toyota	20.6	18.8	21.5	21.7	18.8
Total	89.0	83.0	80.2	78.4	73.9
Overseas					
manufacturers	11.0	17.0	19.8	21.6	26.1

[a]Nissan ceased local manufacturing in 1991.
Source: Bamber and Lansbury (1997: 85).

TABLE 13.4 The location of the Australian motor vehicle industry, 1998

Ford:
- engine, stamping and casting plants at Geelong;
- passenger and commercial vehicle assembly at Broadmeadow, Melbourne;
- engine reconditioning at Ballarat;
- assembly plants at Homebush (Sydney) and Eagle Farm (Brisbane).

GMH:
- plants at Fishermens Bend and Dandenong (Melbourne) and Elizabeth (Adelaide).

Toyota:
- plant at Altona (Melbourne).

Misubishi:
- two plants in Adelaide.

bly plants have closed, including GMH plants in Sydney, Queensland and South Australia; two Toyota plants in Melbourne have been replaced by one (in Altona). By 1995 there were just four producers and seven production plants (Table 13.2). Exports have trebled in the 1990s but imports take one-third of the domestic market.

By the mid-1990s only five models of car were produced in Australia: the Ford Falcon, Holden Commodore, Mitsubishi Magna, Toyota Camry and Toyota Corolla. Despite the contraction in production lines and component manufacturers, and therefore also reduced employment, other outcomes have been a relatively cheaper and better range of Australian cars, more exports and increased investment. For example, a new A$500 million Toyota plant has been built in Melbourne and the other manufacturers are engaged in substantial investment programmes to preserve their share of the market (Table 13.3). The four manufacturers still producing in Australia account for 70 per cent of the market, though this includes some vehicles imported by these companies. Indeed, in 1995 imports accounted for 44 per cent of the market; between 1988 and 1994 vehicle and component imports more than doubled, from A$4.33 billion to A$9.2 billion (Bamber and Lansbury, 1997: 85). Domestic production fell from 370 000 vehicles in 1990 to 312 000 in 1995. A minimum production of 100 000 vehicles per model is usually considered necessary to achieve the benefits from scale economies, but only two plants in Australia operate at this level.

In terms of location, vehicle manufacture and assembly has been dominated by Victoria, with Ford and GMH concentrating activities there from the 1920s. This remains the case today, as shown in Table 13.4. The industry may be an excellent example of a highly flexible, globally sourced and globally integrated form of production, but the local market, without the cushion of protective tariffs, is proving too small to sustain more than a few key operators.

New Zealand

As in Australia, vehicle manufacturing in New Zealand was a classic example of an import substitution industry surviving only because of government intervention. The tiny domestic market made it unfeasible for the kind of mass production normally associated with this industry and has restricted it to vehicle assembly and production of some components. Even within this more limited character of the industry there has been significant rationalisation and changes in production emphasis since the 1960s. The key developments have been:

(a) elimination of the smallest production runs;
(b) reduction of the number of models assembled;
(c) sourcing cheaper inputs;
(d) introducing more efficient and flexible production systems, favouring assembly of fully made-up vehicles rather than kits with original equipment componentry;
(e) allowing substantial import of second-hand vehicles from Japan, thereby reducing prices for domestic used cars and forcing rationalisation of new car production (Britton *et al.*, 1992: 136).

The result has been to shift the dominant market supply to imports from Australia and Japan rather than home-assembled vehicles. Domestic production has increasingly concentrated in Auckland.

Before the Second World War, the first assembly plants were located in Wellington by General Motors (1926), Ford (1936) and a New Zealand-owned firm, Todd Motors (1935). With the post-war growth of private car ownership, new plants were opened in Christchurch and Nelson; there was expansion in Wellington; and by the mid-1960s Auckland was also a centre for car assembly. There was significant expansion in the 1970s when new companies entered the market and there was dispersal of production to provincial centres, including Thames, Waitara and Wanganui. The dominance of domestic production by the Wellington–Hutt Valley fell from 73 per cent in 1955 to 44 per cent in 1972. This also reflected closure of some plants in Wellington in favour of locations in Auckland and the provinces. By the 1970s the New Zealand Motor Co., a subsidiary of the British Motor Co., had joined

the three above-named pioneers of the major assemblers, importing own-brand kits and supplies of made-up vehicles for subsequent sale to franchised retail outlets. By 1980 there were also six New Zealand-owned franchise holders of overseas marques assembling kits.

The inefficiency and over-capacity of the industry was tackled by the fourth labour government in the mid-1980s, as part of the implementation of closer economic relations (CER) with Australia. Industry protection levels were reduced or eliminated, excise tax on vehicles was removed, as were quantitative import restrictions, and forced use of local components was phased out. These changes in policy had dramatic impacts upon the industry (Britton et al, 1992: 138–40):

(a) Changes in ownership. Between 1984 and 1988 seven overseas parent companies took full control of their New Zealand franchisors: Daihatsu, Honda, Mazda, Mitsubishi, Nissan, Suzuki and Toyota. By 1990 all assembly production was foreign-owned and reflected the broader context in which TNCs operated. Hence the tendency has been to construct some models locally and import others from either Australia or Japan.
(b) Contraction of locally based production. The numbers of assembly lines, plants and employees have been cut substantially. Over half the assembly plants were closed between 1985 and 1991, with four of the 10 companies ceasing operation in New Zealand. Mitsubishi owns the only assembly line remaining in Wellington (at Porirua, a satellite town for the capital established post-1946); Nissan are in Auckland (at Manukau) whilst the others are Toyota in Thames, and Honda in Nelson. The total number of assembly jobs fell from 7800 in 1976 to 2947 in 1991, supported by 1087 administrative workers and a further 5100 in component industries. However, there has been a reduction in the use of home-produced components, with more reliance on imported parts, especially from Australia. Against this trend the Ford

alloy wheel plant at Wiri (Auckland) has expanded to meet export orders from Australia and North America. In March 1997 the Ford–Mazda assembly joint venture in Auckland closed (under the trade name of Vanz). Around 39 000 passenger and light commercial vehicles were produced in 1996 compared with 73 000 in 1989. A further review of the industry in 1997 will lead to a move to zero tariffs at the end of 2000, after which it is unlikely that the remaining assemblers will continue to produce in New Zealand.

(c) Alteration of local production practices. Japanese-style labour practices have been introduced by the Japanese firms as well as just-in-time supply practices, whereby suppliers deliver parts to the assembly line exactly when they are required (Linge, 1991b). This reduces the need for storage of parts.

(d) Increased reliance on imports. Removal of most of the barriers to importing ready-built new cars and second-hand cars has put the viability of the local assembly industry in doubt. There was a five-fold increase in the number of cars imported in just three years at the end of the 1980s (Britton *et al.*, 1992b: 140). This influx (now near to 100 000 vehicles per annum) was encouraged by the retention of an export facilitation scheme by Australia under which assemblers received an export subsidy per vehicle. Despite the agreement in CER to remove such subsidies, those for export vehicles were not phased out until the mid-1990s. Nevertheless component manufacturing for the Asian market has greatly increased, with 40 firms employing around 4300 people in 1997, netting NZ$400 million per annum of which NZ$180 million are exports.

Towards economic reform in the 1980s

Despite the operation of protective tariffs, post-1945 there have been long periods with few barriers to foreign investment in Australia. This enabled a large influx of overseas funds into the mining industry in the 1960s and early 1970s. For example, there was A$1.5 billion capital inflow in 1971. This large investment had destabilising effects and raised concerns over the extent of foreign ownership. The Whitlam Labor government of the early 1970s attempted to institute controls on this capital (in a policy known as 'Connorism'), but the financial mismanagement that led to this government's downfall effectively ended these attempts. Subsequently, foreign investment has also been high in the land market, banking and other services, notably tourism where Japanese money has been prominent.

In summarising this growth and impact of foreign investment and control of important sectors of the economy, Crough and Wheelwright (1982) referred to Australia as a 'client state'. They estimated that, by the late 1970s, foreign control extended to over 60 per cent of the minerals industry and over 30 per cent in manufacturing, non-bank finance and general insurance. However, for some particular industries effective control was higher: for vehicles (nearly 100 per cent); oil refining (90 per cent); lignite mining and petroleum extraction (84 per cent); basic chemicals (78 per cent); silver, lead and zinc (75 per cent); coal (59 per cent); and iron ore (47 per cent). In Queensland 80 per cent of the minerals sector was in foreign hands.

Subsequently, legislation such as the Foreign Takeovers Act and the establishment of the Foreign Investment Review Board had little impact upon the growth of overseas investment, especially as deregulation of key sectors fostered more foreign involvement in banking, property, financial services and tourism, concentrating their focus upon Sydney, the other state capitals and holiday resorts. Between 1982 and 1987 the level of foreign investment trebled, though foreign investment in mining had peaked in the early 1970s. This has further prompted domestic concerns that, despite Australia's greater involvement in the circuit of capital within the Asia–Pacific region, this has been at the expense of Australian control over its own economy. Nevertheless, such concerns have not stopped substantial economic reforms.

In the mid-1980s significant moves were made by the federal government to release the protective measures applying to a wide range of Australian industry and to foreign investment. By this time decades of protectionism had generated particular characteristics in the economy. Industry was dominated by small firms and a few large enterprises which had a disproportionate share of overall production: in 1975, 200 businesses out of 30 000 produced half the output, and with high levels of foreign ownership: of these 200 businesses, 87 were overseas-controlled (Catley, 1996: 58). Nearly all domestic industry operated at what would be considered well below the minimum efficient scale in the US, despite being substantially more concentrated. Only nine Australian industrial corporations were in *Fortune* magazine's 1997 list of the top 500 global industrial firms, BHP coming first, in position 125.

Manufacturing industry produced almost entirely for the domestic market, with limited reinvestment, but it also supplied the primary sector which therefore suffered from utilisation of poorer, high-cost inputs than would have been the case if Australian manufacturing had been competitive internationally. A similar situation applied in New Zealand, though close government regulation of the economy reached new heights under the Muldoon government between 1975 and 1984.

In contrast with the individualism of the United States, state enterprise has tended to be much more important in Australasia, and especially New Zealand, where the state initially acted as a landlord and also provided infrastructural needs. State enterprise in New Zealand was extended during the twentieth century to embrace post office savings, insurance, coal mining, organisation of statutory primary producer boards, a significant role in forestry, and nationalisation of the Bank of New Zealand and domestic and international

airlines. Mixed state and private enterprises were developed in the pulp and paper, steel, oil and development finance sectors. As a result, by 1984 state sector employment dominated the communications and electricity industries and accounted for over half the employment in forestry and transport. The Post Office was the largest single employer in the country with 45 000 employees. The value of goods and services produced by central and local government was around one-quarter of GDP (Rudd, 1990).

Despite this substantial role of the state within economic activity, the returns on the state's investments were often very poor and this was one of the factors that helped encourage the Labour government in 1984 to introduce sweeping reforms. Another was a general acceptance by key individuals in the Treasury that the ideas of the new right in Britain and North America regarding the need for reduced state involvement in the economy were well founded. These ideas were also pursued in Australia under the Hawke Labor government, elected in 1983. Therefore both countries have experienced substantial economic reforms dating from approximately the same starting points in the mid-1980s.

Globalisation

Since the late 1970s, economic restructuring has been observed at all geographical scales from global to local. It has involved rapid movements of money and other resources into and out of industries and sectors. These have been accompanied by dramatic geographical shifts of capital investment by firms and governments, both globally and inside countries (Fagan and Webber, 1994: 6).

During the post-1945 boom, American and Japanese investment in Australia increasingly replaced that from Britain, with more of the world's major corporations opening branch plants in Australia. Initially, as described in

TABLE 13.5 Central characteristics of the global economy

(a) Economic transactions transcend nation-states through global circuits of production, trade and finance capital.
(b) Globalisation is underpinned by information technology which can compress space and time.
(c) Older industrialised countries experience automation and restructuring of production in situ, impacting on industrial production (e.g. the Australian steel industry), whilst economic growth has been closely associated with high-productivity, knowledge-intensive production in the West and high-productivity, low-cost production in newly industrialising countries (NICs).
(d) Rapidly changing global flows of money permitted by the global financial system and widespread abandonment of financial regulation by the state have had dramatic impacts on production.
(e) The composition of the workforce has changed, as has the nature of work, related to deregulation of labour markets, increased female participation and international labour migration.

Based on Fagan (1997a: 199).

Chapter 8, investment by TNCs concentrated on the minerals and energy sector. However, the nature of this investment has changed to encompass globalisation and functional integration across national boundaries between banking, investment, production and trade. Globalisation processes include the liberalisation of financial markets, increased international commodity flows, transnational production systems, and enhanced transport and communications technology (Gertler, 1997) (Table 13.5). However, this has often been accompanied by a countervailing localisation process involving intensive linkages between small and medium-sized enterprises (SMEs) which encourages the geographical clustering of productive activity. The combination of processes has been termed global localisation

or glocalisation, and has been associated with new business strategies. Glocalisation has increased regional disparities, especially those between metropolitan and non-metropolitan areas, though differential specialisation between cities has led to variable economic fortunes. The automobile industry, discussed above, is a good example of glocalisation in which a particular market is supplied by local assembly and component plants, many of which will be tied to an overseas-based TNC whose local operations will reflect decision-making taken at the company's headquarters overseas. For example, in Australia the top 50 vehicle component supplier firms are foreign-owned TNCs that control 86 per cent of the component market in the country. The small Australian-owned suppliers are fast disappearing (Marceau, 1997: 46).

TNCs with a globalising strategy centralise strategic assets, resources, responsibilities and decisions, but operations, which are based in several countries, are aimed at tapping a unified global market (Bartlett and Ghoshal, 1989). The home base or country of origin of a TNC may remain important to the nature of its operations and affects the character of globalisation (Bonnano, 1993). Fagan (1997a) distinguishes between TNCs operating globally, by obtaining finance capital from multiple sources and often investing domestic surpluses internationally or repatriating profits, and those called 'multidomestics' who sell mainly to the national markets in which they are located. The latter are typical of foreign-owned branch plants of the TNCs in the Australasian food and clothing industries.

Australian governments have used the globalisation concept as a crucial ideological construct in policy formulation, emphasising the need to dismantle protectionism through deregulatory and economic liberalisation measures. Globalisation has been assisted by new institutional mechanisms of control and co-operation in the global movement of capital, as illustrated by moves towards freer trade during the Uruguay Round of the GATT negotiations in the early 1990s, in which both

Australia and New Zealand strongly supported worldwide removals of tariffs, subsidies and other forms of protectionism (see Chapter 17). This mirrored their own policies, pursued from the early 1980s, in which measures based on 'new right' economic liberalism were substituted for previously dominant protectionist policies.

AUSTRALIA: DEREGULATION, COMPETITION AND THE ASIA–PACIFIC REGION

The economic reforms, begun by the Hawke Labor government from 1983, have been based on three main premises:

(a) the need to reduce the balance of payments deficit by expanding the export sector through a focus on industries with a demonstrable competitive advantage (Parker, 1990), such as resource-based manufacturing, tourism and skill-intensive services;
(b) the need for closer integration with the dynamic production, trade and financial systems of the Pacific Rim;
(c) the need to embrace the 'growth culture' of the 'Asian way'.

In short, neo-liberalism, closer integration with Japan and the east Asian newly industrialising countries (NICs), and globalisation have been at the heart of political discussion and policy-making in Australia since the early 1980s. Reforms have included deregulation of the financial sector, flotation of the Australian dollar, creation of an Industries Commission (from 1989) to develop economic reforms in specific industries, and a gradual reduction of protectionism for specific industries (Linge, 1991a). The latter meant tackling the 'complex but wide-spread system of interfaces between government and business, involving taxpayer subsidies by discounted inputs of land or energy, tariff protection from import competi-

tion, price support schemes, licensing arrangements, a heavily subsidised transport system, and wage regulation' (Catley, 1996: 87). These policies have 'internationalised' or globalised the Australian economy, but have not cut the chronic deficits in the current account and have had only limited success in raising manufacturing output, the most notable successes being in processing related to primary production, e.g. food processing and energy-intensive processed metals (Fagan, 1991).

Deregulation and restructuring

The principal characteristics of deregulation and the associated restructuring processes are summarised in Table 13.6. Deregulation of the financial marketplace was one of the first reforms pursued by Treasurer Paul Keating in the mid-1980s. Tariff protection cutbacks were introduced more gradually towards the end of the decade and then sharply from 1991, accompanied by the privatisation of public sector assets and public sector downsizing (Henderson, 1997: 119). State governments also played a part in the privatisation process, as illustrated by the privatisation of the electricity industry in Victoria which brought the state government much needed revenue following the recession of the early 1990s. By mid-1996 it had realised A$13.6 billion through sales to multinational consortia including a number of UK companies (e.g. PowerGen, Scottish Power) and the US utilities, Pacificorp and Destec Energy. A further A$8 billion should be realised by sales of five more power stations by the end of the millennium, following receipts of A$2.35 billion in 1996 for the oldest and least efficient state-owned coal-fired station at Hazlewood, 110 km east of Melbourne.

More recent reforms include the 1993 Industrial Relations Act aimed at deregulating the labour market, though this did not go as far as similar legislation in New Zealand (Harbridge, 1993). It retained a strong role for trade unions in enterprise and workplace bargaining whilst facilitating the introduction

TABLE 13.6 The characteristics of deregulation and restructuring in Australia since the early 1980s

Deregulation

(a) Removal of entry barriers to allow other local or overseas companies to establish operations in hitherto restricted markets.

(b) Permitting foreigners to become members of national exchanges (e.g. for stocks, bonds, futures and commodities).

(c) Removal of restrictions on activities previously heavily deregulated or disallowed by law.

(d) Forcing local producers from all sectors to compete internationally because reduced border protection has stimulated demand for services.

Restructuring

(a) Financial capital moved rapidly into some sectors of the economy, producing growth of large conglomerates. Service industries were fastest growing whilst manufacturing stagnated.

(b) There were record levels of foreign capital inflow after 1980, but Australian investment overseas also increased.

(c) Manufacturers placed emphasis upon lowering production costs and introducing new working practices in order to meet fiercer competition in the domestic market (e.g. Linge, 1991b).

(d) Pressure grew on the federal government to extend deregulation, stimulated by the chronic current account deficit on the balance of payments which increased dramatically during the 1980s, especially through a deficit on services (including payments for freight and insurance, port services, overseas business services and travel services) and income loss on dividends to foreign operators of businesses in Australia. By the early 1990s Australia's net foreign debt was over A$150 billion, one of the highest per capita in the world.

Based on Britton *et al.* (1992: 145) and Fagan and Webber (1994: 49).

of enterprise bargaining in non-union workplaces – a means of workers gaining wage increases. The Act was introduced against a background of high unemployment (with nearly one million or 11.2 per cent of the workforce unemployed at the end of 1992) but also sustained economic growth that characterised the first half of the 1990s.

The reforming measures have helped open the Australian economy to the world market (Box 13.1). One dramatic effect of this has been an increase in the proportion of GDP that is traded internationally. Exports as a percentage of GDP rose from 13 per cent in December 1983 to 20.5 per cent ten years later and, during the same period, the proportion for imports rose from 14 to 19 per cent. The rise in exports reflects a response to Australian producers being exposed to markets growing faster than the Australian domestic market. For nearly three decades from the late 1960s this overseas market was most spectacularly represented by Japan and the rest of south-east Asia. Hence it was their demand for 'elaborately transformed' manufactures (ETMs), ranging from auto components to computer software that presented some of the greatest opportunities for Australian-based manufacturers.

In some sectors, for example food processing (Fagan, 1997b), policy reforms may have disadvantaged Australian production relative to producers elsewhere controlling the major markets. However, there is some dispute as to the extent to which the policies were detrimental to particular sectors of the economy. Indeed, there are three major contextual changes which may have reduced negative impacts (Marceau, 1997):

(a) economic growth throughout the Asia–Pacific region, which has provided new or growing markets (sometimes referred to as the 'third wave of capitalism');

(b) globalisation processes, generating a broader range of locational possibilities through greater specialisation and the opening-up of new markets where quality rather than price is crucial;

(c) the spread of new technologies, including micro-electronics, telecommunications and

Box 13.1 Deregulating sugarcane production

The growing and processing of sugarcane was one of the most tightly regulated industries in Australia until the introduction of a series of deregulating measures from the late 1980s (Hungerford, 1996). These measures have contributed to the restructuring of the industry, favouring large growers, penetration of the processing sector by foreign capital, and interstate rivalry amongst processors.

Australia produces one-seventh of the sugar traded on the world market, with the crop grown in a relatively narrow strip running discontinuously for over 2000 km in coastal plains and river valleys from Mossman (Qld) to Grafton (NSW) (Figure 13.1). Long-term regulations have controlled the price paid for sugarcane, the quality of cane grown and the sugar produced. Individual growers and processors of cane were unable to make decisions on their level of output and marketing arrangements if these decisions conflicted with the regulations. Controls were first established in the late nineteenth century through the principal processor, the Colonial Sugar Refining Co. Ltd (CSR), with growers required to sign contracts stipulating the types of cane that could be grown, the method of payment and penalties for damaged cane. A system of production developed comprising three distinct horizontally integrated stages: grower, mill, refinery, with varying degrees of control over each stage exerted by groups based within one of the three production stages (see Robinson, 1995a).

Following Federation in 1901 protective import duties were placed on sugar, followed later by prohibition on raw sugar imports. In Queensland all raw sugar produced was acquired by the state government's Sugar Board. All domestic raw sugar was placed in a marketing pool from which it was sold to overseas markets or to refineries for the local market. As agent under contract to the Queensland state government, CSR controlled the sales of all raw sugar exports and refined 95 per cent of Australia's domestic refined sugar needs. A board adjudicated on the division of monies between mills and growers; it set individual acreages for growers and the tonnages which CSR undertook to acquire from the mills; and another board set a uniform wholesale price for refined sugar for consumption in Australia.

Under these controls, and with favourable international quotas negotiated in international sugar agreements, the output, area under cane and value of production increased dramatically post-war. Area increases were especially marked in the North and Herbert-Burdekin regions, but pressures for reform grew because of the limiting constraints imposed on growers which restricted production, the system's inability to respond swiftly to the changing world market, and well-recognised inefficiencies in the milling and refining sectors.

The Hawke government championed deregulation as one method of generating increased production to take advantage of rising world sugar consumption, and in July 1989 the 73-year-old embargo on sugar imports was lifted. Subsequently, a sliding scale of tariffs has restricted imports, but other deregulatory measures have brought about changes to the industry. Greater flexibility has been introduced for growers and millers, with the former allowed to produce beyond previous quota levels. This has especially benefited larger growers, and the area under sugarcane has risen by over one-third since the early 1980s. However, the reforms have not entirely swept away government controls, with a new body, the Queensland Sugar Corporation, empowered to acquire and sell all raw sugar produced in Queensland, partly as a means of extracting premiums from overseas buyers.

In NSW the grower co-operative that runs the state's three mills has built its own refinery (at Harwood) in order to refine and sell its sugar on the domestic market rather than selling to state authorities in Queensland. Elsewhere, mill closures have continued since the early 1970s reflecting a long term decline in world sugar prices and the need to realise scale economies. Strong grower co-operative mills have emerged in NSW and Mackay, the latter accounting for over one-fifth of the country's raw sugar production and utilising the town's bulk handling facilities which make it the largest bulk sugar terminal in the world (Figure 13.2).

Deregulation has weakened CSR's dominant position in sugar refining, through the growth of competition. Although the company has diversified into a wide range of food products, notably into non-alcoholic beverages, its milling operations have been reduced and it now only controls half of the refined sugar market. The growing impact of TNCs in the food-processing sector is apparent in the industry in the form of British companies entering both the milling and refining sectors. For example, Tate and Lyle has taken over Bundaberg Sugar Pty, thereby acquiring a range of products, including the Famous Bundaberg Rum, industrial alcohol, textiles, machinery manufacture for the sugar industry,

mining, and power generation (Figure 13.3). Most of the cane supply to the mills in the Bundaberg district comes from large company-owned plantations, making Tate and Lyle the largest cane-grower in the country as well as processing 20 per cent of Queensland's cane sugar. Another example is the joint venture between E.D. and F. Man and Mackay Sugar to build a new refinery to supply Sydney, New Zealand and the Asian market.

biotechnology. These have created a series of new productive possibilities, including decentralised organisational forms and new management strategies.

Nevertheless, one negative outcome has been the rise in unemployment to levels last recorded in the great depression. Rates have tended to be higher in non-metropolitan areas with narrow economic bases. However, significant pockets of high unemployment have been associated with the 'fallout' from manufacturing, especially in old industrial areas of the inner city in Adelaide, Melbourne and Sydney. Unemployment rates of over 20 per cent have also been recorded in suburban industrial areas, notably Fairfield–Liverpool in Sydney, Melbourne's industrial west and north-west around Broadmeadows, and north-west Adelaide.

Broken Hill Proprietary (BHP) and the Australian steel industry

As summarised in Table 13.7, there have been various types of corporate strategy as a response to changing internal and external circumstances. Globalisation processes and changes in government policy have meant that Australasian companies face both new opportunities but also new competitors. In order to compete, the companies have had to develop new strategies, often with significant consequences at a local level: for example, affecting the character of jobs, changes to labour

processes, job security, hours of work and rates of pay. One example of this with respect to a 'traditional' heavy industry is steel production.

BHP has had a near-monopoly over the Australian steel industry for the last 60 years. Until 1915 BHP concentrated its activities on mining at Broken Hill. Then it began steel production at Newcastle, and subsequently, in 1935, it purchased its chief competitor at Port Kembla, south of Wollongong, and British-owned fabrication plants that had developed in the 1920s around its own plant in Newcastle.

TABLE 13.7 Changing corporate strategies in the 1980s and 1990s

(a) There were fierce struggles for market power, often leading to the construction of national markets to replace regional ones.

(b) There were shifts into particular sectors, e.g. food production, where control over a particular market share could be obtained.

(c) Some Australian companies reconstructed themselves as TNCs, though company indebtedness could restrict this or lead to greater foreign control as assets had to be realised. Foreign firms have used global sourcing of products to sell locally using 'local' brand names, sometimes with local production bases.

(d) Companies in slow-growth sectors (e.g. heavy industry) attempted to diversify, but often also shed labour, changed management and labour practices, and closed old plant.

Based on Fagan and Rich (1990).

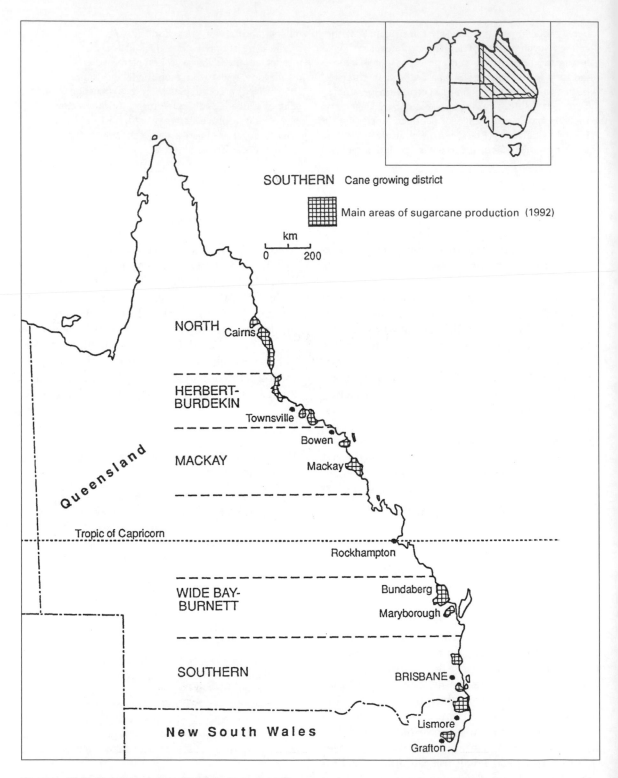

FIGURE 13.1 Cane-growing districts in Australia

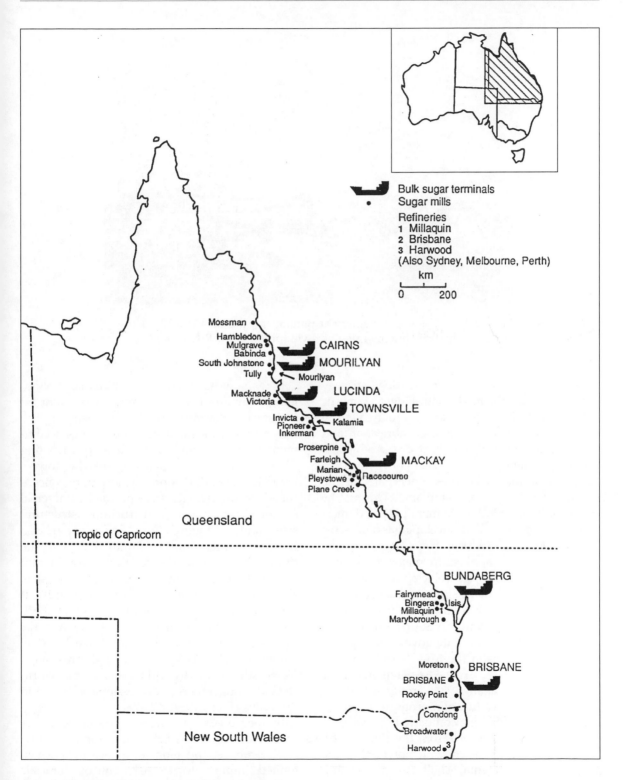

FIGURE 13.2 Processing sugarcane in Australia

FIGURE 13.3 The sugar refinery at Bundaberg, purchased in the late-1980s by the British firm Tate and Lyle (photo G.M. Robinson)

This was part of a consolidation strategy whereby BHP obtained control over domestic iron and steel production through vertical integration: forwards into steel fabrication and backwards into coal and iron ore production. Shipping interests were also developed to supply raw materials from other parts of the country to the processing plants. During the post-war boom, BHP had a monopoly of domestic steel production, and developed as a low-cost fully integrated producer benefiting from the tariff protections applied to major steel-using industries such as motor vehicle manufacture.

The rising demand for steel encouraged expansion, and a third integrated steelworks was opened at Whyalla in South Australia in 1964 (in return for the state government granting iron ore leases in the Middleback Ranges) and a small blast furnace at Kwinana, near Perth, in 1968 (Fagan, 1984; Trengrove, 1975). BHP began a diversification programme in the mid-1960s to enable the company to participate in the export of iron ore from the Pilbara and the exploration of the mineral potential of the Bass Strait continental shelf (in partnership with the American firm Exxon).

Like Newcastle and Port Kembla–Wollongong, Whyalla became heavily dependent on its steelworks and related industries. The development of the steel industry was accompanied by the state housing trust constructing 7000 houses, or 70 per cent of the town's housing stock. BHP's production there concentrated on the heavier steel products, universal sections for major construction, structural steels and rail manufacture. This was part of a horizontal integration of production, with the three main steelworks focusing on different products. Port Kembla manufactured steel slabs for strip mills, including the mill at Westernport near Melbourne, in which BHP took a 50 per cent share. At Newcastle, merchant bars, light structural steels, steel blooms and billets were produced. This made Newcastle especially vulnerable as the world steel industry entered a recessionary period in the 1980s (Rich, 1987: 235–7).

By the mid-1980s BHP was producing 97 per cent of the country's 6.5 million tonnes of raw steel, protected by long term measures which limited imports to 15 per cent of domestic consumption (Donaldson, 1982). In addition to its major steelworks it also had steel fabricating establishments in all the state capitals and in Darwin, Townsville and Geelong. However, protectionism could not completely shield BHP from falling demand worldwide and increased competition from Japan and the Asian NICs. Australia's steel industry suffered in comparison because of the fragmentation of production and lack of sufficient technological innovation. As a result, from 1982 onwards, BHP has been forced to restructure and reduce the scale of its steel-making operations, revealing its previous reliance on a combination of state subsidy and availability of cheap domestic raw materials to offset outdated labour-intensive technology.

Yet the company has diversified successfully into other non-steel products from operations elsewhere in Australia and, increasingly, in south-east Asia and the USA (Tsokhas, 1986). Its oil division grew strongly in the 1970s and by 1982 was generating three-quarters of BHP's profits from just 16 per cent of sales revenue and less than 0.5 per cent of company employment (Fagan and Webber, 1994: 103–4). Therefore BHP has been able to maintain its position as the largest Australian company despite substantial cuts in its domestic steel-producing activities.

The first substantial wave of cuts came in 1982 with BHP's Australian workforce cut by 30 per cent, affecting operations in all four of its production centres, but with a reinvestment programme supported by federal government funding. At this time the steel division accounted for two-fifths of sales and employed 58 per cent of BHP's workforce of 72 000. The Port Kembla plant (Figure 13.4) employed nearly 21 000 people (just under 10 per cent of Wollongong's total population and 28 per cent of the workforce), and BHP claimed it supported indirectly 72 per cent of jobs in the town (Haughton, 1990: 77). In the steelworks 60 per cent of the non-salaried employees were non-English-speaking migrants, especially Yugoslavs, Italians and Turks, willing to accept restrictive workplace conditions and relatively low wages by Australian industrial standards, but with a high level of apprentice training provided (Schultz, 1985).

The newly elected Labor federal government in 1983 introduced a five-year steel plan to assist the struggling industry. This incorporated a system of subsidies to purchasers of BHP steel, which included BHP-owned steel fabrication enterprises, and which meant paying substantial financial aid to BHP, worth A$45 million (Haughton, 1990: 83). There were further guarantees that BHP should have at least four-fifths of the domestic market. In return BHP agreed to avoid further compulsory retrenchments and to invest A$500 million during the operation of the plan. The trades unions agreed to abide by an ACTU–federal government accord so that damaging strikes would be avoided.

The federal government also allocated A$100 million to the regions centred on Wollongong, Newcastle and Whyalla, of which the former received approximately half. This investment was primarily aimed at improving economic and social infrastructure, and funding start-up costs for new industries in the steel towns. The aid to BHP enabled the firm to embark on much-needed technological updating, but this has not saved the Australian steel industry from further decline, nor did the steel plan prevent subsequent wholesale job losses and plant closures, so that BHP's operations at Newcastle and Port Kembla are now a shadow of their former size.

The labour force of BHP's Newcastle steel-making operation was reduced from 11 114 to 3726 between 1981 and 1992. During the same period, mining employment in the area fell from 11 000 to 3500. The closure of the state dockyard in 1987 took a further 400 jobs. The collapses of Carrington Slipways and Hexham Engineering (1280 jobs directly and indirectly) were further manifestations of the demise of heavy industry that had dominated the town's economic make-up (Metcalfe and Bern, 1994). However, numerous other closures could be added to this list as the town came to symbolise industrial failure rather than the strong bedrock of the economy that it had portrayed in the first half of the twentieth century. New opportunities within heavy industry, e.g. producing oil rigs, and

FIGURE 13.4 BHP's Port Kembla steelworks (photo G.M. Robinson)

some diversification into information technology and education services, have not replaced the former employment in the steel industry, leaving the town with consistently high unemployment rates (Dunn *et al.*, 1995). Overall employment in the Australian-based steel production sector has been cut severely: by 100 per cent in Perth, by 75 per cent at Newcastle, 55 per cent at Port Kembla and 40 per cent at Whyalla between 1982 and 1992 (Fagan and Webber, 1994: 107). This had some effect in reducing financial losses in this sector, though returns from the three remaining plants have rested heavily upon their ability to secure export sales, and there has been further rationalisation involving job cuts. Nevertheless, recent upturns in the world market, coupled with BHP's greater efficiencies have led to increased Australian crude steel production from 6.2 million tonnes in 1991 to 8.1 million tonnes in 1997.

BHP has continued its move towards greater focus on export-orientated mineral production and diversification overseas, notably through the takeover of the American firm Utah International in 1984 at a cost of A\$2.4 billion. This added Utah's interests in large coking coal export mines in Queensland so that BHP controlled half of the highly successful coal export trade with Japan. Other Utah interests were purchased, including iron ore mines in Brazil and copper mines in Chile and Papua New Guinea (Fagan, 1996). In 1985 A\$1.7 billion oil and gas interests were purchased in the US, thereby giving BHP more access to the lucrative US energy market and conferring tax concessions in the US for oil exploration costs incurred outside America. In the 1990s BHP has continued to diversify both its interests and its markets, placing more emphasis on the Asia–Pacific region and especially the NICs. This has included coal mining in Indonesia and oil production in the Timor Sea and Papua New Guinea. There has been oil exploration in China and Vietnam. BHP remains Australia's largest single exporter, but one-quarter of sales now come from production outside Australia, one-quarter of its workers are employed overseas, and one-quarter of its shares are in foreign hands (Wiseman, 1998: 53). Steel-making operations in Newcastle have ceased.

ROGERNOMICS AND ECONOMIC REFORMS IN NEW ZEALAND

Until the 1940s New Zealand's role as 'Britain's farm in the South Pacific' meant that exports of primary produce to the UK were complemented by imports of manufactured goods from the UK. The secondary manufacturing that was developed was heavily reliant on imported supplies and hence tended to focus on the four main urban centres at which these imports arrived, and especially around Auckland. However, industry was generally small-scale and, with the exception of food processing, focused on the local market.

The Second World War provided a stimulus to the manufacture of metal goods, from which a domestic steel industry was launched in the 1950s. Post-war there was an expansion in industries that could produce goods more cheaply than overseas-made supplies, notably building materials, fabrics, carpets, small-scale equipment, and general metal-working and engineering. These were part of a growing production of consumer goods which could be produced locally despite the lack of economies of scale consequent on the small size of the domestic market. Given the higher costs of domestic production, strong import controls were applied to enable home-based producers to survive. Even so, a high proportion of New Zealand-made consumer goods relied on foreign investment, with capital from North America and Australia gradually replacing the former British dominance. Manufacturing production was dominated by small-scale operations, with more than half of those employed in manufacturing in the mid-1970s being in establishments employing fewer than 10 people. Ten per cent of all manufacturing production came from just 14 plants. The following reference to the New Zealand white-goods industry (fridges, cookers, etc.) illustrates the general problem faced by manufacturers: 'the small scale of the New Zealand market ... forced New Zealand manufacturers to adopt a range of production techniques that seriously compromised the ideals of mass production' (Le Heron and Pawson, 1996: 173). The demands of the internal market were far smaller than production runs of a single model in other countries.

Regulation of the economy was increased in the 1970s in attempts to combat the destabilising effects of the so-called 'oil crisis'. This and the continuing high expenditure on social welfare contributed to New Zealand developing a higher foreign debt per capita than all but a handful of developing countries. However, when they were introduced, the sweeping economic reforms made by the fourth Labour government impacted most swiftly upon the farming sector which, at a stroke, lost its price supports (see Chapter 6). The reforms affected all parts of the economy, fundamentally altering the context within which economic activity operates. In a very short period the country went from being one of the most heavily regulated economies in the developed world to one relatively free of government controls. This transformation has had some equally far-reaching consequences upon the geography of economic activity.

Rogernomics

The architect of the reforms was the Treasurer in the fourth Labour government, Roger Douglas, whose first name has given rise to the term 'Rogernomics' to describe the wide-ranging set of measures enacted during his tenure of office. As shown in Table 13.8, the sobriquet subsumed a remarkably broad sweep of policies. The particular nature of the relatively small New Zealand economy was one characteristic encouraging this breadth, but the desire of Douglas and like-minded colleagues in the Treasury to eliminate the policies of the previous Muldoon government was another. Perhaps such control of the economic levers of power by a small Treasury team may not have been feasible in other countries, but the fact that these 'new right' policies were being pursued by a Labour

TABLE 13.8 Rogernomics: the range of liberalisation measures

- deregulation of entry licensing into industry
- floating the exchange rate
- removal of price controls
- removal of import licensing
- revision of town and country planning
- abolition of quangoes and quasi-government organisations
- elimination of many special exemptions and lowering levels
- replacement of erratic indirect structure with a comprehensive goods and services (GST) tax at 15 per cent
- review of competition regulation
- reform of local government
- resource management law reform
- removal of concessions for favoured investment
- establishment of Closer Economic Relations (CER) with Australia
- reorganisation of core government departments
- corporatisation of some local authority trading activities
- partial deregulation of occupational licensing
- revision of corporate, personal and direct taxation
- removal of other operating barriers to industry
- significant decrease in import tariffs
- removal of financial controls
- liberalisation of foreign exchange controls
- changes to education and health provision
- more user pays
- partial deregulation of energy sector
- deregulation of ports and waterfront work
- removal of concessions for favoured sectors
- removal of shop trading hours restrictions
- removal of monopoly rights in state trading
- deregulation of transport and financial services sectors

Source: Savage and Bollard (1990).

Box 13.2 Corporatisation and privatisation

One of Douglas's key aims was to tackle government involvement in the production of goods and services. His contention was that the role of government departments was not conducive to the efficient production of goods and services for the market, and therefore the institutional framework needed to be altered. The alteration took the form of separating commercial and non-commercial objectives and functions via the establishment of state-owned enterprises (SOEs) which were expected to operate commercially. This was termed corporatisation, with nine new SOEs established under the State-Owned Enterprises Act of 1986 (see Table 13.9). Similar ideas with respect to accountability were applied to the existing state corporations: Air NZ, Petrocorp, the Shipping Corporation, the Tourist Hotel Corporation and the Railways Corporation. This process of obtaining 'value for money' for government investment and 'cutting costs' to the taxpayer was also seen in the commercialisation of activities in government departments, notably by charging for services previously offered gratis and by raising charges to achieve 'full cost recovery', e.g. charging for recreational facilities on conservation land. However, the existence of some of these SOEs was short-lived, and between 1988 and 1991 SOEs were sold to the private sector in a wave of privatisation that affected other state assets such as the Rural Bank, Government Print and State Insurance. As shown in Table 13.10, the purchasers were a mixture of large domestic and overseas businesses. Of the sales, one of the most acrimonious was that of the Forestry Corporation, which is discussed in Chapter 14.

government was repeated across the Tasman under Hawke and Keating.

Douglas's views were set out in his book *There's got to be a better way* (Douglas, 1980). However, implementation of the key ideas relied on the widespread perception that the economy was inflexible and in terminal decline if reforms were not swiftly implemented. At the heart of Rogernomics was a desire for both market and trade liberalisation, supported by many politicians both in the government and opposition. However, it was the speed and extent of the associated policies that were dramatic and far-reaching. In a short period of

TABLE 13.9 The creation of state-owned enterprises (SOEs), 1986/7

Former government department	Newly created departments and SOEs (*)
Department of Lands and Survey	*Land Corporation Department of Survey and Land Information Department of Lands (dissolved in 1990) Department of Conservation
NZ Forest Service	Department of Conservation *Forestry Corporation{p} Ministry of Forestry
Commission for the Environment	Ministry for the Environment Parliamentary Commissioner for the Environment
Ministry of Energy	*Electricity Corporation *Coal Corporation Ministry of Energy (absorbed by Ministry of Commerce in 1989)
Ministry of Transport	*Airways Corporation Ministry of Transport
Post Office	*Telecom Corporation{p} *Postbank{p} *NZ Post
State Services Commission	*Government Property Services State Services Commission

{p} assets subsequently privatised.
Based on Pawson in Britton *et al.* (1992: 165–9).

time in the mid-1980s, deregulation of finance markets, removal of industry protections (starting with agriculture), trade liberalisation, deregulation of foreign investment, liberalisation of resource management, deregulation of the media, a corporatisation programme (including the welfare state), privatisation (Box 13.2), public service restructuring, the pursuit of a monetarist agenda, labour market deregulation and public sector finance cuts transformed the country. Subsequently, in the 1990s under successive National governments reforms have been carried into new areas, notably labour reforms and reductions in state support for the welfare state (Easton, 1994; Harbridge and Hince, 1994; Stubbs and Barnett, 1991).

Impacts of the economic reforms

Despite the dramatically changed context for economic development in New Zealand, the impacts upon overall economic growth were disappointing initially. Between 1985 and 1992 the average growth across OECD countries was 20 per cent; for New Zealand there was a 1 per cent decline. Only after 1992 was there a strong economic boom when New Zealand had the highest rates of growth of any developed country (Dalziel, 1994). This was eroded by the economic recession in south-east Asia and Japan from 1997, illustrating how much more trade and investment had been developed between New Zealand and this region. The other long term outcomes of the package

TABLE 13.10 The sale of state assets and trading activities, 1988–96

Asset	Cost ($NZmillion)	Principal buyers (NZ unless stated otherwise)
Telecom	4250	Ameritech (US), Atlantic Bell (US), Fay Richwhite, Freightways
State Plantation Forest	2300	Carter Holt Harvey, Fletcher Challenge, NZ Forest Corporation, Timberlands, Juken Nissho (Jap)
Housing Corporation mortgages	2176	Fay Richwhite and Co.
Forestry Corporation NZ	1600	Fletcher Challenge, Brierly Investments, Citifor
Bank of New Zealand	850	National Australia Bank (Aus)
Petrocorp	801	Fletcher Challenge
State Insurance	735	Norwich Union (UK)
Rural Bank	688	Fletcher Challenge
Postbank	678	ANZ (Aus/UK)
Air NZ	660	BIL, Qantas (Aus), American Airlines (US), JAL (Jap)
Fletcher Challenge shares	418	
NZ Timberlands	366	ITT Rayonier (US)
NZ Rail	328	
NZ Steel	327	Equiticorp, Helenus (Aus)
Maui Gas	254	
Synfuels	206	
NZ Liquid Fuel investment	203	
Taranaki Petroleum mining licences	121	
Development Finance Corporation	111	National Provident Fund (UK), Salomon Bros (US)
Works and Development Services	108	
Radio Company	89	
Motonui Synfuel	82	Fletcher Challenge
Landcorp Financial Investments	77	Mortgagees
Tourist Hotel Corporation	72	Southern Pacific Hotels (US)
GCS	47	
Government Print	39	Rank Group
Shipping Corporation	32	Blueport-ACT (UK)
Maori Development Corporation	21	
Export Guarantee	20	
Health Computing Service	4.25	Paxus (Aus)
Wrightson Rights	3.4	
Government Supply Brokerage	3.2	
National Film Unit	2.5	Television NZ
Waikato Regional Airport	2.1	
Communicate NZ Ltd	0.25	DoC Group

NB sale price includes any subsequent purchase price adjustments.
Based on Mascarenhas (1991: 47) and SNZ (1997).

of economic and social reforms introduced since 1984 have been a sharp drop in inflation (from 18 per cent p.a. in 1987 to 2 per cent p.a. in 1994), a reduction in net public debt (from 51 per cent of GDP in 1992 to 30 per cent in 1996/7), a reduced current account deficit, budget deficits (despite asset sales and new taxation measures and until surpluses were attained under the National Party government in the mid-1990s), reduced investment, a widening gap between rich and poor, and higher unemployment rates (peaking in 1993 at an official rate of 11 per cent).

The reforms have also had differential effects upon the individual sectors of the economy. The service sector has been the main beneficiary, with deregulation creating lucrative new opportunities in public relations, economic and management consultancy, legal work and investment advice. By the mid-1990s services comprised one-quarter of exports, with a substantial expansion in financial services, and mounting profits in telecommunications, transport and energy. The move towards a greater concentration of employment in the service sector was accompanied by 'fallout' in older manufacturing industries. In manufacturing there was a reduction of 25 per cent in the workforce by 1991 and a halt in the export growth that had characterised the 1970s. Analysis of changes occurring between 1981 and 1986 revealed the biggest increases in employment were associated with mining and quarrying (notably via expansion of coal-fired power stations and steel-milling capacity), restaurants and hotels, financial services, research and scientific services, and other community services. Declines were recorded in a range of manufacturing industries, notably transport equipment, clothing and chemicals. This net outflow from manufacturing towards services 'reinforced a growing differential between the skilled and unskilled labour forces' (Kelsey, 1997: 254).

The monetary and fiscal policies, followed by labour market reforms and lower interest rates, have helped increase exports of industrial machinery and electrical and electronic equipment. Excluding the food and fibre sector, manufactured exports now exceed NZ$7 billion per annum or one-fifth of manufacturing output and 38 per cent of all exports. This was part of a strong spurt of growth in the 1990s, especially apparent in the main industrial areas of Auckland where new ventures located. This has helped contribute to Auckland's increasing primacy despite a slight fall in the labour participation rate in the early

1990s. The Auckland district has over one-third of national manufacturing employment and 40 per cent of business services. However, analysis of aggregate employment information reveals a trend in favour of consumption-based activity virtually everywhere (referring to the retail and wholesale trade, hotels and restaurants, and personal and household services) (Le Heron and Pawson, 1996; 283–7). Also it is not Auckland that has the highest concentrations of activity in this sector but rather Northland, Otago, Wellington and the Bay of Plenty – emphasising retirement, tourism and contraction in other sectors. The latter has been most evident in Wellington where the impact of the 1984 reforms was to bring job contractions in most economic sectors. Growth has been most apparent for firms employing over 100 people as it is this size of enterprise that has been most closely associated with production for export.

One of the direct consequences of deregulating the economy has been the growth in foreign investment. Strategic purchases have been made in existing industries, with considerable investments in a range of value-added processes, especially where sales of former state-owned assets have occurred. For example, in 1994 NZ$75 million foreign investment in timber processing plants occurred. One of the principal targets for investment has been food processing, as illustrated by the purchase in 1992 of the largest food processor, Watties, from its Australian-based owners, Goodman Fielder Wattie, by H.J. Heinz Ltd, an American-based TNC. By consistently positioning itself to take advantage of growing markets, Heinz has extended its operations outside the United States so that half of its profits are now derived from ventures based overseas. The attractions of Watties were its range of products, cheaper labour costs (up to one-third lower than in Australia) and the ability to use its New Zealand-based production to supply the Asian market. Furthermore, Watties' canning, freezing and food processing facilities all had the capacity to increase production. Within two years of the purchase, Heinz had more than doubled its sales in south-east Asia and Australasia. Within New Zealand it has expanded production across a wide range of convenience foods, including soup, pasta, beans, sauces, 'simmer' meals, fruit, vegetables and baby food. Its network of contract growers has been expanded and new methods introduced, notably organic crops in its three main growing regions, Hawke's Bay, Manawatu and Canterbury. Further examples of increased foreign investment in both Australia and New Zealand are considered in the following Chapter.

FURTHER READING

There are a number of texts that deal with the evolution of economic protectionism in Australia. Good examples are Bell (1993, Chs. 2–4), Capling and Galligan (1992), Catley (1996, Ch. 3), and Maddock and McLean (1987). Protection of the Australian motor vehicle industry is analysed by Bamber and Lansbury (1997) and Schedvin (1976). This industry is one of several analysed from a geographical perspective by Fagan and Webber (1994), who also give a useful guide to the nature of globalisation; see also essays in Burch et al. (1999). For the New Zealand car industry see Britton in Britton et al. (1992: 135–40). The notion of Australia as a client state is expounded in Crough and Wheelwright (1982). A more recent overview is provided by Marceau (1997). There are several good articles on key geographical changes in certain Australian industries by Fagan (1991; 1995; 1997a; 1997b; Fagan and Le Heron, 1994). There are numerous detailed studies of the impacts of Rogernomics, including Castles et al. (1996), Easton (1989) James (1992), Kelsey (1997), Massey (1995), Roper and Rudd (1993), and Savage and Bollard (1990). Subsequent labour deregulation is dealt with by Harbridge (1993). Much of Britton et al. (1992) and Le Heron and Pawson (1996) deals with the geographical impacts of Rogernomics, globalisation and associated economic and social restructuring, using detailed case studies. The restructuring of the Australian sugarcane industry is analysed by Hungerford (1996) and Robinson (1995a). Work on BHP and the Australian steel industry includes Haughton (1990), Fagan (1986), Metcalfe and Bern (1994), and Rich (1987).

14

CHANGING CONTEXTS FOR SERVICES AND THE PRIMARY SECTOR

This Chapter complements the previous one by concentrating on different parts of the economy, primarily the rapidly expanding service sector. As indicated in Table 13.1, this encompasses a range of different activities in which employment has grown dramatically in recent decades, thereby contrasting with the decline in manufacturing jobs. Again, the impacts of deregulation and globalisation are analysed, with key sectors examined. Special attention is given to the growth of tourism, in which both Australia and New Zealand have developed as international destinations, generating substantial revenue and attracting investment to new locations. The second half of the Chapter returns to consideration of the primary sector. In particular, new opportunities for primary production and processing in Australia are discussed, emphasising the impacts of the expanding Asia–Pacific market and new relationships between land-based production and the processing of primary produce. The focus on the Australian agro-commodity sector complements that on its New Zealand counterpart in Chapter 7, whilst the forest products industries of both countries are also examined here.

DEREGULATING AND GLOBALISING SERVICES

One of the most significant changes to the Australasian economies post-1945 has been the expansion of the service sector. In Australia, employment in services, retailing and whole-saling reached 50 per cent for the first time in the late-1950s; by 1980 it was 66 per cent; and by 1996 over 70 per cent, contributing more than two-thirds of GDP (excluding ownership of dwellings). In New Zealand there was a similar development, with employment in the service sector rising from 41.9 per cent in 1956 to 65.9 per cent in 1996. 'Services' comprise a broad range of activities, including some that have grown faster than the rest of the sector, such as finance and business services, tourism and community services (including education and health). Employment in finance, property and business services increased by over 90 per cent between 1971 and 1991, whilst that in community services and public administration doubled, to account for nearly one-quarter of the total workforce in both countries.

Employment in services has drawn upon female labour as part of a growing participation in the labour market by women, though some of this has been only part-time employment (Stimson, 1997). Indeed, women comprise three-quarters of all part-time workers. In both countries, women account for around 45 per cent of the labour force (sometimes referred to as the 'feminisation' of the workforce), though over 35 per cent of women's work is in jobs involving working for less than 30 hours per week (i.e. part-time).

The most rapidly growing services have often been affected by the process of deregulation since the early 1980s (see Table 14.1), which has increased their responsiveness to global forces, and also by privatisation measures which have introduced greater commercialisation, e.g. affecting services provided by the principal utilities, power, water and telecommunications (Crabb, 1991; Langdale, 1991). In most industries domestic producers have been exposed to overseas imports and foreign investment. This has placed greater pressure on businesses to succeed as exporters or international investors (Perry, 1991). In turn local businesses have been forced to place more emphasis on services, adding value to products and on service-sector products, e.g. design, research, marketing, packaging, franchising and branding. The impacts of globalising processes can be seen in increased foreign-direct investment (FDI). FDI in producer services in Australia increased from 43 per cent of all inward FDI in 1975 to 56 per cent in 1991. This is not one-way traffic, though, as during the same period Australian-based firms increased their producer services FDI in the Asia–Pacific region from 33 per cent to 57 per cent (Clegg, 1996).

Another impact of globalisation can be seen in the retail sector where retailers have been able to look beyond their country of operation for their produce. Hence, in Australia, key retailers like Coles, Jewel's and Franklins, have sourced their own-label products globally whilst also offering a bigger range of products drawn from around the world. This has exerted

TABLE 14.1 Deregulation of the service sector

(a) Removal of entry barriers to allow local or overseas companies to establish operations in previously restricted markets.

(b) Allowing foreign nationals/firms to become members of national exchanges, involving dealings in stocks, bonds, futures and commodities.

(c) Removal of restrictions on activities previously disallowed by law or heavily regulated.

(d) Reducing tariffs and thereby generating international competition.

Based on Britton et al. (1992: 148).

pressure on domestic suppliers, e.g. food processors, despite their attempts to develop value-added products as a means of maintaining product differentiation and minimising direct competition with low-priced imports (Burch and Pritchard, 1996). The growing power of the major retailers can be seen in numerous sectors, but a good example is the clothing industry where the Coles Myer group has become the twelfth largest retail chain in the world, and the largest private employer in Australia.

The substantial changes in the macro-environment of service provision have contributed to the continuing expansion in office-based employment from the early 1980s (Table 14.2). Some of the changes can be subsumed within the dominant contribution of globalisation processes that have generated demand for a range of sophisticated producer services. These have affected the banking, accounting and legal sectors, both through in-house units internal to large enterprises, and industrial producer service firms occupying particular 'niches' from which they can exploit new market conditions. Through deregulation specialist suppliers of business services have been able to increase their operations substantially, servicing both primary and manufacturing sector producers. Good examples of

TABLE 14.2 Factors influencing the expansion in office-based employment

(a) Corporate takeovers, reflecting economic restructuring especially within manufacturing, but generating jobs in legal services, accounting, real estate, advertising, currency trading and insurance services.

(b) Deregulation of banking and the general growth of financial transactions.

(c) Transformation of share and commodity trading, e.g. new futures markets.

(d) New investment opportunities, especially involving assurance companies and superannuation funds.

(e) Computing and communications technology which has helped large corporations maintain control over geographically dispersed operations.

Based on Walmsley and Sorensen (1993: 127–9).

changes linked directly to deregulation can be seen in the banking and aviation industries.

Banking

Various deregulatory measures were adopted in and were applied to the banking industry throughout the 1980s, enabling banks to develop beyond the core borrowing and lending functions to which they had been restricted, and also removing distinctions between banks and non-bank financial institutions (such as merchant banks and finance companies). Deregulation promoted greater competition, blurred the distinction between local and global markets, stimulated product innovation and encouraged the entry of overseas banks to compete with local ones (Fagan and Webber, 1994: 116–21).

Technological changes have also enabled the introduction of new services such as electronic self-service systems and greater centralised control over a wider range of services, including new functions such as sales of insurance, personal pensions and travel services.

Locational changes stimulated by deregulation have included a 30 per cent increase in the number of retail bank branches in Australia, but a 70 per cent cut in the number of agencies run by banks. Specialist products have been offered in some of the larger branches or in new complexes. Greater concentration of assets has occurred, including expansion of the four main banks: ANZ, National Australia Bank, Westpac and Commonwealth, but some new small banks have also been created. However, head office control functions have remained concentrated in Sydney and Melbourne with hardly any significant banking headquarters elsewhere. Retail banking remains largely in the hands of Australian-owned banks, with less than 100 branches of foreign banks, nearly half of which are in Sydney. Changing technology has provided more opportunities for part-time employment in the banking industry, and this has been associated with a growth in the number of banking jobs occupied by women.

One of the central elements in Rogernomics in New Zealand was the comprehensive deregulation of the financial sector, including removal of exchange controls. The financial institutions in the country expanded their services rapidly, developing new products and niche markets, with concomitant growth of merchant and investment banking, the futures market, property trusts and unit trusts. However, high domestic interest rates promoted movement of funds offshore. Financial sector expansion involved mergers and takeovers as well as streamlining, whereby branches of major banks were closed, staff cut and unprofitable retail services curtailed. Greater concentration and overseas control were other consequences, so that by 1995 of the country's 16 registered banks 14 were entirely or substantially foreign-owned. For example, the Bank of New Zealand was sold to the National Australia Bank in 1992. For a few years after deregulation speculative finance capital rather than productive investment led to a brief share price boom before it was ended by the worldwide financial slump at the end of the 1980s.

Aviation

Despite the removal of the economic regulation of domestic interstate aviation in Australia in 1990, after over 30 years of control, there has been relatively little change in the dual domination of domestic airline services. Ansett and Qantas remain the two groups operating these services, but now within a market without federal government regulation of fares, market entry, capacity and aircraft importation. Deregulation did prompt the emergence of new operators, notably Compass Airlines, but having commenced operations in December 1990, this company went into liquidation within 12 months. Its second launch in August 1992 was followed by collapse within eight months. Another domestic airline, Eastwest, had operated as a separate entity within Ansett, but was formally amalgamated in 1993.

There has been a growth in the numbers of regional carriers, competing on some of the routes previously served by the domestic airlines. Even here, though, the Qantas and Ansett groups are significant, owning the two largest regional airlines, Eastern Australia Airlines (Qantas) and Kendall Airlines (Ansett) (Quinlan, 1998). A more lasting impact of deregulation has been lower average fares, improved quality of service and a sharp rise in passenger numbers: 25 per cent in the first four years of the 1990s as prices fell by the same proportion, related to increased availability and discounted fares.

In June 1992 the Australian government merged and sold as a single entity its international and domestic carriers, respectively Qantas and Australian Airlines. British Airways purchased one-quarter of the newly merged airline and there was a public flotation of the remainder. A memorandum of understanding was signed between Australia and New Zealand in 1992 to implement a single aviation market, phased in over three years. In November 1996 the Trans-Tasman single aviation market (including cabotage rights) came into force, replacing the previous restrictive arrangement in which only Qantas and Air New Zealand were designated carriers and both governments had to agree on fares, frequencies and capacity (Kissling, 1998; O'Connor, 1998). Australia imposes a limit of 25 per cent on ownership by one foreign airline of an Australian airline. In contrast, New Zealand since 1988 has allowed 100 per cent foreign ownership of a domestic airline, as is the case with Ansett New Zealand, now a wholly owned subsidiary of the US-based News Corporation, which commenced operations in 1987. Subsidiary regional carriers, such as Mount Cook Airlines and Eagle Air have been linked through ownership to either Air New Zealand or Ansett.

INTERNATIONAL TOURISM

Australasia as an international tourist destination

One of the most rapidly growing and economically significant aspects of the service sector has been tourism, and it has increasingly been international tourism that has been vital to the two economies as Australasia shows the growing buying power of Pacific Rim tourists. Today tourism often generates more revenue than existing trade links.

Between 1980 and 1995 the average annual rate of increase in international tourists to Australasia was nearly double the rate of growth in world international tourist numbers, and expanding much faster than domestic tourism. For Australia, international tourism has been the leading industry export sector since 1987, and by the mid-1990s it was generating A$14 billion in foreign exchange, equivalent to nearly 13 per cent of total export earnings and accounting for between 6 and 7 per cent of all jobs in the country (BTR, 1995). Between 1985 and 1995 international visitor arrivals to Australia increased by an average of 13 per cent p.a., well above the rate of growth of domestic and outbound tourism, as Australia became a signif-

TABLE 14.3 Percentage share of short-term arrivals to Australia (1975–97) and Zealand (1980–97) by country/region of origin

(a) to Australia

	USA	NZ	UK/Ireland	Other Europe	Japan	Other Asia	Rest of world
1975	14	28	14	11	5	10	18
1980	12	34	15	12	5	10	11
1985	17	21	14	12	9	14	11
1990	11	19	13	12	22	16	8
1997	9	14	11	10	21	22	13

(b) to New Zealand

	USA	Australia	UK/Ireland	Other Europe	Japan	Other Asia	Rest of world
1980	18	43	11	7	9	4	8
1985	19	43	11	5	8	7	7
1990	15	35	8	7	13	6	16
1997	10	29	12	8	11	20	10

Sources: Australian Tourist Commission and New Zealand Tourist Board.

icant international tourism destination. It has benefited from its close proximity to east and south-east Asia, the fastest growing region in terms of both world tourism arrivals and departures (Payne, 1993). In 1996 there were four million overseas visitors, staying on average just over 20 days, an indication of the 'long-haul' nature of many visits. Nearly two-thirds of visits were for recreational purposes, and 99 per cent arrived by air. Tourism represents 12.7 per cent of the country's current account credits.

Just over half of the tourists to Australia come from Asia, with Japan (21 per cent) being the principal supplier followed by New Zealand (14 per cent) and the UK (9 per cent). Initially, Japanese tourists led the Asia–Pacific market, overtaking New Zealand as the single greatest source of visitors in 1990, but other Asian nations have supplied sufficient tourists recently so that the Asian market as a whole now outstrips the traditional suppliers from New Zealand, the UK and North America (Table 14.3). Japanese tourists include specialist market segments, such as honeymooners, but also a broad spectrum from students to family

groups and the elderly (Opperman, 1994). Yet, with an average length of stay of just over one week, the Japanese spend a shorter amount of time in Australia than tourists from other destinations (Griffin and Darcy, 1997; Mules, 1998).

The number of tourists grew rapidly during the late 1970s, and again in both the late 1980s and mid-1990s (Figure 14.1) as the proportion of tourists from Asia exceeded 40 per cent for the first time in 1992. This reflected the growth in both long-haul tourism from Europe and North America and the rise in tourists from Asia (though the recent devaluation of Asian currencies has cut the numbers of Asian visitors to Australasia). Tourism plays a significant role in the economies of the principal 'gateways', and has led to large-scale resort developments, especially in Queensland (Box 14.1), though with some significant inputs from domestic tourism, e.g. the Gold Coast (Mullins, 1992; 1994). During the 1990s new forms of tourism have emerged, such as ecotourism, which Australia has been well placed to exploit (e.g. Orams, 1999), though not without environmental consequences (e.g. Craik, 1987).

Box 14.1 Daikyo Incorporated and tourism in Cairns, northern Queensland

Geographically, most of the direct impacts of tourism in Australia are concentrated on the three principal destinations: Sydney, the Gold Coast and Cairns. In each of these locations most of the tourist-related infrastructure, and especially luxury hotels, has been funded through FDI, with the Japanese being the leading investors in the first 'boom' of the mid- to late 1980s, followed by more direct sources in the 1990s, including other parts of Asia (Daly et al., 1996; Rimmer, 1992; 1994).

In northern Queensland, international tourism has expanded considerably since the opening of Cairns International Airport in 1984. Tourism in Cairns accounted for 14 700 full-time jobs directly and indirectly in 1993, a total likely to rise to 35 000 by 2001 (OCGQ, 1994). With its rapidly expanding tourist base, the economy of the Cairns region nearly doubled between 1986 and 1994 (in terms of gross regional product), with an annual growth rate of 5.5 per cent and with tourism's contribution exceeding 25 per cent (AHURI, 1995). Visitor numbers during this period rose from 410 000 to 1.2 million, with 48 per cent of the latter representing international visitors, a 24 per cent per annum growth in these numbers (Stimson et al., 1998: 165). The state government has invested substantially in infrastructure projects, including A$412 million on transport between 1992 and 1997, but FDI has been a significant factor in the expansion of the tourism sector, dominating the provision of luxury hotels (Table 14.4) and playing a smaller role in the development of duty-free stores.

With respect to tourist-related investment in Cairns, Stimson et al. (1998) have highlighted the role of the Japanese corporation, Daikyo. Daikyo's investments have been focused on Cairns, the Gold Coast and Brisbane, and they also own the Park Royal Hotel in Christchurch. In Cairns the company owns three of the top four hotels and resorts, a four–star hotel, the largest of the Great Barrier Reef tour companies, two island destinations, a scuba-diving company, a golf course, several major development sites, and real-estate management and development services. It owns over one-third of all foreign-owned land in the

Cairns area and has continued to expand its operations despite the economic downturn in Japan, redeveloping the Green Island resort on an offshore cay and developing a mainland integrated tourism resort with golf course, residential sales and recreation facilities. By 1996 its overall investments in Queensland exceeded A$1.6 billion. At this time its enterprises in Cairns employed 978 people, of whom 628 were full-time and 53 per cent female.

In fact, the character of this labour force typifies the emergent tourist sector. In part-time and casual labour, women outnumber men two to one, with a much larger proportion of men than women in the sales and personal service staff. In the high-wage occupational categories of managers and professionals, there is a great dominance of males. Over 40 per cent of the workforce are labourers, relatively poorly paid, often on a casual or part-time basis with little prospect for career development. As these jobs offer little security, they can be hard to fill and are associated with transience and a high proportion of migrant workers. A proposed joint Daikyo and state government programme to convert casual employment into permanent jobs may increase the numbers of full-time employees. This highlights a problem generally overlooked by casual observers of the tourist industry, namely that it is associated with transient workers, underemployment and low-paid jobs, and it does not necessarily help local people, in contrast to its impact on the business community or overseas investors.

Nevertheless, in contrast to many analyses of tourism based on FDI, Stimson et al. (1998) argue that Daikyo's operations in Cairns have had a very positive effect: gross output of A$140 million in 1995; a substantial proportion of wages and salaries remaining in the town (A$43 million); major contributions to local trade, community services and transport; and A$75 million of value-added effects for the local region. Despite the recession of the early 1990s in Australia, and the 1997/8 recession in Japan, Daikyo has been a stable force in the Cairns economy, at the heart of tourist development.

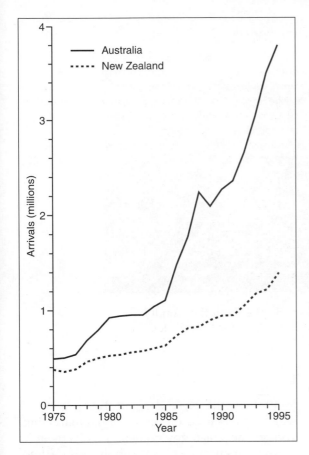

FIGURE 14.1 International visitor arrivals to Australia and New Zealand, 1980–96

New Zealand had nearly 1.6 million overseas visitors in 1997, of whom 55 per cent were 'on holiday' whilst a further 23 per cent were 'staying with friends and relatives'. Tourism now accounts for about 15 per cent of the country's international exchange, comparable with earnings from dairying and superior to forestry and horticulture. Total foreign exchange earnings from international visitors (excluding airfares) in 1997 was NZ$3232 million. International tourist arrivals in New Zealand doubled in the 1980s and rose again by two-thirds between 1990 and 1997. Even more than Australia, New Zealand relies on long-haul tourists and hence the growth of tourism originating in Asia has been highly significant (Table 14.3).

One million international tourists per year were recorded for the first time in 1992, forming part of the doubling of arrivals between 1984 and 1994. As for Australia, there was a surge of growth in international arrivals in the mid-1980s followed by a levelling off and then a further spurt from the early 1990s. Australia still provides the highest single source of international tourists (and the only short-haul visitors), but Japanese visitors have grown steadily since the early 1980s, as have numbers from other Asian countries, particularly Taiwan and South Korea (Waitt, 1996). As in Australia, the Japanese account for a relatively greater share of expenditure and a lesser proportion of total nights. Europeans generate a larger number of bed-nights and are more likely to be independent travellers as opposed to taking a package holiday. Many of the latter are characterised by circuit tourism 'consisting of entry through either Auckland or Christchurch international airports followed by a sightseeing tour, at first by coach and increasingly by air or independently (e.g. rental car or campervan), of the country's major natural and scenic attractions such as the geothermal area of Rotorua, Mount Cook, the Southern Lakes, Milford Sound and perhaps the glaciers of the West Coast' (Pearce and Simmons, 1997, 203–4). Hence there is a concentration of tourist bed-nights and expenditure in Auckland, Canterbury and Otago, with some spread into the West Coast, Nelson and the Coromandel (Opperman, 1994; Page and Piotrowski, 1990; Pearce, 1990).

With the main international airport, Auckland has the greatest share of person-nights by all arrivals and also accounts for nearly one-quarter of all person-nights amongst holiday-makers. The 'circuit' tourism undertaken by the Japanese is reflected in the fact that 60 per cent of Japanese tourist person-nights are spent in Auckland and Christchurch and a further 20 per cent in Queenstown (Figure 14.2) and Rotorua. In contrast, only half of the person-nights spent by Australians are in these four locations. Australian and British tourists are proportionately more important in secondary

FIGURE 14.2 Queenstown and the Remarkables from the Gondola restaurant; the main resort in the Southern Alps for 'circuit' tourists (photo: G.M. Robinson)

regions such as Northland, the Bay of Plenty, West Coast and south Canterbury. Auckland, Christchurch and Wellington account for half of the total expenditure of international tourists but only one-quarter of domestic tourists.

There has not been the same degree of foreign investment in hotels and resorts in New Zealand as in Australia. Nevertheless, all but one of the major hotels in Queenstown now belong to offshore interests, mainly in Asia. Moreover, Hotels New Zealand Ltd, the largest hotel owner-occupier in the country, is a subsidiary of CDL Hotels International, the hotel arm of the largest public-listed property developer in Singapore. Foreign investment has generally been in existing property rather than newly built and has accompanied a pronounced reduction in direct government intervention in tourism (Pearce, 1990). For example, the government's Tourist Hotel Corporation (THC) sold off its hotels, whilst the Tourist Department withdrew from its commercial operations before being replaced

in 1991 by the New Zealand Tourism Board which has focused on international marketing. Today, the chief government involvement is through management of supply-side 'attractions', notably through the Department of Conservation's management of national parks and other protected areas which cover 30 per cent of the country's land surface.

The dramatic changes in the economic climate of the 1980s and 1990s may have favoured the creation of employment and income in the service sector, but there have also been opportunities for innovation and growth in other areas. Some of these have been discussed with respect to Australian mining (Chapter 8), manufacturing industry (Chapter 13) and New Zealand agriculture (Chapter 7). However, this omits certain key aspects of change associated with both Australian agriculture and the broader Australasian primary processing sector. This is now addressed, with an emphasis upon the responses to the new challenges posed by the changing context within which production is operating.

TABLE 14.4 Levels of foreign ownership in the Queensland tourism industry (early 1990s)

Facility	Extent of Foreign ownership (%)			
	Gold Coast	Sunshine Coast	Cairns and the Far North	Total
Five-star	66	66	64	65
Four-star	31	1	29	28
Three-star	12	2	0	7
All three- to five-star	29	12	27	25
Tourist attractions	25	5	0	8
Coach tours	8	4	3	5
Boat tours	4	0	6	4
Duty-free stores	70	0	20	53

Source: Stimson *et al.* (1998: 166).

THE RESTRUCTURING OF AUSTRALIAN AGRICULTURAL PRODUCTION

Globalising the agro-food chain

Food processing in Australia employed 180 000 people in 1995, making it the largest sector within manufacturing industry, with 17 per cent of the national manufacturing workforce. Foodstuffs, both raw and processed, represent one-quarter of merchandise exports, though processed foods accounted for only one-fifth of the food exports. This emphasises the continuing importance of agricultural activity whilst providing a reminder of the growing significance of forward linkages from agriculture to processing and retailing as part of the agro-food chain.

This chain has increasingly been influenced by large-scale capital investment. For example, the key food sectors are dominated by large companies, often subsidiaries of TNCs, and acting as multidomestics. In 1992 the 20 largest food manufacturers accounted for two-thirds of total industry turnover and most of the processed food exports. Of these 20 firms, eleven were subsidiaries of foreign-owned

TNCs. Yet much of this sector has not kept pace with growth rates in other areas of manufacturing. Indeed, food imports have grown, not only through more 'exotic' foods such as high-value internationally branded products, but also goods competing with domestic production, e.g. canned vegetables and orange juice. There have been some strong domestic growth sectors in the 1990s, notably wine and processed cereals and vegetables, and hence overall this is a sector with variable performance, in which some of the traditional domestic products have come under increasing pressure in recent decades.

Some food processing conglomerates have disintegrated (e.g. Adelaide Steamship) and their food production activities have been split up; there has been some increased market concentration (e.g. chicken production); and elsewhere small producers have increased in number, pursuing market niches (e.g. 'boutique' wine production). These developments reflect not only globalisation and national economic reforms, but also changing patterns of food consumption favouring 'fast' convenience foods, reduced meat eating, 'healthy' foods and preference for named brands (which may enable globalisation processes to be cloaked by use of a local brand).

Some of the strategies adopted by processors and growers to deal with the new economic

FIGURE 14.3 Export-based wine production in the Hunter Valley (NSW) (photo: G.M. Robinson)

and regulatory climate can be seen in Burch and Pritchard's (1996) study of the tomato growing and processing industry following the deregulation of tomato production in 1992. This industry is concentrated in the Murray River basin, especially in Victoria where two-thirds of the country's tomatoes for processing are grown. Deregulation has introduced great flexibility into grower–processor relationships, favouring larger growers, who are in a better position to secure more favourable contracts from processors. This has been reinforced by the tendency for the processors (in this case Heinz and Unifoods) to prefer dealing with fewer and larger growers. Reduced government aid to producer co-operatives has also been a feature of recent policy, with the result that the future of some of these organisations is now in doubt. However, domestic producers have faced strong overseas competition as a result of the policies of global sourcing introduced by the major retail chains during the last two decades, and increased availability of subsidised products from the European Union.

Since the early 1980s, the policy emphasis towards land-based production has been placed on deregulation and tariff removal (see Chapter 13). This has contributed to the substantial restructuring of both agriculture and food processing, involving diversification from its traditional mainstays, large-scale wheat, beef and sheep production. Moreover, the terms of trade for wheat and sheep products have fallen since the 1950s, helping to stimulate new enterprises, especially in the horticultural sector. This has contributed to Australia being the only country in the OECD to increase the absolute size of its agricultural workforce in the 1980s (Ferguson and Simpson, 1995: 9). Yet the new economic conditions have not encouraged widespread rural prosperity. In terms of 'broadacre' livestock production (extensive livestock and crops), only cattle grazing properties in the remote heart of the continent generated substantial cash surpluses in the early 1990s. Elsewhere many farm families were sinking into poverty, though without the levels of indebtedness facing cropping (Taylor, 1996). This has contributed to increased farm bankruptcies and to 'flight from the land'.

A declining pastoral base?

Despite the changes affecting the growing, marketing and retailing of foods, much remains of the traditional patterns of land-based activity. Australia remains the world's largest exporter of wool, producing one-third of the world's output, though production has fallen since the peak in 1989/90. It is the second-largest exporter of meat, the third-largest exporter of wheat and a major trader in cotton, dairy produce, fruit, grain, rice and sugar. It has 15 per cent of the world's sheep, is second only to New Zealand in the export of sheepmeat, and is first for the export of live sheep, beef and veal. Nearly two-thirds of beef production is exported, primarily to the US, compared with one-third of dairy production, fourth-fifths of wheat and one-sixth of wine (Figure 14.3). Nevertheless, the relative importance of agriculture in the Australian economy has fallen steadily post-1945: from 20 per cent of GDP to 4 per cent. In contrast, as described above, the processing of agricultural produce has continued to be a vital component in the manufacturing sector, with new opportunities (and threats) posed by globalisation and new marketing opportunities.

As indicated in Table 14.5, production of crops, fruit and vegetables now exceeds that from the livestock sector, a significant change to the long-established pattern whereby wool and beef were the largest sources of farm revenue. Wheat currently generates more output by value than either beef or wool, the latter now falling behind milk production as falling wool prices and drought have affected trade. This reverses the trend in the late 1980s when grain prices were depressed following policy-induced distortions in world grain markets. However, export-led production of wheat and increases in the amount of grain fed to livestock have contributed to its area more than doubling in the 1990s. As with barley production, the main concentrations of wheat growing are in south-west Australia, southern parts of South Australia, extending through the 'fertile crescent' of Victoria northwards into New South Wales (Figure 14.4).

Of nearly 17 million tonnes of wheat produced per annum, 13.7 million tonnes are exported. However, the fastest growing sector of the industry is production for domestic feedgrains (12 per cent of output). Dairy cattle are increasingly being fed on feedgrains and there has been rapid growth in the commercial beef feedlot sector. The latter is being penetrated by US and Japanese companies seeking to export to South Korea and Japan. Thus production has encouraged a diversified grains market, with output of barley, oats and sorghum rising in response. The Australian Wheat Board (AWB) remains a significant player in the domestic market but grower co-operatives have been established to take advantage of the changing situation.

This development of grower co-operatives has also been stimulated by the removal of guaranteed minimum prices for wheat since 1989. Until 1989 the AWB had a monopoly on wheat purchase from growers. It has retained a statutory monopoly on export trading, but commercial ventures such as the food conglomerate, Goodman Fielder Ltd (now New South Wales' largest grain buyer), have entered the field. As suggested in Table 14.6, the importance of grain production within the rural economy is greatest in South Australia and Western Australia, though often in conjunction with sheep or beef (Figure 14.5). Sales of livestock are most important in New South Wales and Queensland, with livestock products in New South Wales and Victoria.

Despite the surge in grain production and diversification into viticulture and fruit and vegetable production, the products of livestock grazing continue to play a major role in the Australian rural economy and hence the problems faced by both sheep and beef producers have affected large areas of the country. Sheep and cattle still represent over 35 per cent of the country's rural production and two-fifths of merchandise exports by value (Figure 14.6). Only 17 million ha of Australia are under broadacre and intensive crop production, a further 26 million ha are under improved (sown) pasture, but around 467

TABLE 14.5 Australian agriculture, 1996

(a) Value of output

	Gross value A$m	%		Gross value A$m	%
Crops			*Livestock*		
Barley	1347.0	4.9	Cattle and calves	3474.3	12.6
Oats	311.0	1.1	Sheep and lambs	1005.5	3.6
Wheat	4602.0	16.7	Pigs	589.2	2.2
Other cereals	683.5	2.5	Poultry	964.6	3.5
Sugarcane	1319.7	4.8			
Fruit/nuts	1421.6	5.2	All livestock	6066.4	21.9
Grapes	680.6	2.5			
Vegetables	1465.0	5.3	*Livestock products*		
All other crops (includes			Wool	2686.8	9.7
pastures and grasses)	3752.1	13.7	Milk	2965.8	10.7
			Eggs	250.9	1.0
All crops	15582.5	56.7	All livestock products	5937.9	21.4
			Total	27595.9	100.0

(b) Area, yield and output

	Area million ha	Yield million tonnes		Output 000 tonnes
Agricultural area	463.3		Beef	1742
grazing/fallow	414.0		Veal	38
sown crops	16.9		Lamb/mutton	589
wheat	9.8	17.0	Pigmeat	349
sugarcane	0.4	37.4	Poultrymeat	462
barley	3.2	5.5	Wool	865
oats	1.2	1.9	Milk	8206 million litres
rice	0.2	1.0		
Forestry	105.3			
native	40.7 (of which 11.3 privately owned)			
woodland	63.4			
plantations	1.2 (of which 0.4 privately owned)			
coniferous	1.0			

From 62 048 ha of vineyards, 531 million litres of wine were produced and 777 373 tonnes of grapes.
NB: no. of holdings = 116 193; with cattle = 71 803 (61.8 per cent); with sheep = 56 026 (48.2 per cent).

(c) Tasmanian Apple production

	Output tonnes
Red delicious	24 539
Golden delicious	8881
Democrat	4128
Fuji	3619
Granny Smith	2995
	52 400

Note: Worth A$34.8 million; three-quarters comes from the Huon Valley

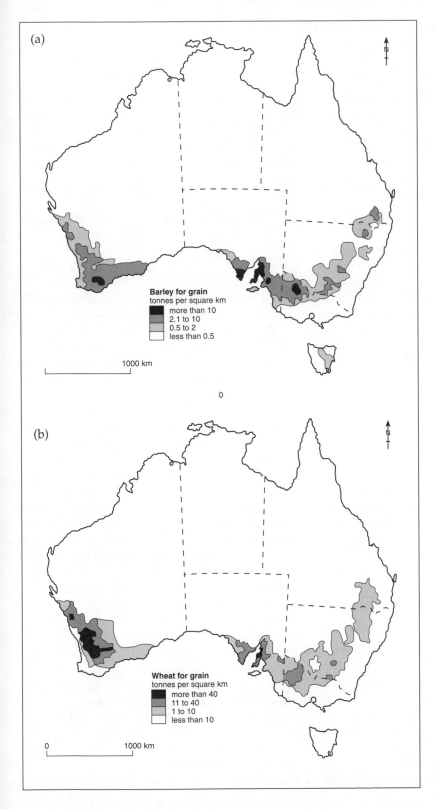

FIGURE 14.4 Distribution of (a) barley and (b) wheat in Australia

TABLE 14.6 Financial aggregates of Australian farm businesses, 1996

	NSW		Vic		Qld		SA		WA		Tas		NT		Total
	A$m	%	A$m	%	A$m	%	A$m	%	A$m	%	A$m	%	A$m	%	A$m
Crop sales	2103	21.4	1307	13.3	3028	30.9	1206	12.3	1938	19.8	193	2.0	29	0.3	9804
Livestock	2029	32.3	1073	17.1	1875	29.9	502	8.0	575	9.2	125	2.0	100	1.5	6279
Livestock products	1688	30.2	1886	33.7	556	9.9	455	8.1	807	14.4	201	3.6	3	0.1	5596
Turnover	6321	26.9	4585	19.5	6005	25.5	2354	10.0	3530	15.0	571	2.4	150	0.7	23 516
Purchases/ expenses	3835	28.4	2574	19.0	3406	25.2	1279	9.5	2034	15.0	320	2.4	69	0.5	13 517
Gross indebtedness	5143	28.2	3026	16.6	4924	27.0	1605	8.8	2988	16.4	462	2.5	120	0.5	18 268

Source: Australian Bureau of Statistics (1998d).

million ha of natural and semi-natural vegetation are used for cattle and sheep grazing (Figure 14.6). It is this production of cattle and sheep that played a vital role in economic development in the nineteenth century and which is still the economic mainstay of huge parts of rural Australia.

Only 3.7 per cent of the agricultural area is under crops; a further 0.6 per cent is fallow and 5.3 per cent in use as sown pastures. Nevertheless two-fifths of holdings are cropping-based, albeit often in conjunction with sheep (Figure 14.5). Sheep and cereals predominate in South and Western Australia; beef cattle in Queensland and Northern Territory; dairy cattle in Victoria and Tasmania; and sheep in New South Wales.

Intensive cropping is concentrated in three areas:

(a) Vegetable growing occurs around all the metropolitan centres.
(b) Specialist fruit and vegetable and wine production occurs in:
    (i)   the Barossa Valley, South Australia (wine);
    (ii)   Tasmania (apples and pears);
    (iii) along the Queensland coast and the northernmost coastal zone of New South Wales (sugarcane).
Total production of fruit and vegetables has

increased by 25 per cent since 1986, mainly for domestic consumption; Western Australia is the leading exporting state.

(c) Inland irrigated agriculture. This has been government-sponsored, most notably in the Murray–Darling Basin, e.g. the Murrumbidgee Irrigation Area.

There is little extensive cropping as beef and/or sheep have been more profitable in the drier areas.

Extensive pastoral land use is associated with beef cattle in the drier areas of Queensland and the Northern Territory, which have half of the country's 13 million beef cattle and where four-fifths of properties carry only beef cattle (Figure 14.6a). Elsewhere, 40 per cent of properties carry exclusively beef. Initially British cattle breeds, such as the Shorthorn and Hereford, dominated. The latter have remained important and account for 40 per cent of all cattle. However, tropical breeds, notably the Brahman and Santa Gertrudis, have increased in popularity and have helped further the development of cattle rearing in Queensland and the Northern Territory.

The majority of sheep rearing is still geared to wool production from merinos and merino crosses (accounting for over 70 per cent of all sheep). Two-thirds of the sheep are in large flocks of over 2000, almost entirely in areas to

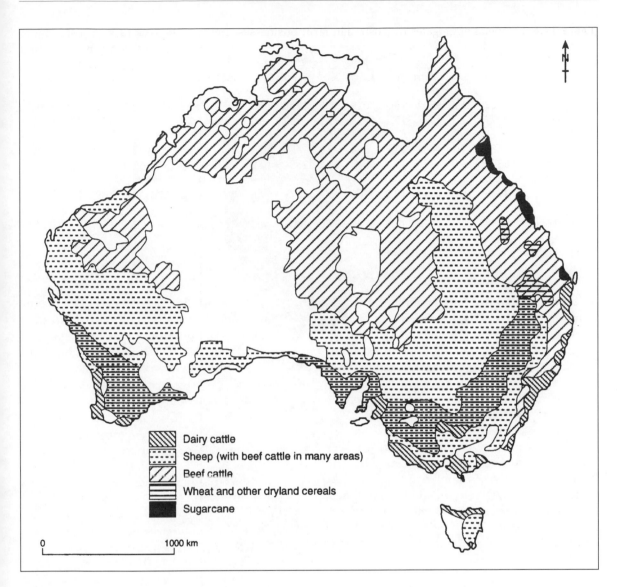

FIGURE 14.5 Farming types in Australia

the west of the Great Dividing Range (Figure 14.6b). These flocks also produce mutton, whereas lamb is produced from smaller flocks in the 'wheat and sheep' zone of south-east Australia (Figure 14.5). Thus wool production is associated with leaseholders, low capital investment and low output per unit area.

In contrast, the smaller holdings for fat lamb production are owner-occupied. Many of the largest holdings are now under company control, often with multiple properties operated by a single company who transfer stock from one property to another. The largest cover over 450 000 ha, but tend to concentrate on cattle rather than sheep. For example, when sold in the late 1980s, the Sherwin Pastoral Co. owned 73 454 km^{-2} of Queensland and the Northern Territory, with 300 000 cattle, of which one-quarter would be slaughtered each year. This was the biggest 'station' in the

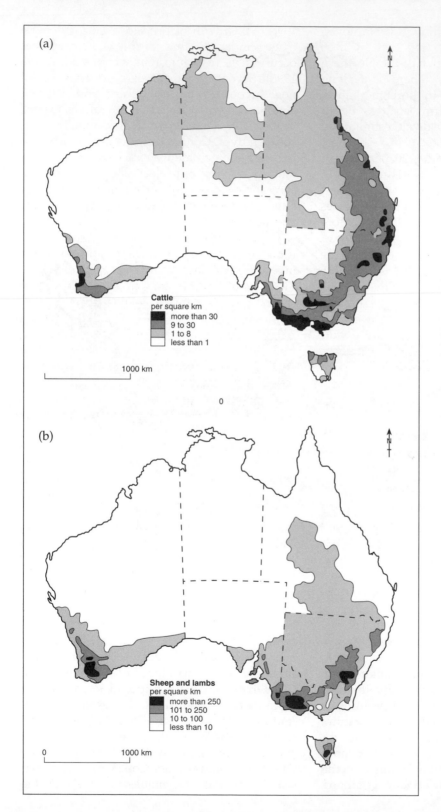

FIGURE 14.6 Distribution of (a) cattle and (b) sheep and lambs in Australia

(a)

Cattle
per square km
more than 30
9 to 30
1 to 8
less than 1

1000 km

(b)

Sheep and lambs
per square km
more than 250
101 to 250
10 to 100
less than 10

0   1000 km

country in terms of area, though it employed only 200 people.

Pastoral leases cover over half the Northern Territory (Rawling, 1987). There is a long history of these leases in foreign hands, initially through British firms such as Bovril and Vestey, but with major investment by American companies post-1945, so that American firms accounted for 20 per cent of the leases by the mid-1980s (Bauer, 1984). Subsequently both south-east Asian and domestic investment through multiple leases has grown (e.g. the Hooker Corporation). More intensive forms of management of herds have only been applied where stock can be marketed in good condition and so a more remunerative return is more readily guaranteed. This is the case in parts of eastern Queensland and those areas near to the coastal meatworks that were established as export outlets. However, not all of these meatworks have proved profitable, the plant in Darwin having closed in 1975.

Dairying has been a strong growth area since the early 1980s, with Australian dairy exports accounting for 70 per cent of the world dairy trade. Under the federal government's Kerin Plan in 1986, domestic product prices were directly linked to international returns, boosting exports so that 45 per cent of production is exported in a manufactured form. From the 1930s the pattern of fresh milk supply was controlled by government regulation to ensure consistency of supply. In the 1970s more attention was given to assistance for low-income farming, starting with the Dairy Farms Reconstruction Scheme. This contributed to a 50 per cent increase in milk production per dairy farm and a 40 per cent rise in production per cow. Deregulation since 1986 has enabled market forces to dictate certain changes, with less profitable production regions losing out to higher productivity regions (Robinson, 1996b). Economies of scale reduced the number of dairy co-operatives from 44 to 27 and the number of dairy proprietary companies from 65 to 31 between 1983 and 1993, whilst overall production rose by 40 per cent (Pritchard,

1998: 68). Growth of producer dairy co-operatives similar to that in New Zealand has occurred in Tasmania (Grosvenor and Wood, 1996). Seventy per cent of dairy exports are now accounted for by Bonlac and Murray-Goulburn, two Victorian co-operatives formed in 1986/7 by the merger of 13 regionally based co-operatives. Contract production and branding agreements have become far more important in efforts to protect and build market access. New forms of finance for the industry have also been introduced, e.g. share issues and debentures.

There has been some penetration of agricultural production by TNCs and overseas investors, notably through Japanese-based TNCs and trading companies (e.g. Nippon Meats) and their involvement in beef feedlotting using grain (Le Heron, 1992: 170). This has followed similar Japanese policy towards using multiple suppliers of raw materials, including Australian coal and iron ore, with production based in Australia fostered by a highly valued yen, Japanese trade liberalisation, Australia's desires for foreign investment, and pressures from farmers seeking alternative outlets for farm products. Half of the feedlots are in Queensland and one-third in New South Wales, with export output increasing dramatically since the first Japanese investments in 1988. Overseas ownership/ shares in abbatoirs now account for half the cattle kill of the two states. New beeflots in the Riverina can raise 130 000 cattle at a time and have a throughput of 250 000 animals per annum. This marks a shift from small-scale beef enterprises, though the region's farmers supply stock and feed to the feedlots under contract. Le Heron (1992: 171) notes the environmental hazards associated with this industrial style of production, and the susceptibility of such operations to international competition.

By the mid-1990s all five of Australia's largest beef exporters were foreign-owned, including ConAgra Inc. of the USA who purchased the agrifood divisions of the largest Australian agribusiness conglomerate, Elders IXL, in 1992. This has enabled the foreign enterprises to supply exports to world markets from a range of sources, reducing the market share actually produced in Australia. A similar trend of foreign investment is apparent in the forest products industry, but with increased production within Australasia for overseas markets.

NEW OPPORTUNITIES IN THE FOREST PRODUCTS INDUSTRY

New Zealand

In 1907, when a major report on the state of the New Zealand timber industries was published, the output of sawn timber from native forests was 432 million superfeet from 411 mills (McKinnon, 1997: 47). This marked a concerted clearance of 'bush' that helped transform the country's landscape. Much of the timber was used for building purposes and in railway construction, but some was simply burnt off in forest clearance for agriculture. Some was exported to Australia with specialist trade in kauri gum from Northland significant at the turn of the century. Nevertheless, of the country's 8.6 million ha of indigenous forest, 5.3 million ha remains undisturbed.

The forest products industry developed under the Industrial Efficiency Act of 1936, which awarded three domestic monopolies protected by full import licensing, though neither of the first two made adequate returns on capital invested:

(a) New Zealand Forest Products (NZFP) held the monopoly for fine paper, packaging papers and paper board;
(b) Tasman, initially a joint venture between the government, Fletcher Construction and offshore manufacturers, held the monopoly for newsprint;
(c) Caxton had the monopoly for tissues and lightweight wrapping. Carter Holt Harvey (CHH), a New Zealand-based firm, purchased Caxton in 1988.

The state-owned New Zealand Forest Service (NZFS) supplied raw materials to the processors if they were not already self-sufficient. In particular, this meant supplying up to 54 per cent of its harvests to Tasman, small amounts to Caxton, and also to the sawmilling industry. The NZFS was not permitted to compete with pulp and paper producers through forward integration, but it played the primary role in forest management for catchment protection and conservation, and also acted as an instrument for regional development in some rural areas.

Tasman processed timber from the Kaingaroa Forest on the North Island's Volcanic Plateau, establishing the town and mill of Kawerau. NZFP, centred on Kinleith and the adjacent town of Tokoroa, 60 km west of Rotorua, owned 54 per cent of the privately owned exotic forests on the Plateau. In 1989, though, Tasman, by then a subsidiary of Fletcher Challenge Ltd (FCL), agreed to conserve permanently tracts of its native forests, notably in the Bay of Plenty, largely because they had limited commercial value.

The reforms introduced under Rogernomics were intended to privatise the state's plantation forestry assets and to separate conservation from production objectives. As a result, the Forestry Corporation (FC), as the trading arm of the NZFS, became one of the first nine SOEs established on 1 April 1987. The FC was given the responsibility for making a profit from operating 170 040 ha of state plantation forests, representing 13 per cent of the country's planted forests. One immediate consequence of their need to operate at a profit was their reduction of the workforce from 8070 employees in 1987 to 2597 in 1990 and 1300 in 1994 (Duncan et al., 1994). The FC moved into the export log trade, targeting Asian markets through its Red Stag brand and exporting mouldings and millwork to the USA through a short-lived joint venture arrangement with the Californian-based firm, Fibreform Wood Products (Le Heron and Pawson, 1996: 167). The FC is still the country's leading exporter of logs, accounting for around one-third of logs exported.

In addition to the creation of the FC, there has also been the sale of the state forests in the form of cutting rights worth NZ$1.3 billion. This has facilitated the entry of overseas firms (Roche, 1991), e.g. Juken Nissho, Ernslaw One Ltd and Royonier Ltd, the latter now controlling 8 per cent of the planted forest. Since the early 1980s there have been a series of mergers as companies have sought to benefit from scale economies and as overseas capital has been attracted by the potential offered by forestry privatisation. In the 1990s investment by foreign capital has dominated the sector at a rate of approximately NZ$75 million per annum invested by both foreign and domestic sources. There have been some implications for Maori ownership and/or compensation to Maori in the sales of state forests, with 23 claims being made to the Waitangi Tribunal between 1991 and 1994 regarding forest lands (see also Taiuepa *et al.*, 1997).

Both NZFP and CHH had clearance from the Commerce Commission to purchase state forests in different regions, and the newly merged company was allowed to proceed with their purchases on the grounds that regional monopolisation would not be the outcome. As a result CHH and Elders Resources (which had merged with NZFP) controlled 53 per cent of the North Island and 49 per cent of national sawn timber production. This compared with 19 and 15 per cent respectively for Tasman.

In 1991 the American-based International Paper Ltd (IP), the world's second-largest pulp and paper company, purchased a one-third stake in CHH (which also owned 325 000 ha of forest and NZ$454 million of operations in Chile). Rayonier and IP, through its majority ownership of CHH, now own nearly three-quarters of the forest plantations in Northland. This gives them a significant amount of control over employment in a region with both a large Maori population and high unemployment rates. However, the latter have risen as a result of continued labour shedding by IP and other foreign enterprises.

New Zealand (and Australian) timber has been used as a substitute for that formerly provided to industrial concerns around the Asia–Pacific region by the forests of south-east Asia (suffering from depletion), British Columbia and the US north-west (facing environmental control regulations) (Barnes and Hayter, 1998). This has led to substantial increases in exotic afforestation and in output of forest products, providing opportunities for both existing New Zealand-based firms and foreign firms new to New Zealand. Firms in this sector have also taken advantage of economic deregulation to expand and diversify their operations.

In the case of New Zealand-based FCL, this has helped them become the eleventh-ranked pulp and paper company in the world, and to be included amongst the world's 500 largest companies by *Fortune* magazine (Le Heron and Pawson, 1996: 161). By the mid-1990s a concerted period of expansion had enabled them to develop interests in seven different countries in the forest products industry, with this sector of their operations contributing substantially to the firm's global assets of NZ$14 billion. Meanwhile, CHH had become the largest forest owner in New Zealand by the mid-1990s, with one-quarter of the country's forest plantations. To develop this timber resource, CHH and the other large forest owners have embarked on substantial planting and investment programmes. The scale of the afforestation in the 1990s has far exceeded that carried out in the previous 'boom' of the 1970s, with both sales of land by graziers and the growth of farm forestry. In the case of Rayonier (NZ) Ltd, a local subsidiary of an American company, this has included a NZ$2 million upgrading of a mill near Gisborne, and NZ$120 million for the construction of a medium-density-fibreboard (MDF) plant in Southland. For CHH there was a NZ$75 million investment to double the capacity of its Canterbury Timber Products MDF plant, NZ$8 million on sawmilling equipment for its Nelson mill and NZ$1.3 million on drying kilns at Tokoroa (Le Heron and Pawson, 1996: 164–5).

New Zealand is currently the ninth largest exporter of woodchip, accounting for 2 per cent of world exports. The majority goes to Japan, the world's largest importer, for whom Australia is the main supplier. Around half of New Zealand's exported woodchip comes from indigenous forest remnants on private land, principally from kamahi (*Weinmannia racemosa*) and silver beech (*Nothofagus menziesii*) used for production of fine writing paper (G.A. Wilson, 1994). There are three principal chipmills: Newman's Nelson Pine Forests Ltd (50 per cent owned by Sumitomo Forestry of Japan); South Wood Export Ltd, located near Invercargill; and NZFP (formerly Elders), based on the Kinleith Pulpmill. The first two operators have accounted for over two-thirds of indigenous forest destruction since the late 1960s.

Australia

In Australia the forest resource has experienced major pressures in the last three decades through the start of the woodchip export trade, the restructuring of the wood industries and the rise of environmentalism as a social and political force. One aspect of the various conflicts associated with forestry has been considered in Chapter 8, with respect to preserving forests in Tasmania. However, this is only one aspect of a multifaceted story. Overall, there are 3.2 million ha of native forest remaining in Australia from the 69 million ha that existed in 1788 (Figure 14.7). In addition there are one million ha of forest plantations. In terms of the economic exploitation of these forests, Table 14.7 gives an indication of the complexity of the management of exploitative regimes. It should be noted, though, that demands on the forests have increased post-1945, first through the growth of sawmilling and the new pulp and paper industry which was part of the 1950s and 1960s economic expansion and, more recently, through the emergence of the woodchip export trade.

In 1969 no woodchip were exported from Australia. By 1980 there were nearly 4 million m³ being sent largely to Japan and South Korea. The 4 million m³ mark was exceeded in the late 1980s, representing at least three-quarters of the total cut of hardwood pulpwood. This substantial consumption of forest resources has usually occurred near locations with deepwater anchorages, as at Eden, Twofold Bay, south-east NSW, supporting the Harris-Daishowa woodchip concession. State governments have granted various concessions over forests to encourage the lucrative trade, with the Tasmanian government providing five concessions for Associated Pulp and Paper Mills (APPM), Tasmanian Pulp and Forest Holdings (TPFH) and Australian Newsprint Mills (ANM) (Figure 14.8). Clear-felling associated with these concessions has exceeded the size and concentration of domestic pulp mills, causing widespread concern amongst environmentalists.

In the other sectors of the Australian forest products industry there have been developments similar to those in New Zealand. The pulp and paper sector has developed as an oligopoly. The dominant firms, APPM, Australian Paper Manufacturers (APM) (now renamed Amcor) and ANM have continued to divide the market, enabling each company to specialise in particular products. However, there has been substantial investment by these companies overseas, Amcor, for example, now operating in 14 countries. This is not a one-way flow of investment, though, with CHH and Japanese companies entering Australia, either in partnership with Australian firms or buying out existing operations. Increasingly, therefore, Australia's forest products industry is becoming part of the Asia–Pacific web of capital with unpredictable impacts upon those regions most reliant upon the industry, e.g. the 'paper company towns' such as Burnie and New Norfolk (Tas), Traralgon (Vic) and Millicent (SA). Of the 19 local government areas in 1986 with 10 per cent or more of their workforce in the industry, six were in Tasmania and four in New South Wales. Only two have since recorded an increase in this workforce (because of afforestation programmes), a reminder of the continued labour-shedding associated with this industry.

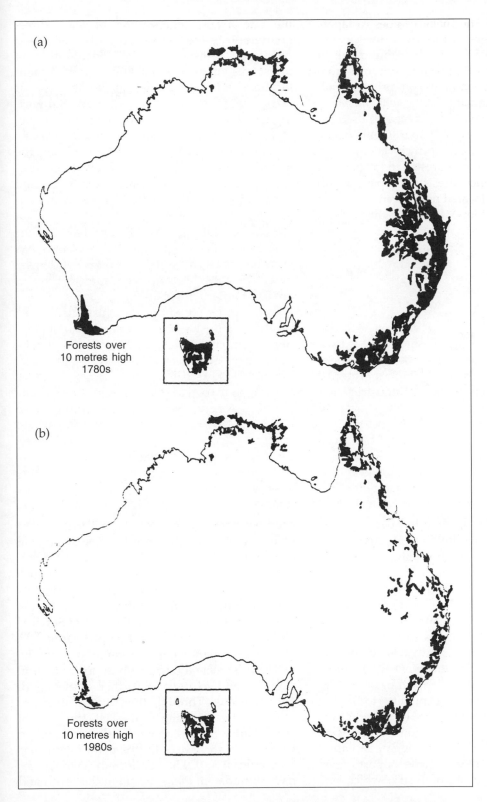

(a)

Forests over
10 metres high
1780s

(b)

Forests over
10 metres high
1980s

FIGURE 14.7
Forests and
woodlands in
Australia in (a)
1780s and (b) the
1980s (Dargavell,
1995: 9)

TABLE 14.7 Major forest resources regimes in Australia

Regime	Capital	Labour	Products	Form	Extent ha	Duration years	Quantity m³/year
Small sawmills	Competitive: family firm, private company	Owner, family, wage	Sawlogs	Exclusive licence	100 to 1000	1–10	1000 to 10 000
Large mills (sawmills, board mills)	Competitive: private or public company	Wage	Sawlogs, now some pulpwood	Special licence, agreement	20 000 to 200 000	15–42	50 000
Exclusive concession (pulp mills)	Monopoly: public company	Wage	Pulpwood, some sawlogs	Legislated agreement	up to 500 000	88	1 000 000
Integrated concession (pulp mills, woodchips)	Global monopoly: public company, multinational corporation	Wage	Pulpwood, some sawlogs	Legislated agreement	up to 1 000 000	15–88	1 500 000
Logging contractors (which supply all regimes)	Contract, sub-contract: family firm	Owner, family, piece-wages	No resource rights in public forests				

Source: Dargavel (1995: 11).

As in New Zealand, there has been a rapid shift in favour of exploitation of non-native softwoods rather than natural forests. Softwoods and other materials, such as steel and aluminium, have increasingly replaced sawn hardwoods in the building industry, thereby reducing the demand for timber from native forests. Whilst this has encouraged more planting of non-native softwoods, there has been public resistance to these plantations because of their alien appearance in the landscape and their inhospitality to native fauna. Eucalypt plantations are one alternative, more favoured by conservation groups who view them as a substitute for logging in native forests. Eucalypts may provide opportunities for the expansion of agroforestry, which has

been occurring in both Northern Territory and Queensland through combinations of grazing and timber production.

Less than one-fifth of the harvest of saw logs is now derived from native trees. Experiments with introduced conifers first began in the 1880s and by the 1920s all state Forest Services had an active pine afforestation programme. In the 1960s a national programme was implemented to create a resource base for industrial supplies and to prevent further depletion of native forests. Through this policy, aimed at self-sufficiency by the year 2000, the area of pine plantations has now exceeded one million ha. For over three decades the planting rate has been at least 35 000 ha per annum. Two species have dominated: *Pinus radiata* in the south and

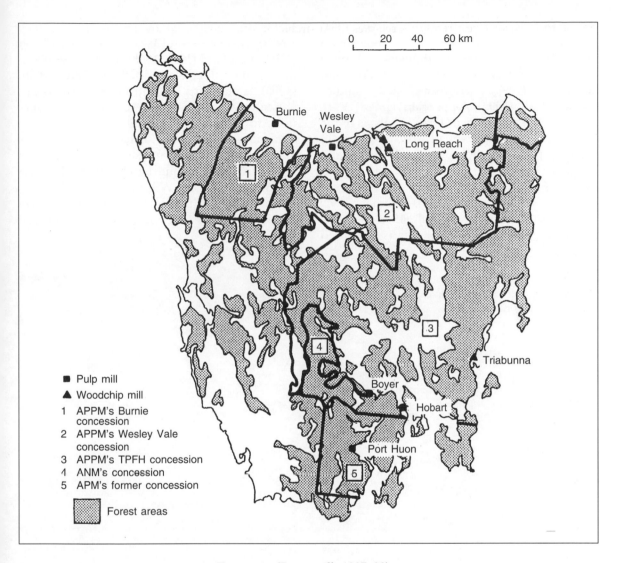

FIGURE 14.8 Forest concessions in Tasmania (Dargavell, 1995: 98)

south-east, *Pinus elliotti* in Queensland. Of these plantations, 70 per cent are owned by state Forest Services and 28 per cent by large private forest industry companies. The expansion of private-sector interests has been restricted by continuing low log prices, reflecting the long term influences of low-price timber from native forests.

FURTHER READING

Various aspects of the impacts of globalisation and deregulation of services are covered in Fagan and Le Heron (1994), Britton *et al.* (1992) and Le Heron and Pawson (1996). See also Perry (1991). Examples of privatisation are provided by Crabb (1991), Duncan *et al.* (1994) and Langdale (1991). Overviews of the growth of the international tourist industry are given by Griffin and Davey (1997) and Pearce and Simmons (1997). Tourism-related studies include Hall (1995), Opperman (1993; 1994), Payne (1993) and Pearce (1990). Analysis of the growth of tourism-dominated urban centres in Queensland is provided by Craik (1991) and Mullins (1990; 1992; 1993; 1994). The significance of Japanese investment in resorts is covered by Daly *et al.* (1996), Rimmer (1992; 1994; 1997a) and Stimson *et al.* (1998). Various developments in the Australian agro-food

sector are considered by Fagan (1990a; 1995; 1997a), Grosvenor and Wood (1996), Le Heron (1992), Pritchard (1996; 1998), Taylor (1996) and in a series of essays edited by Burch *et al.* (1996). A standard text introducing Australian agriculture is Williams (1990). Good introductions to the forest products industry in New Zealand are given by Roche (1990) who has also contributed some excellent material in Britton *et al.* (1992) and Le Heron and Pawson (1996); see also G.A. Wilson (1994) on the New Zealand woodchip industry. Consult Dargavel (1995) for a fine overview of the Australian timber industry.

15

REGIONAL DEVELOPMENT AND PLANNING

SPATIAL PATTERNS OF ECONOMIC GROWTH

Within the set of global processes affecting Australasia, new types of 'economic spaces' have emerged at both local and regional levels as a response to more flexible production systems associated with globalisation and the post-Fordist economy (Storper, 1992). There has emerged a complex set of rapidly changing relationships between global processes and responses at local level which, in turn, affect global developments (Barnet and Cavanagh, 1995). Within Australasia there have been distinctive changes in the spatial patterns of economic development, especially associated with the differential performances of services and manufacturing. This Chapter examines some of the key elements in the regional patterns of economic growth, focusing on measures adopted specifically to manage the space economies at a regional and local level.

Changes to the Australasian space economies

As suggested in the previous two Chapters, macro-level economic changes have included increased wealth creation through the devel-opment of a multifaceted service sector, accompanied by a shrinking of the employment base in certain arms of manufacturing and primary production. Hence major cities have become key sources of employment in services, especially Sydney, Melbourne and Auckland with their high levels of producer services. Different services are prominent in the state capitals: for example Adelaide, Brisbane and Perth have high levels of consumer services and public administration employment, but all have experienced substantial growth in personal services. Adelaide and Melbourne have the highest concentrations of employment in manufacturing, partly because their industries have traditionally been heavily protected (especially the textiles and machinery/equipment sectors) (Beer, 1998).

The 'mega metro' regions centred on the five largest Australian cities continue to dominate both aggregate population growth and growth in economic activity, followed by the coastal non-metro areas (Hugo, 1995; Walmsley et al., 1998). Auckland has become more dominant within New Zealand, reflecting the way in which globalisation and economic restructuring have favoured large centres through agglomeration processes (O'Connor, 1993; Paris, 1994). However, there has also been dispersal of certain activities associated with the attractiveness of some environments and development of new industries located in non-traditional manufacturing centres. This dispersal has been complex, involving both old and new industries. It is a pattern representing a greater differentiation between economic landscapes of production and consumption, with the latter favouring employment generation in the 1990s. Various processes are reinforcing the dominance of Sydney, Melbourne and Auckland whilst also dispersing population to their outer suburbs or beyond (Table 15.1) (Murphy, 1995).

In terms of employment the largest increases in their national share of employment since 1981 have been in the south-east Queensland 'sun-belt metropolis' stretching from the Sunshine Coast through Brisbane to the Gold Coast and the border with New South Wales (Figure 15.1). Significant growth has also occurred in Perth and Canberra whilst the greatest losses have been in Sydney, Melbourne and heavy industrial regions, notably the Hunter. For example, whilst the number of jobs in Sydney rose by one-quarter in the 1970s and 1980s, those in manufacturing fell by one-third.

This fall has impacted most on the unskilled and poorly educated, especially young males with poorly developed English language skills (Yeates, 1997). Meanwhile, most growth was accounted for by financial, business, health and community services. The retail, recreation and leisure sectors also expanded. Immigrants and greater labour force participation by women significantly increased the size of the city's workforce. Throughout Australasia the highest increases in employment have been in producer services, community services and public administration, communications, recreation, personal and other services (Stimson, 1997; Stimson and Taylor, 1997). This has also been accompanied by a significant effect of deregulation, namely the growth of small to medium enterprises (SMEs) oriented towards exporting value-added products. These are most concentrated in the large urban centres and fast-growing smaller centres where there are pools of highly skilled labour.

TABLE 15.1 Key factors contributing to concentration and dispersion of population, employment and investment in the largest Australasian cities

(a) International migration continues to swell the populations of Auckland, Sydney and Melbourne, contrasting with the main trend of internal migration to coastal regions of the Bay of Plenty, New South Wales and Queensland, e.g. the Gold Coast. International migrants stimulate local economic activity in the two main Australian cities whilst internal migrants have significant effects on retailing, personal services and some manufacturing activity.

(b) Tourism has helped to promote new foci, e.g. Cairns (Qld) and Queenstown (Otago) for international visitors, and Hervey Bay (Qld) for domestic tourists and retirement migrants. However, the concentration of international tourists on the leading destination, Sydney, has expanded facilities, hotels and management functions which have reinforced the city's CBD. Similar but smaller effects can be seen in Melbourne and Christchurch.

(c) Sydney has grown as a 'global' city for financial services, with attendant agglomeration of activities and concentration of employment. However, dispersal of producer services to the suburbs, following out-migration from the inner suburbs, has brought growth along the city's north shore, including St Leonards, Chatswood and Hornsby, and in the west around Parramatta.

(d) Industrial restructuring in manufacturing has helped concentrate activity in Melbourne, using its international links for exporting and domestic links for overnight delivery (e.g. O'Connor, 1989). This is most apparent in the south-east suburbs. For example, Melbourne still has over two-fifths of employment in the Australian clothing industry despite substantial change in the labour market (encouraging relocation to non-urban sites: O'Neill, 1991), labour processes (e.g. outworking) and product markets (Peck, 1990; Zhang and Webber, 1992).

Based on O'Connor *et al.* (1998).

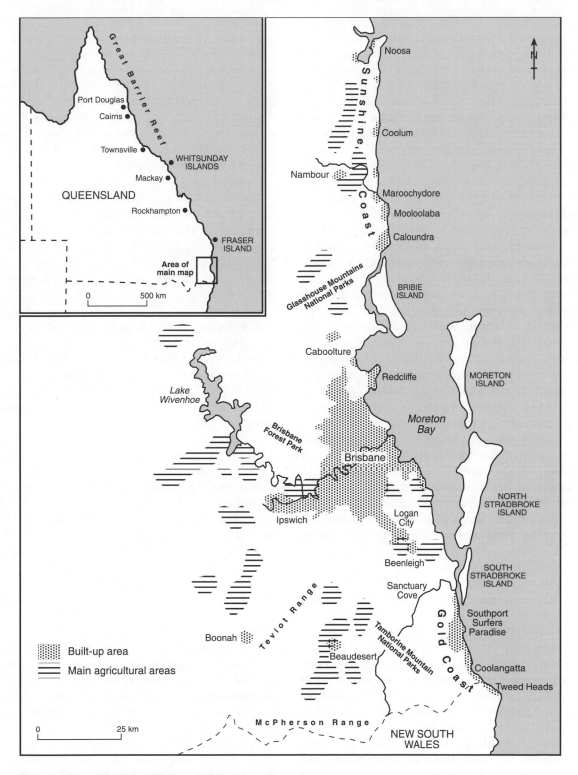

FIGURE 15.1 South-east Queensland

The decline in manufacturing employment was reflected in general in Australasia by the downturn in the clothing, footwear, white goods and metal fabrication industries. In contrast, there has been strong growth in pharmaceuticals, telecommunications and computer equipment, emphasising the importance of elaborately transformed manufactures (ETMs). Often these are represented by small firms with fewer than 10 employees, but a strong export orientation. They are high users of financial and intellectual capital, capable of competing in Asia–Pacific markets, but are not located in traditional industrial areas.

Locations near major research institutions, such as universities and government laboratories, are attractive. Hence in 1994 the New South Wales government announced support for a long-promised Australian Technology Park on a former industrial site in inner Sydney. This will develop ideas originating from Sydney's three largest universities, symbolising the way in which economic growth is becoming more closely linked to synergies created between research, technology and entrepreneurship, in this case focusing on telecommunications and information technology. Another illustration is at North Ryde, 10 km north of the Sydney CBD, supported by the Commonwealth Scientific and Industrial Research Organisation.

Employment gains and losses have also been associated with significant intra-urban changes, with shifts from metropolitan cores to the suburbs. In general the loss of manufacturing jobs has most affected the inner cities and some of the suburban post-war industrial estates. Distributive activities have tended to decline in the inner cities but have grown elsewhere, e.g. Melbourne's outer south-eastern area. Retailing, personal and social services jobs have also grown strongly in outer areas. Producer services have been widespread throughout the cities, but growth has been strongest in CBDs, especially central Sydney.

Most growth has been associated with suburban locations through consumer-related (consumption) economic activity. Outer suburbs and adjacent fringes accounted for 72 per cent of national population growth between 1991 and 1995 whilst there were slight decreases for the city cores and inner suburbs. Non-residential construction activity is increasingly being concentrated in the outer suburbs, though attempts to revive or maintain CBD activity have fostered some construction in CBDs – seen especially in Sydney in preparation for the 2000 Olympics.

Despite losing population through outmigration, Sydney is attracting around 45 per cent of international immigrants (especially from Asia) and 58 per cent of migrating businesses (Jones, 1992; 1998). Meanwhile, Melbourne is experiencing a resurgence in its specialised activities, specifically in transport, manufacturing (especially food processing), high technology and communications. O'Connor et al. (1998) hypothesise that much of the growth and development of Brisbane, Adelaide and Perth is closely linked to key functions in the two major cities which supply markets in these lower-order cities. This may restrict the economic impact of population growth outside Sydney and Melbourne.

Sydney has become the prime city in Australasia in terms of its role within the global economy. Rimmer's (1998) analysis shows that it has been the leading recipient of investment from Japanese service sector firms since 1985, placing it alongside cities such as Beijing, Bangkok, Kuala Lumpur and Jakarta but well behind Hong Kong, Singapore and Los Angeles. Melbourne is in the next 'tier' (with Seattle, Vancouver and Honolulu) whilst Auckland and Brisbane occupy a lower rung (with Panama City and Ho Chi Minh City). Within this activity by the Japanese service sector multinationals, real-estate firms are dominant in Sydney, Brisbane and Auckland, though this concentration has weakened during the 1990s. Airlines and travel agents have a relatively high share in Brisbane and Auckland (Box 15.1).

Sydney's status reflects its role as Australasia's main international air gateway, its selection by the first Japanese and American

---

Box 15.1 Regional performance in New Zealand

In 1986 the Ministry of Works classified New Zealand's regions according to their economic characteristics and performance based on a cluster analysis of indicator variables relating to changes between 1981 and 1985 in population, employment, unemployment, house prices and wages (Table 15.2). In the following decade economic development has reinforced the dominance of Auckland, which has had the highest growth rate of the major urban centres during this period: over 36 per cent of national manufacturing employment, 37 per cent of transport and communications employment and 39 per cent of business services. Meanwhile, the fortunes of Wellington and Christchurch have fluctuated. The latter has benefited from tourist-related development and its diverse economic base, whilst Wellington has been more reliant on white-collar employment which has suffered from cuts in the major government departments. Despite the expansion in the service sector overall, Wellington has lost headquarters office functions to Auckland and is becoming more reliant on consumption-

based activities, as symbolised by the redevelopment of its waterfront (Morrison, 1990).

Population growth in the 1990s has been most rapid in association with tourist development, retirement migration and consumption-related employment. This has been most notable in the Bay of Plenty and Northland, but with some effect in Marlborough and Tasman. Reliance on agro-commodities and food processing remains high in a number of regions (Table 15.3), but has contributed to some economic stagnation, e.g. Hawke's Bay, the West Coast, Manawatu–Wanganui, and Southland. Some 'heartland' regions have been more successful in developing new products or exploiting niche markets, e.g. Canterbury, Marlborough, the Bay of Plenty. Since 1986 Gisborne, Hawke's Bay, Wellington, Manawatu–Wanganui, Southland, Taranaki and the West Coast have lost population through out-migration, but only the last three actually lost population between June 1996 and June 1997, suggesting that these remain the main 'problem' regions.

TNCs locating in the region, its high level of amenity, and possession of icons recognised worldwide (notably its Harbour Bridge and Opera House), and both language and cost advantages for TNC administration within the growing south-east Asia economic zone (Searle, 1996). Producer services are concentrated in the central core of the city, but with a spread to North Shore locations during the past decade and a smaller concentration in the west at Parramatta (Figure 15.2). One-third of all employment in the central core and the inner north is in finance, property and business services (Searle, 1998: 240). Key sectors have been management consultancy, general insurance, graphic design and data processing. Globalisation has reinforced the core focus of many producer services in the CBD. However, technological change has been a decentralising force, encouraging home-based businesses and office locations in professionals' residential areas. Growth in the CBD has been favoured

by the finance sector, law and accounting for whom face-to-face contact and city-centre prestige are important.

REGIONAL PLANNING IN AUSTRALIA

In both countries the patterns of regional development have taken place largely through the workings of the private sector, and increasingly through the impacts of processes and investment originating overseas. Macro-level government policy has provided a critical framework within which these processes operate, but it can be argued that for several decades the impact of 'meso-level' regional policies has been relatively weak. This assertion can be tested by considering the respective regional policies.

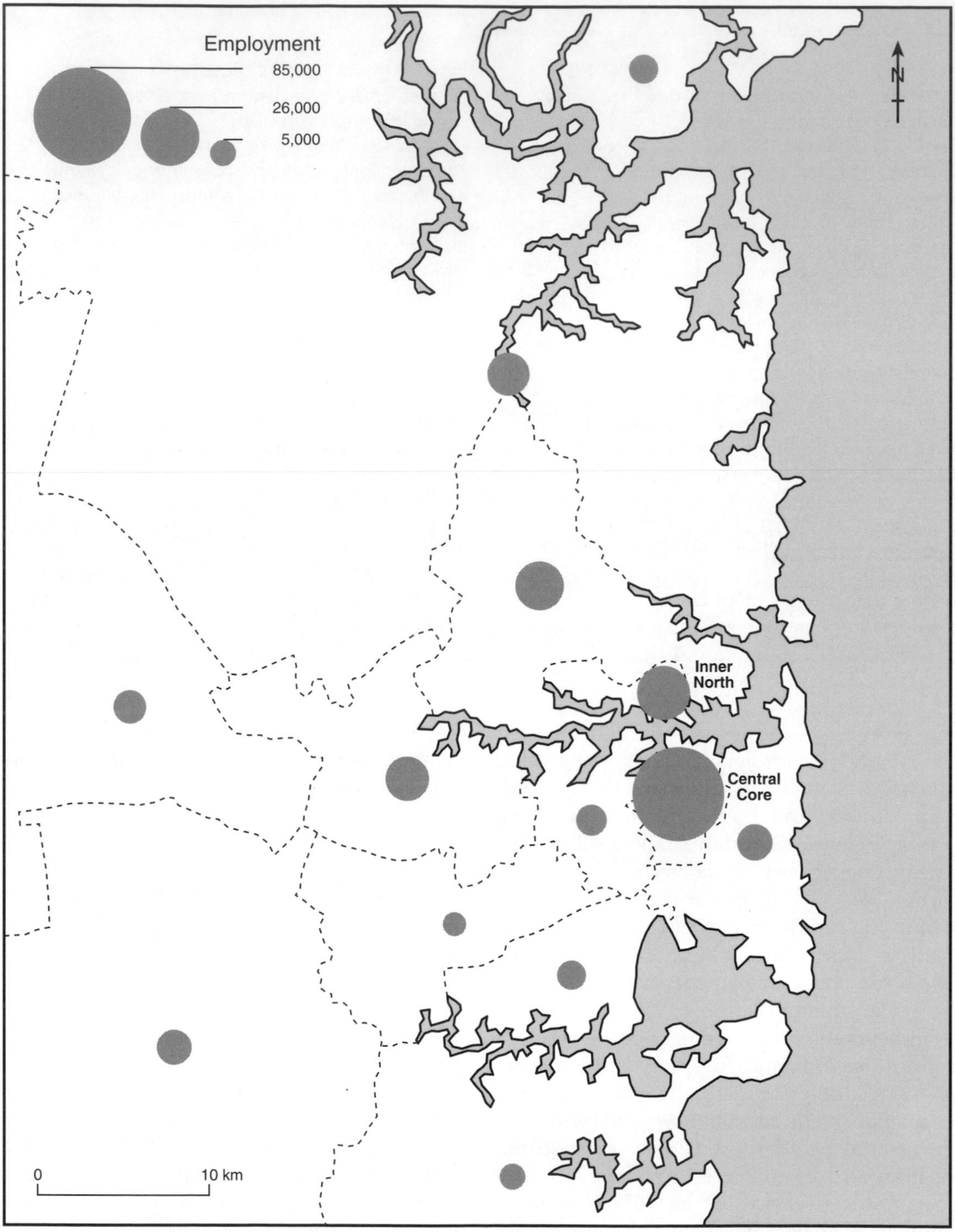

FIGURE 15.2 Finance, property and business services employment by sub-region, Sydney, 1991 (Searle, 1998)

TABLE 15.2 A classification of New Zealand's regions, 1986

(a) Metropolitan centres
1. Auckland, with a large manufacturing sector and a smaller business services sector, high growth and low unemployment. With its good port facilities, command of the largest domestic market and availability of suitable land for industrial development, Auckland has provided the largest magnet for new manufacturing industry post-1945, contributing one-third of the country's manufacturing output by the mid-1970s, compared with one-quarter in Wellington–Lower Hutt, and one-eighth in Christchurch.
2. Wellington, an administrative and business services centre, with a sizeable manufacturing base but low growth despite being the seat of government.

(b) Heartland regions
The rural regions that have traditionally driven New Zealand's resource-based export production. These have the most productive farmland and mixed economies: Waikato, the Bay of Plenty, Manawatu, Nelson, Marlborough and Canterbury. These were experiencing moderate growth and indicators around the national average;

(c) Peripheral regions
These include regions in long term decline (West Coast and Wanganui), highly specialised and vulnerable economies (Otago and Southland), and those facing difficult physical conditions (Aorangi, Wairarapa and East Coast). In all cases GNP was declining, though only Southland, Aorangi and Wairarapa were losing population.

(d) Transitional regions
Northland, Taranaki and Hawke's Bay: mixed performances with growth deemed 'unsustainable' due to over-capacity in key industries (e.g. meat processing in Hawke's Bay) or recent heavy government investment in petrochemicals (e.g. Taranaki).

Source: Ministry of Works (1986).

Prior to the 1970s specific regional policy measures in Australia were largely limited to the Commonwealth Grants Commission's efforts at equalising fiscal capacity and sporadic attempts by state governments to encourage decentralisation, or growth beyond the confines of the state capitals (Matthews, 1981; Robinson, 1993c). The federal government's involvement in regional policy became more marked during the Whitlam government of the early 1970s when an interventionist approach to the creation of social equity and justice was practised. The Department of Urban and Regional Development (DURD) encouraged various forms of economic growth and balanced regional development. DURD aimed to increase economic efficiency by co-ordinating public investment and initiating policies dealing with intra-urban development. Among these schemes was a growth centres strategy to encourage economic development away from dominant state capitals, and especially from Sydney and Melbourne.

Several areas were regarded as potential growth centres, but only one federally funded project was declared: that for Albury–Wodonga, straddling the New South Wales–Victoria border. Here a regional planning authority was established in 1974 as a joint Commonwealth–state statutory authority. Federal funds were also provided, though on a different basis, for Bathurst–Orange and Campbelltown (NSW) and Monsanto (SA). For the first two, development corporations were established to promote economic growth, with incentives to attract new firms, but with a high reliance on public sector employment. For Campbelltown an additional aim was to improve the living conditions for residents in what was a poorly serviced outer suburb of Sydney. Subsequently, this aim was extended to other poor outer suburbs through Area Improvement Plans.

These modest initiatives were followed by the Australian Assistance Plan, designed to form 34 regional social planning authorities, but this was not fully implemented by the time Whitlam lost power. Similarly, financial assis-

TABLE 15.3 Regional employment in manufacturing in New Zealand, 1997

Region	Full-time equivalent jobs		Major types of industry (%)								
	1994	1997	1	2	3	4	5	6	7	8	9
Northland	5374	5715	26	6	14	5	10	5	12	17	5
Auckland	85 965	91 116	14	11	6	12	11	3	14	21	8
Bay of Plenty	11 626	12 618	19	3	35	5	5	3	7	16	7
Gisborne	2002	1770	44	7	23	7	1	3	6	8	1
Waikato	20 102	20 361	25	7	21	5	6	2	12	19	3
Taranaki	7139	8132	39	2	7	4	9	1	21	14	3
Man-Wang	13 184	12 830	30	17	9	6	7	2	7	17	5
Hawke's Bay	11 260	10 063	50	9	11	5	3	2	5	12	3
Wellington	22 031	21 550	14	8	8	14	14	3	11	22	6
Tasman	1768	2374	46	1	31	1	2	6	5	16	2
Nelson	3640	3625	53	5	7	6	1	2	5	18	3
Marlborough	2446	2791	57	2	7	7	2	1	5	14	5
West Coast	1612	1555	40	1	17	5	2	12	2	16	5
Canterbury	35 702	36 912	26	13	7	8	7	3	7	22	7
Otago	12 246	10 839	39	13	8	8	2	2	8	17	3
Southland	7825	8681	53	4	7	4	2	2	19	7	2
NZ	243 922	250 932	25	10	10	9	8	3	10	19	6

Key: 1 – Food, beverages and tobacco; 2 – Textiles, leather and apparel; 3 – Wood processing and wood products; 4 – Paper and paper products, printing and publishing; 5 – Chemicals and chemical, petroleum, coal, rubber and plastic products; 6 – Concrete, clay, glass, plaster, masonry, asbestos and related mineral products; 7 – Basic metal industries; 8 – Fabricated metal products, machinery and equipment; 9 – Other manufacturing industries.
Source: Statistics New Zealand Business Demography Statistics (1998).

tance to local authorities through regional groups of councils was provided in the 1973 Grants Commission Act, but although 68 regions were delineated, this regional organisation was never fully developed and the role of the regions remained limited largely to an advisory capacity.

Regional policies were not a priority for either the Fraser or Hawke governments, with DURD being closed and adjustments to macro- and micro-economic policies replacing specific focus on uneven spatial development. However, the decentralisation of industry to non-metropolitan areas was maintained by offering certain forms of assistance to industry outside the major urban centres, including exemptions from payroll tax, freight rate subsidies, and low interest rates (Taylor and Garlick, 1989). In the 1980s some measures were introduced to compensate peripheral areas not in receipt of major industrial and job-creating developments. They included the Community Employment Program, Labour Adjustment Training Arrangements, the Relocation Assistance Scheme, the Local Employment Initiatives Scheme and various rural development measures, including the Rural Adjustment Scheme, the Farm Household Support Scheme and the National Drought Policy (Barr and Cary, 1992; Malcolm *et al.*, 1996: 321–5).

Many recent measures emphasise a new approach to regional development based on

FIGURE 15.3 Designated country centres

local initiatives, and with government as guides towards new development rather than the overwhelming driving force (Fagan, 1987). A good example of this new approach is the Country Centres Project, a A$500 000 federal pilot programme introduced in 1986 to assist targeted rural communities and rural industry. Eleven centres in six states were selected (Figure 15.3) to test whether local communities could adopt self-help strategies and manage-

ment systems 'to identify feasible options for development and adjustment, and facilitate their implementation within a competitive market framework, with maximum private sector involvement' (Taylor and Garlick, 1989, 93). In 1988 the programme was extended to a further eight centres, with additional federal budgets appended as part of the assistance to specific sectors. Table 15.4 summarises the key elements within the project. Weaknesses were

TABLE 15.4 Australia's Country Centres Project

Goals
- to improve local economic performance by identifying viable local economic opportunities and coordinating means for their realisation;
- to improve federal and state government awareness of local needs so that better co-ordinated and more effective targeting/delivery of their programmes and services could occur at the local level
- to develop effective models for the participation of community groups in self-help local development.

'Bottom-up' components
- local liaison committees
- community consultation to identify opportunities for economic development including interviews, workshops, seminars and public meetings);
- provision of advice and information to state and federal governments on local needs (including constraints to development and provision of government services);
- preparation of statements of opportunities and implementation of viable developments.

Key problems
- the inability to assess the economic viability of identified local opportunities;
- insufficient local infrastructure to support new initiatives;
- lack of risk finance or venture capital;
- anti-development attitudes.

Source: Robinson (1993c).

the over-reliance on primary industry and ancillary manufacturing industry in the new ventures supported by the project, and the limited base from which to work in some of the selected centres (Taylor and Garlick, 1988). The deepening recession of the late 1980s impacted severely on some of the centres, limiting the initiative's impact, though the leading 20 business opportunities in the project attracted investment of A$100 million and generated export earnings in excess of A$20 million (Robinson, 1993c).

After Labor had been re-elected to federal government in 1993, it paid more attention to regional development issues, and conducted a series of enquiries into regional problems and prospects. These included a Taskforce on Regional Development (1993), an Industries Commission (1993) enquiry into regional industrial adjustment, and an expensive consultancy by McKinsey and Co. into regional business investment. In May 1994 the federal government published a White Paper entitled *Working Nation*, focusing on employment and growth, which attracted considerable interest from geographers because of its content on regional development (Australian Government, 1994a; 1994b; BIE, 1994).

The content of *Working Nation* reinforced the Labor government's aim to promote international competitiveness and the need to move away from protection of industry and markets. A central idea was to use subsidies to encourage private sector employers to hire the unemployed. This was termed a 'Jobs Compact', but there was little emphasis on new job creation beyond recourse to traditional economic management. Specific initiatives on regional development were promised, stressing regional leadership and vision, encouragement of international best-practice in local economic activities and improved efficiency in using local infrastructure. Area Consultative Committees were to support new work opportunities schemes to deliver the Jobs Compact.

However, nearly 30 per cent of the A$263 million allocated was to be spent on Badgery Creek Airport in outer south-west Sydney. The consensus amongst geographers was that insufficient was being done to deal with the distributional effects arising from the impact of international/global processes (e.g. O'Neill, 1994). Indeed, some argued that the roles of regional and local processes were being ignored despite the spatially uneven development occurring (e.g. Bolam, 1994; Fagan, 1994; Johnson and Wright, 1994). Weaknesses in *Working Nation* included lack of understanding of the impacts of changing work practices and the dynamics of unemployment, and an ignoring of the potential of local communities to take initiatives and develop solutions to local problems (O'Neill and Fagan, 1994). Several of the proposals were stillborn, being shelved following the subsequent election of the Howard Liberal–National government.

REGIONAL PLANNING IN NEW ZEALAND

In New Zealand the idea of a regional tier of government languished for 89 years after the provinces had been abolished by Prime Minister Vogel in 1876. Therefore after just 19 years of provincial government, New Zealand moved to a bicameral parliamentary system of government, though the upper chamber, a nominated Legislative Council, was dispensed with in 1951. Vogel's achievement was the establishment of a centralised colonial economy and a drastic alteration of the balance of power between the centre and the provinces. The provincial governments, which had been central to the advance of colonisation in the 1860s, were surplus to the requirements of the new policies and were also resistant to the greater degree of control being exerted by the national government.

In the contest that ensued, Vogel utilised the ultimate weapon against the provincial governments by abolishing them. In their place a network of county councils on the British system was instituted, which could pose no threat to the national government. With lesser powers, borough councils, road boards and harbour boards were the units of local government. In addition regional education boards, hospital boards and land boards were created. The next major reform of local government was not made until 1989 (see below).

Formal regional policy was not evident until the 1960s when concerns were expressed about the growing economic dominance of Auckland. This led to two initial policy developments, one concerning the management of the Auckland region (see below) and the other a concern to halt population loss from the South Island and smaller North Island centres. In the early 1970s attention was given to 'priority' regions. Here new or existing manufacturing firms could receive investigatory grants or suspensory or low-interest loans in order to encourage private investment and diversification of the rural base. However, limited resources were allocated, with a total of just NZ$32 million between 1973 and 1985, contrasting with the size of subsidies to the primary sector (see Chapter 6). This regional programme was dismantled in 1985. Northland received 11 per cent of the grants under the scheme, with an emphasis upon small firms, reflecting the character of manufacturing in the region, the average size of firm being 11.3 workers. One-fifth of the grants went to the growth sector of wood and wood products, and one-fifth to the paper and printing industry.

'Think Big'

In the 1970s and 1980s, whilst there was limited funding directed at regional policy, significant regional impacts were intended from a policy pursued by the Muldoon government, known as 'Think Big' (Le Heron, 1979; 1987a). Intended to create 400 000 jobs and to develop showcase industrial projects based primarily on developing new sources of energy supply, it had the subsidiary agenda of stimulating growth in selected regions. Its rationale was based on the country's reliance on imported oil supplies, a vulnerability highlighted during the 1973 oil crisis, and the lack of large domestic supplies of coal and oil. Sub-bituminous coal mined near Huntly in the Waikato was first exploited commercially in the 1860s and now accounts for around half of the country's coal production, the majority of the rest coming from Westland. However, only 3.6 million tonnes per annum are produced, of which over one-third is premium grade for export. 'Think Big' projects included an oil refinery at Whangerei

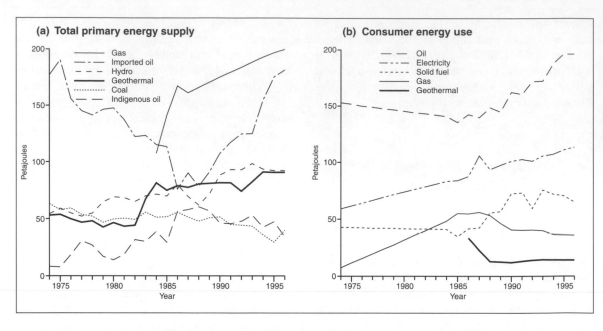

FIGURE 15.4 (a) Primary energy supply, (b) Consumer energy use, New Zealand, 1974–96 (Ministry of Commerce)

(Northland), the New Plymouth power station (Taranaki), a synthetic fuels plant at Motunui (Taranaki), the Clyde Dam (HEP) project on the river Clutha (Otago), the Bluff aluminium smelter (Southland), offshore oil and gas explorations (Taranaki) and the Marsden B (oil-fired) power station (Northland). Much of the investment for these came from overseas. Whilst job creation was well below target, there was a demonstrable success in terms of increasing energy self-sufficiency by two-thirds and so reducing reliance on imported oil (Figure 15.4). Developments in HEP, oil and natural gas have been central to this.

HEP has long been more important in New Zealand than Australia because of the former's relative dearth of other energy supplies. Nearly three-quarters of the country's electricity is generated by HEP, the first scheme opening in 1885. Today the Electricity Corporation of New Zealand has 40 power stations, of which nine generate electricity by thermal means and 31 from HEP. These have over 7000 MW capacity,

accounting for nearly all the country's electricity. The remainder comes from 23 small plants owned by electricity supply authorities.

Initial investigations rejected central Otago as a source of HEP because of the potential for interfering with gold-dredging operations. So HEP was first developed in the Waikato prior to the First World War, to provide power for crushing low-grade ores in gold-mining operations. More substantial developments for generation of electricity with a regional grid did not occur until the 1950s. The first site in central Otago, at Roxburgh on the Clutha, was selected in 1949, with the first of this station's generators commissioned in 1956. Developments here were completed in 1962, producing a lake of 6 km² extending 2 km along the Roxburgh Gorge, contributing an annual average of 1500 GWh. The majority of HEP generation today comes from this and other sites in the South Island so that electricity is the South's biggest 'export', transferring electricity via a national grid to the North Island.

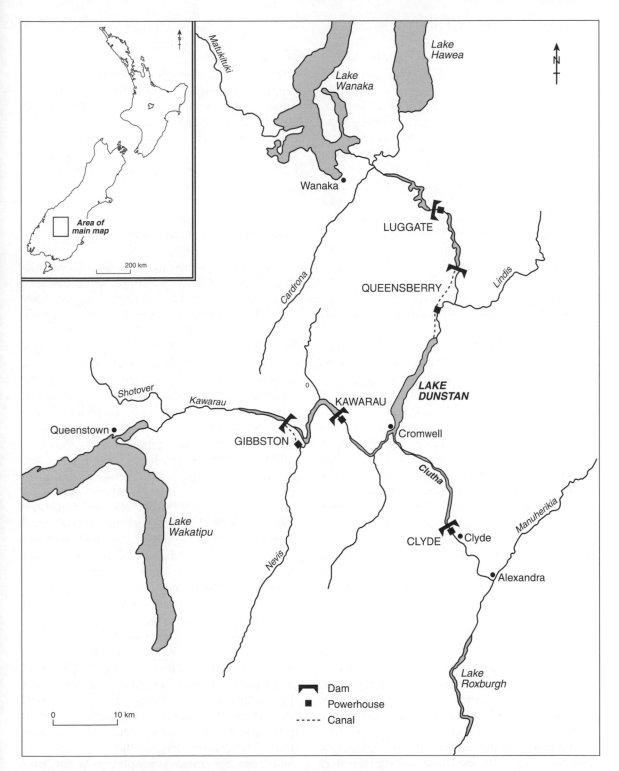

FIGURE 15.5 The Clutha Valley (Otago) hydro and irrigation scheme

Expansion of HEP facilities continued in the 1970s and 1980s, as part of the 'Think Big' programme, which included the Clutha Valley development in Otago. This was part of a scheme to produce sufficient electricity to support Christchurch, Dunedin and Timaru, and involved the construction of the Clyde Dam, costing NZ$1573 million and flooding part of the town of Cromwell and some of the most productive apricot orchards in the country. Several new lakes were created as part of the scheme, notably the 26 km² Lake Dunstan, extending through the Cromwell Gorge and the Lowburn Flats (Figure 15.5). A smaller arm stretches up the Kawarau River to Bannockburn and the entrance to the Kawarau Gorge.

The project has constructed five separate power stations, one at the Clyde Dam, two on the Upper Clutha and two on the Kawharu. The Clyde Dam is the country's largest: 64 m high and producing electricity equivalent to the burning of two million tonnes of coal per annum. The scheme also provides irrigation waters much needed for arable farming in central Otago and the Canterbury Plains where water 'races' have been created in association with the rivers crossing the Plains. The scheme links the races of the Clyde and Fraser rivers to irrigate 7750 ha of the Earnscleugh area.

At the time of the 1973 'oil crisis', 58 per cent of New Zealand's energy consumption came from imported oil. This proportion had been cut by nearly half a decade earlier, and today imported oil accounts for less than one-quarter of energy consumption. In its place, there have been increases in natural gas, especially from the Maui field off the coast of Taranaki which supplies the power stations at New Plymouth and Stratford. Crude oil production is currently 40 petajoules, with 200 petajoules of natural gas produced from five gasfields. The Maui field has two-thirds of domestic oil and gas reserves, but its economic life is estimated at less than 20 years.

An additional element in the country's energy budget is provided by geothermal power. There are two geothermal power stations in the North Island's Volcanic Plateau, at Wairakei and Ohaaki. The former was a pioneer development in the tapping of water and steam held at pressure at least one kilometre below the surface. Bores release about 1200 tonnes of steam per hour, sufficient to generate an annual output of 1100 GWh. After cooling, the condensed steam is released into the Waikato River. The same process is used at Ohaaki, also on the Waikato, commissioned in 1989.

New regional initiatives

There have been two major state policy initiatives from the late1980s with respect to regional development (Le Heron and Pawson, 1996: 289–92). The first is the 'reinvention' of regional policy in the form of local economic development (LED). Although the Lange government did not pursue a regional policy when it came to power in 1984, the regions received greater attention as unemployment rates began to rise at the end of the decade. Nevertheless, it has been localities rather than regions that have had direct policies aimed at them.

These policies include small business programmes to foster enterprise development and community-led programmes aimed at employment generation (Le Heron and Pawson, 1996: 108–9). Together these have been termed LED programmes, and in many respects are the antithesis of regional policy familiar in Western Europe where governments have used incentives to foster business relocations and major public investment decisions to influence regional development. This European approach has been viewed by the 'new right' as both interfering with the operation of market forces and as having been of extremely limited benefit both to the target regions and to society as a whole. In its place, LED targets the supply side of local economies in the spirit of 'enterprise culture' as espoused under Thatcherism in the UK.

Perhaps the most far-reaching of the programmes has been the Regional Development

Investment Grant Scheme, introduced in 1986 to stimulate innovative, economically viable and self-sustaining development new to regions. Available throughout the country, the maximum grant available was NZ$50 000 for 50 per cent of costs. A related scheme introduced in 1989 was the New Business Investigation Grant Scheme to assist the unemployed or recently redundant to investigate possibilities for establishing viable, self-sustaining businesses. The maximum grant was NZ$20 000 per project for 80 per cent of costs. In 1988/9 NZ$2.3 million were allocated to the scheme.

To tackle the problem of rising unemployment, a training programme called ACCESS was established under the 1988 Access Training Scheme Act. This was managed by a network of regional employment and ACCESS councils (REACs), to match training to the needs of local labour markets. A separate but complementary Maori ACCESS scheme has been established which provides approximately 20 per cent of ACCESS training administration through the *iwi* authorities. In July 1990 a new Education and Training Support Agency took over responsibility for the

ACCESS training scheme, the apprenticeship system and other aspects of vocational training. The intention was for the agency to improve links between education and work, with its prime responsibility being to train individuals for the labour market.

The second initiative is local government reform (see Box 15.2) which established regional councils as a middle tier of government, whilst also reorganising a lower tier of city and district councils. These reforms paralleled resource management law reform and so provided the framework within which the 1991 Resource Management Act (RMA) could operate (see Chapter 16). The creation of a tier of regional government established regions based partly on water catchments. This has helped the sustainable management of natural and physical resources under the RMA but without specific promotion or planning for these activities (Memon and Gleeson, 1995) as this has not been encouraged at ministerial level. Therefore a more conventional managerial role has been pursued via issuing consents for resource development and providing services. Meanwhile, though, there has been the growth of enterprise culture within local

---

Box 15.2 Reforming local government in New Zealand

The first unit of regional government was established in 1963 when the Auckland Regional Authority was created as a directly elected regional authority with responsibility for urban passenger transport, planning, parks and reserves, urban water supply, drainage, refuse collection and disposal, roads, community development, civil defence, beach patrol and rescue services, the regional orchestra and water catchment. Regional councils with similar responsibilities were created for Wellington in 1980 and Northland in 1987. However, the 1974 Local Government Act had made provision for the creation of 22 regional or united councils. These were established between 1977 and 1983 as united councils for a redesigned 20 areas, to provide a form of regional government for regions unwilling to establish, or not warranting the expense of, a

regional council. The united councils had three mandatory functions: regional planning, regional civil defence and petrol rationing planning. Subsequently, they have subsumed a range of other functions, notably responsibility for regional reserves, roads and community development.

In 1989, as part of the ongoing transformation wrought by the Labour Government, there was a major reform of local government (Table 15.5). This reduced the number of regions from 22 to 14, making them directly elected regional councils, and giving them extra powers. Subsequently the 14 were reduced to 12, plus four unitary authorities which act in a similar fashion to the territorial authorities, but have some regional powers (Figure 15.6). The main functions of the regional councils are set out in Table 15.6.

TABLE 15.5 Reforms to local government in New Zealand, 1989

The Local Government Amendment Act 1989 recognised two principal *types of local authorities*:

• directly elected *regional councils* with a major role in resource management functions;
• directly elected *territorial authorities* responsible for broadly the same range of functions as at present.

Other key features were:

• the number of regions was reduced from 22 to 14, with directly elected *regional councils*. There are now 12 of these;
• the number of territorial authority districts was reduced from 204 to 74. This replaced the previous system in which territorial authorities comprised borough and city councils (minimum population 20 000), county councils, district councils or town councils. From 1989 only two types of councils remained: *city councils* (15) and *district councils* (59), all headed by mayors;
• local authorities are required to conduct their affairs in a more open and transparent manner, with clear lines of accountability and opportunities for corporatisation;
• provision was made for 159 *community boards*, usually relating to a ward or group of wards. There are now 154 boards.

The purposes of *community boards* are:

• the consideration of and reporting on all matters referred to it by the territorial authority or any matter of interest or concern to the board;
• the overview of road works, water supply, sewerage, stormwater drainage, parks, recreational facilities, community activities and traffic management within the community;
• the preparation of an annual submission to the budgetary process of the territorial authority for expenditure within the community;
• communication with community organisations and special interest groups within the community.

Source: Statistics New Zealand.

TABLE 15.6 The functions of New Zealand's regional councils

• the functions under the Resource Management Act (see Chapter 16);
• the functions under the Soil Conservation and Rivers Control Act;
• control of pests and noxious diseases;
• harbour regulations and marine pollution control;
• regional aspects of civil defence;
• overview transport planning;
• control of passenger transport operators;
• some councils have other functions, such as those formerly undertaken by land drainage boards.

Source: Statistics New Zealand.

government, through amendments to the Local Government Act 1974 and small grants from the Department of Labour amongst various initiatives. This culture has often involved public–private partnerships, including combinations of local councils, businesses and business agencies, educational institutions, labour organisations and *iwi*. A weakness of this has been that benefits tend to accrue only to specific sectors of society, e.g. tourists or professionals, rather than to society as a whole, whilst poorer groups may be excluded or disadvantaged. A specific exception to this undesirable trend is the Jobs Skills programme of Christchurch City Council, which is operated in conjunction with the Canterbury Development Corporation and the New Zealand Employment Service.

PLANNING FOR CITIES AND CITY REGIONS

In Australia, Melbourne and Brisbane are excellent examples of how the differing concerns of state, metropolitan and local government have been manifested in the

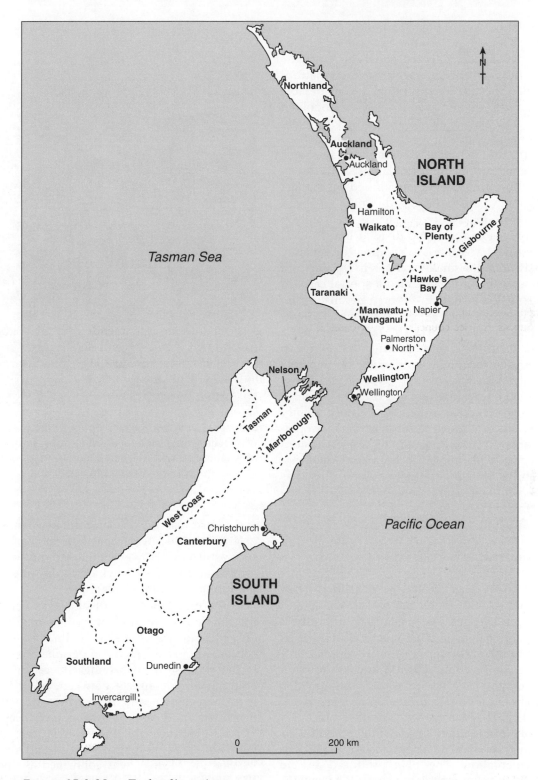

FIGURE 15.6 New Zealand's regions

formulation and implementation of metropolitan plans. Planning for these two cities and for the other state capitals also shows how, at state level, the domination of a single large metropolis has the effect of turning a considerable amount of state planning into metropolitan planning, with regional planning effectively comprising planning for the city region.

With the possible exception of Brisbane, problems of policy formulation and implementation have been compounded by the fact that metropolitan areas do not comprise unitary political entities. Hence the built-up area of Melbourne is sub-divided into 56 local government areas, not to mention the separate jurisdictions for smaller centres in the adjacent region. Melbourne presented a special case for much of the twentieth century because, from 1890, the existence of the Melbourne Metropolitan Board of Works (MMBW) provided a public instrument through which metropolitan Melbourne acquired a degree of separate identity for purposes of physical land use planning (McLoughlin, 1992). Meanwhile, Brisbane is unique in having the only unitary local authority covering much of the metropolitan area. In New Zealand, the longest history of regional planning has been for Auckland, which, as the largest metropolitan area, has some problems comparable with those of the Australian state capitals.

Melbourne

The MMBW was created in 1890 to provide Melbourne with water supply, drainage and sewerage systems. As a statutory authority it was subject to state parliament's control of its constitution and functions, but it was accountable to the local municipal councils through a commissioner system in what was termed a 'parliament of the suburbs' (Gardner, 1989). Post-1945 the MMBW took on other functions, notably preparation of planning schemes for the metropolitan area, and, from 1954, plan implementation. The plan produced in the mid-1950s relied on zoning controls to encour-

FIGURE 15.7 A tram running down the middle of a major inner city road – an enduring image of Melbourne and marketed as a tourist attraction (photo: G.M. Robinson)

age centrifugal growth at a time when such growth was widely regarded as the mark of a successful city. However, the rapidity of this growth and urban sprawl led to disputes over the future direction of planning.

In the mid-1960s this dispute was between the state's Town and Country Planning Board (TCPB) and the MMBW, with the former concerned over the focus on peripheral growth at the expense of a thriving city centre. This also mirrored conflicts between, on the one hand, developers and landowners who were interested in developing commercial centres beyond the inner city and, on the other, those who advocated more growth in the city centre. In 1968 the MMBW was confirmed as one of Victoria's Regional Planning Authorities, charged with the task of preparing regional plans under the controls and guidelines set by the superior TCPB. The resultant master plan of 1971 maintained support for centrifugal growth, but to be channelled into specific areas via a series of urban growth corridors and enlarged satellite towns to the north at Melton and Sunbury. Eight corridor zones were identified, to be based on transport routes separated by open 'wedges' (Robinson, 1994c: 64). Population growth within a 95 km radius of the city centre was projected to expand by a further three million by the end of the century. Yet the eight innermost suburbs had suffered a population decline of 18 per cent between 1947 and 1971, and the local authority for the city centre, Melbourne City Council (MCC), responded with their own plan proposing revitalisation of the CBD and inner suburbs (Logan, 1985).

The resulting conflict over who should plan the development of Melbourne and its region and what form development should take eventually led to a loss of functions for the MMBW, though not before its new strategy plan had been formally adopted in 1983. This now recognised the continuing loss of population in the inner suburbs (a decline of nearly 100 000 during the 1970s), the emergence of some inner city gentrification by young professionals, growth of suburban retailing at the expense of the CBD (Edgington, 1982: 234), CBD reliance upon public sector office employment, and rising unemployment amongst blue-collar workers.

There was a move away from the previous championing of the decentralised city region (Logan, 1986), as illustrated in the attempt to focus retailing, space-using industry and certain types of offices in 20 suburban 'activity centres' in a ring around the inner suburbs. Peripheral development was to be channelled into four designated corridors where there were centres to act as foci for administrative, commercial, retail, cultural and entertainment facilities. However, this plan was strongly criticised (Robinson, 1994c, 67), and developers demonstrated their ability to use cheaper sites

and to expand in non-designated suburban centres that they deemed more suitable than the activity centres. Nevertheless the strategy did introduce certain equity considerations into metropolitan planning, with a focus on incremental growth, more local participation in decision-making, private sector initiatives, and encouragement of higher-density housing to restrict sprawl (Bunker, 1987).

Significant outcomes of urban sprawl and decentralisation have been the creation of successive rings of new outer suburbs, with large new estates in the 1980s over 30 km from the city centre (e.g. Berwick; Carrum Downs), use of green-field sites by space-using industries moving from the inner city, and major out-of-town retail and office complexes. For example, between 1980 and 1987 the suburban share of office employment rose from 10 per cent to 27 per cent (Edgington, 1988; 1989), as part of the growing trend towards employment concentration in the eastern and southern suburbs.

In the mid-1980s a new state Labor government restructured regional planning in Victoria so that responsibility for metropolitan planning was removed from the MMBW, which reverted to its original function of providing water, sewerage and drainage services. In its place came an enlarged state Ministry for Planning and Environment (MPE) geared towards creating a climate of investment confidence by streamlining the processing of development applications, by releasing government-owned land for development, and by upgrading the environmental and cultural image of the city. These measures concentrated closely on the city centre where the MCC was stripped of its planning powers. This was a reflection of the state government's desire to shape the character of the CBD and to manage business investment there to enhance the reputation of both the state and its capital within the context of the ongoing competition with neighbouring New South Wales and Sydney. There was a clear recognition of a need for capital investment in the city if it was to compete effectively with Sydney and other state capitals for jobs and economic growth opportunities, with concerns voiced that it was Sydney and not Melbourne that was benefiting from the growth of producer services, international finance, and increases in trade in

minerals and energy (Edgington, 1989). Nevertheless, Melbourne was gaining more investment in manufacturing and distribution activities, providing jobs in both new and traditional industrial areas. The MPE endeav-

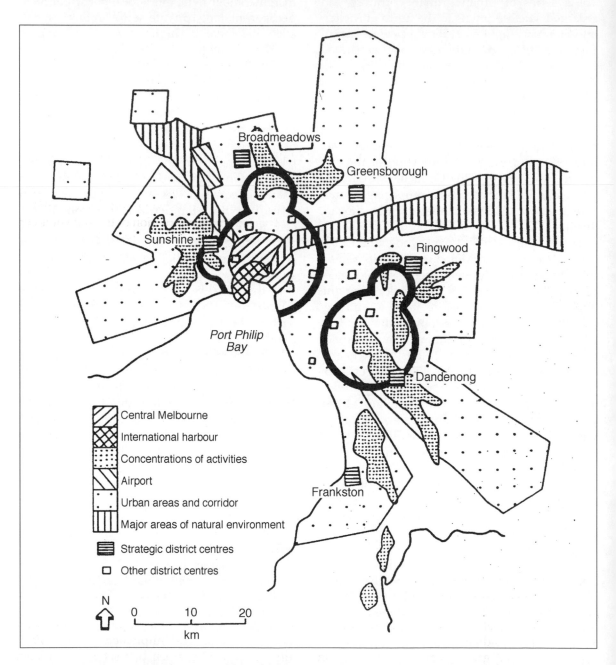

FIGURE 15.8 Shaping Melbourne's future 1987

oured to manage this via a planning framework that designated five key areas in which economic activity would be stimulated, all including extensive holdings of government-owned land. It also introduced plans for new tourism, recreation and leisure opportunities and environmental improvements prior to publishing its long term aims in 1987 in *Shaping Melbourne's Future* (*SMF*).

SMF addressed imbalances within the metropolis associated with the concentration of factories, offices, services and transport nodes in the east and south of the city. Here suburban expansion policies had really taken off, creating areas of growth adjacent to fashionable suburbs stretching along Port Philip Bay. The new plan sought to compensate the continuing growth here with complementary developments in the west and north (Figure 15.8). Federal government support was sought to promote job opportunities in depressed western and north-western suburbs, with various suggestions regarding job creation, notably targeting the area around Tullamarine Airport as a focus for manufacturing and the service sector.

Effectively, these measures extended a Western Suburbs Action Program (first established in 1982) which focused on environmental and infrastructure improvements worth A$17 million. Building on the general outlines of *SMF*, the following year the state government designated five technology precincts to combine the type of facilities and developments associated with science and technology parks. Subsequently, plans were also implemented for a teleport and regional communications facility near the city's docklands where other improvements in cargo handling have occurred. Various transport improvements have also been introduced to assist the growth corridors through several key arterial road projects, and more flexible planning controls have been applied to high-tech industry which has been deemed less incompatible with residential development than traditional manufacturing industry.

One result of greater state intervention in planning the Melbourne region has been the succession of high-profile schemes focused on the inner city and city centre, including the National Tennis Centre, redevelopment of the South Bank and the area around Museum station, the World Trade Centre Hotel complex and a A$2 million transformation of retailing property. This has been part of 'fast-track' planning encouraged from the mid-1980s by successive state governments, stimulating building activity in the city centre, which some likened to the 'marvellous Melbourne' of the speculative land and finance boom of the 1880s (Bunker, 1987). However, the 1980s' equivalent was adversely affected by the end of the construction boom which contributed to A$5.5 billion worth of losses in government-related institutions and the A$2.7 billion failure of the State Bank's merchant banking arm, Tricontinental (Armstrong and Gross, 1995).

This led to the resignation of the state premier, but new regimes in the 1990s have pursued various fast-tracking schemes in the guise of 'market' solutions to development issues. In the 1990s the metropolitan and regional levels of governance have been neglected by the state in favour of voluntary co-operation between groups of local councils who are required to produce municipal strategy statements under the 1992 Local Government Act, setting out developmental visions and planning strategies for their areas (Huxley, 1999). Various options are still being considered to deal with the forecast of 1.5 million additional people and 845 000 extra dwellings to be accommodated in the Melbourne region by 2031 (Fincher, 1998).

Brisbane

The City of Brisbane Act in 1924 provided the city with a single authority controlling a range of services and town planning. Its jurisdiction was extended to cover 1220 km^2, but conflicts between the state parliament and the city council prevented a metropolitan plan from being prepared and implemented until 1965, though in 1952 a green belt scheme had been proposed for the city, to prevent urban sprawl.

As a result it was low-density expansion, devouring small hamlets and much farmland, that accommodated the population growth from 220 371 in 1921 to 593 668 in 1961.

The 1965 Town Plan was basically a statutory zoning plan with accompanying ordinances which systematised the zoning of land use that had developed organically. The main zoning provision was the designation of a Future Urban Zone which emphasised expansion along major routeways and allowed for infill in the area that had been suggested as green belt in the early 1950s. This plan was revised in 1971 when provision for higher-density residential zones was introduced around the inner city. In part this was a

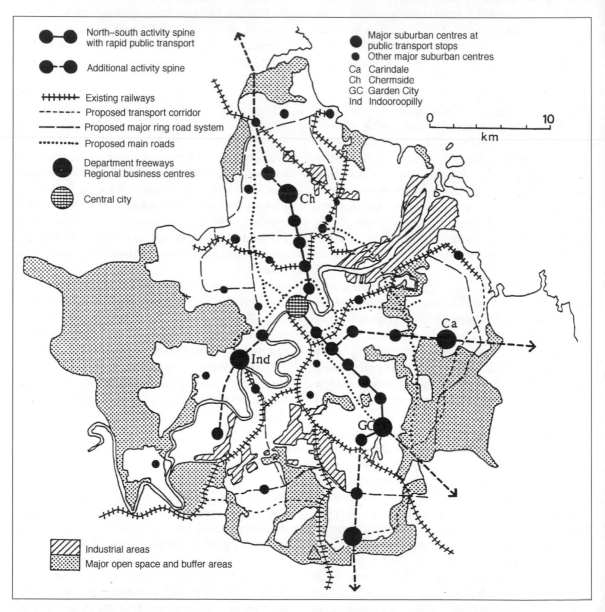

FIGURE 15.9 Brisbane's regional business centres (based on the *Brisbane Town Plan*, 1979)

FIGURE 15.10 Brisbane's CBD from the Brisbane River (photo: G.M. Robinson)

response to the dramatic urban sprawl of the 1960s, which, in the south-east, stretched towards Logan City, 25 km from the city centre and outside the city limits. So great was the development here that Logan City was gazetted as a separate local authority in 1978 when it had a population of 120 000.

New plans in 1975 and 1979 acknowledged the strong suburbanisation of both retailing and office development, by designating five regional business districts (RBDs) (Figure 15.9) whilst recommending that steps be taken to ensure the maintenance of a vibrant city centre retail complex, e.g. through pedestrianisation and investment in public transport including a rapid mass transit system. The RBDs were to be the foci for stores offering opportunities for comparison shopping in a total provision of retail floorspace in the order of 50 000 m², accompanied by offices for professional services, government branch offices and financial and commercial services. They were also to provide community and recreation services and facilities. Designation of the RBDs recognised the emerging concentrations of commercial development in the outer suburbs, but of the five designated only three (Chermside, Garden City and Indooroopilly) have grown to a substantial size. The CBD has remained as the largest retail centre but with only around

one-fifth of the retail turnover in the metropolis and one-quarter of retail floorspace (Robinson, 1993b). The CBD and immediate inner city have retained two-thirds of metropolitan office space, partly through capturing the majority of new office development during the 1980s' service-sector boom. This process changed the face of the city, producing at its core a cluster of high-rise office blocks running along the north bank of the Brisbane River (Figure 15.10).

The ambitions of both city and state politicians have been expressed in the region by attempts to raise the national and international profile of Brisbane, in part to attract tourism and foreign investment. In the 1980s the most obvious returns to the city for these policies were the securing of the 1982 Commonwealth Games and Expo '88, the world exposition which attracted 15 million visitors to a 44 ha site directly across the Brisbane River from the CBD. Subsequently, this site has been redeveloped by a corporation established by the state parliament in a dramatic 'instant' creation of a new urban environment on what was formerly an old warehousing and disused dockside area (Robinson, 1994b).

The site, renamed Southbank Parklands, now has 180 800 m² of commercial space, 84 000 m² of residential accommodation,

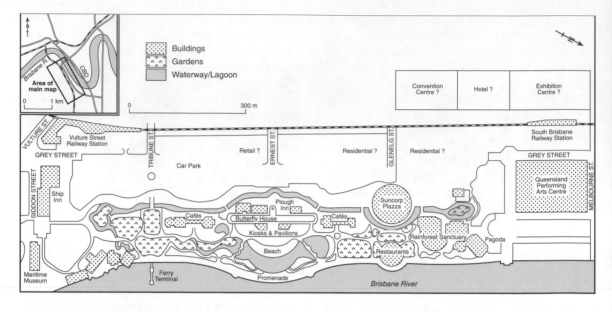

FIGURE 15.11 Southbank Parklands, Brisbane

14 400 m² of retailing, a 45 000 m² A$170 million convention and exhibition centre, 10 000+ m² for food and drink outlets, nearly 40 000 m² for public purposes, a 68 250 m² A$40 million hotel, apartment blocks and a A$25 million home for the Queensland Conservatorium of Music (Figure 15.11). Artifical lagoons, beaches, waterways and a rainforest experience centre provide attractions for tourists just a five-minute ferry ride from the CBD. Total investment to create the new urban environment on the site was in the order of A$1 billion, largely from private companies, though most of the initial A$130 million was from the public sector. Opened in June 1992, there were an estimated 4.5 million patrons within the first eight months.

Further planning of the Brisbane region since the mid-1980s has taken place in the context of rapid population growth, linked to the 'northward drift' of the Australian population. For Brisbane itself there have been further political conflicts that have restricted the impact of the major planning initiative, the 1990 *Brisbane Plan: A City Strategy*, developed under project director Robert Stimson (1991; 1992), a geographer. This was the most substantial planning exercise in the history of both city and state, with a major consultation exercise (Caulfield and Minnery, 1994) and publication of numerous ancillary discussion papers. It refers to Brisbane as the 'magnet city', symbolising its potential to act like a magnet as the centre of gravity of population and economic activity 'shifts from the Sydney–Melbourne axis to a Sydney–Brisbane–East Coast axis' (Stimson, 1991: iii).

One problem with the *Strategy* has been that some of its proposals regarding the role and status of various types of suburban centres are contrary to the larger regional planning exercise, South East Queensland 2001 (SEQ2001), which was initiated at state level towards the end of the Brisbane City exercise (Bunker, 1995: 161). This project is aimed at dealing with the extra one million people or up to 450 000 additional households expected to be resident in the region between 1991 and 2011. This champions the production of 'bottom-up' sub-regional strategic plans prepared by regional organisations of councils, but with directions from above towards possible 'solutions', notably the possibility of 'edge city' development reminiscent of western Sydney and north-west Melbourne.

The preparation of SEQ2001 is a significant development in a state where regional planning has been neglected, in part because local authorities rather than state government have the major responsibility for land-use development and physical services (Caulfield, 1992). Moreover, the rapid growth of southeast Queensland presents a particular set of imperatives for planning to manage the rapid development.

The Gold Coast has been one of the fastest growing urban areas in the country during the last two decades, yet the planning responsibility has largely been that of a former rural shire council, Albert Shire. This helps to explain the extent of urban sprawl, encroaching on to fragile coastal ecosystems, producing major problems of environmental destruction and pollution, notably erosion, flooding and silting, coupled with high costs of servicing the linear settlement (Robinson, 1994a).

However, the role of the state government under Premier Joh Bjelke-Petersen, in the 1970s and 1980s, was also highly important in encouraging the rapid growth of tourist-related development along the 250-km coastline south of Noosa Heads and incorporating the Sunshine Coast, Moreton Bay and the Gold Coast (Figure 15.1). The ruling National Party drew upon its close links with business, and especially property developers, to maximise capital investment into expanding tourism. Coincidentally, federal government deregulation of finance and banking encouraged overseas investment in this and other parts of the service sector. The state government manipulated the planning process, often through 'creative legislation' such as rezoning and use of special legislation to permit resort development. On the Gold Coast the interests of the local council, entrepreneurs and state government coincided to produce a property developer's heaven, though there has been more resistance by local citizenry to similar

development on the Sunshine Coast (Craik, 1991; Mullins, 1990).

Many aspects of the *Strategy* have been affected by changes in the political composition of state government and new initiatives developed both at state and federal level. One of the latter is the federal government's Better Cities Programme (Paris, 1992). Launched in 1991, this allocated A\$816 million to be spent over a five-year period, with 20 area strategies to provide and upgrade existing infrastructure, reduce urban development costs, improve urban land use, reduce costs associated with traffic congestion and pollution, improve urban planning, and increase housing choice and affordability.

This has been described as a 'renovation agenda' for the cities (Peel, 1995), with the aim being to demonstrate how 'higher density, planned urban development, integrating housing, services and employment will help the three aims of economic efficiency, environmental sustainability and social justice in Australian cities' (Orchard, 1995: 71). Area strategies are intended to deliver more coherent regional centres, improved rail services (especially in the outer suburbs), upgrading of old public housing estates, integrating medium-density redevelopment and sales of public housing to private owners, inner city housing redevelopment, and improvement of infrastructure. It also encompasses a more flexible approach to Commonwealth/state relations rather than focusing on agendas dominated by Canberra-based politicians.

Nevertheless, there are clear ideals underlying the programme, notably a perceived need for more compact cities through denser urban development, better-integrated public transport and strategic investment in infrastructure.

In the Brisbane region the programme includes extensions to the Brisbane–Gold Coast rail line, erection of new low-cost housing, and substantial improvements in the poorer outer suburban area south-west of the city towards Ipswich. In the city itself the focus is on the inner north-east, with plans for 14 000 new dwelling units, including 3500 affordable housing units for low-income earners.

Auckland

Comparisons can be made between the solutions adopted for dealing with urban growth in Brisbane and Melbourne and those for Auckland. In particular there are similarities involving initial use of a green-belt policy, followed by a relaxation of this measure in the

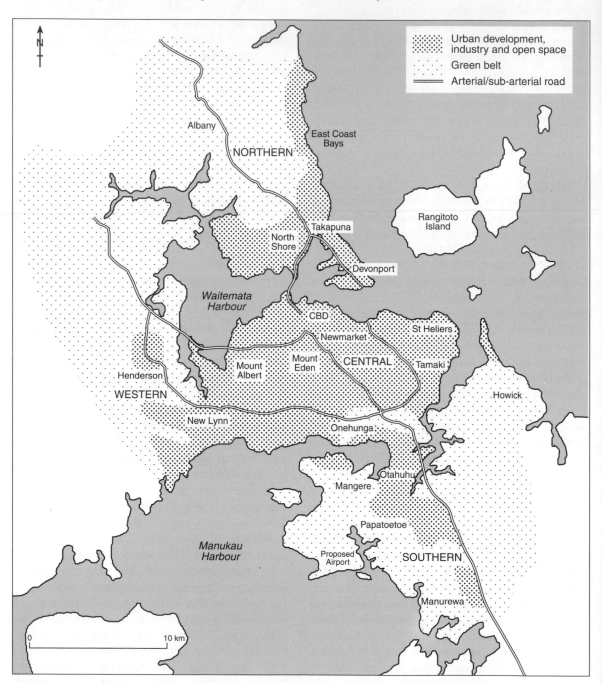

FIGURE 15.12 The Auckland Development Plan, 1949

FIGURE 15.13 Auckland Harbour Bridge (photo: G.M. Robinson)

face of urban sprawl, and more recent plans to reinvigorate the city centre. The 1949 Auckland Development Plan introduced land-use zoning and prescribed limits to the city's population and spatial growth. It envisaged a maximum metropolitan population of 600 000 which would be contained within encircling green belts (Figure 15.12) and with the application of strict land-use zoning. However, the population projections were swiftly exceeded by rapid population growth in the 1950s and 1960s, and there was urban sprawl associated with rising car ownership which produced a decentralised residential pattern breaching the green belt.

The Master Transportation Plan of 1955 advocated the construction of motorways to serve Auckland's future transport needs. At the heart of this was construction of the Harbour Bridge (Figure 15.13), connecting the city with the north shore, then sparsely settled. Once opened in 1958 the bridge was a catalyst in encouraging settlement on the north shore, with the growth of suburbs such as Birkenhead and Takapuna. However, this was also encouraged by state housing schemes, e.g. Tamaki, and with low-interest loans for new houses. There was assistance for private builders, such

as Fletcher Trust, to develop new suburbs served by shopping malls which catered for car-borne shoppers. Population growth was partly linked to in-migration of Maori from the early 1950s. Initially, they concentrated in inner city neighbourhoods such as Freemans Bay, but in the 1960s there was out-migration to state housing in the outer southern suburbs, notably Mangere and Otara.

By the mid-1960s the city's population had risen to half a million. In the next decade there was a further 25 per cent increase, but then growth levelled off, especially in central areas, whilst urban sprawl continued. Within the dominant centrifugal tendencies clear social differentiations emerged, with middle-class Pakeha suburbs dominating the north shore, working-class Pakeha in the western suburbs, and Maori and Pacific Islanders having significant concentrations in the south (McCall and Connell, 1993). From the mid-1970s, redevelopment of the city centre was accompanied by a process of gentrification, as discussed in Chapter 12. Outward extension of the urban area has not ended, with outlying coastal townships enjoying the most rapid growth during the last two decades, followed by outlying rural districts, e.g. the Albany Basin, 25 km north of the CBD.

In 1970 the government acquired land in Albany for a second university in Auckland, with the expectation that the Basin would house 40 000 people. Yet the Basin has been partly by-passed by urban growth which occurred instead along the coast in East Coast Bays and the Hibiscus Coast. Instead it has become a service centre for more distinctive centres both to the north and south, and with a campus of Massey University. Meanwhile, Auckland University has established a second campus in the south-east of the city (Tamaki Campus, operational from 1992, but not on the site originally reserved for a university).

In 1988 the Auckland Regional Planning Scheme proposed limits to the spatial expansion of urban sprawl in order to protect farmland and to make more efficient use of existing infrastructure and services. These ideas have been given a new framework by the 1991 RMA, with a focus on the environmental impacts of urban sprawl. However, there have been disputes between four local authorities in the region and the regional council regarding the latter's desire to impose metropolitan urban limits.

Local planning schemes at the northern edge of the urban limits have zoned land for development in both the Albany Basin and the Okura–Long Bay area which could mean substantial infill settlement between East Coast Bays and the Hibiscus Coast. Although development is supported by North Shore City Council, which projects a population increase of 28 000 in the city's population by 2001, there is strong local opposition. Meanwhile, major new developments have occurred in the city centre with a 'sky tower' now dominating the skyline, the Britomart complex development (see Chapter 12), and anticipation of the America's Cup yacht race.

FURTHER READING

Recent articles on the geography of economic change in Australia include those by O'Connor *et al.* (1998), Stimson (1997), Morrison (1997) and Stimson and Taylor (1997). Details of changes occurring in Sydney are in Searle (1996; 1998). For New Zealand see Le Heron and Pawson (1996: *passim*), which also contains studies of new regional initiatives and planning in the Auckland region. For a detailed consideration of employment changes in the four main New Zealand cities in the 1980s see Morrison (1989). The development of regional planning in Australia is discussed by Robinson (1993c). An issue of *Australian Geographer* (vol. 25) contains several essays on *Working Nation* (Jacobs, 1996). There are several references to the 'Think Big' policies in New Zealand in Britton *et al.* (1992). Memon and Gleeson (1995) refer to the implementation of New Zealand's Resource Management Act. Background to urban planning is given in Troy (1995). Case studies include McLoughlin (1992) and Robinson (1994a) on Melbourne; Peel (1995) on Better Cities; Caulfield and Minnery (1994) and Robinson (1993b; 1994a; 1994b) on Brisbane/south-east Queensland.

16

ENVIRONMENTAL HAZARDS AND ENVIRONMENTAL MANAGEMENT

Environmental management is largely a reaction to perceived hazards, because hazards threaten human well-being and the future of the planet. Hazards may be geophysical or human-induced. The former include earthquakes, volcanic activity, cyclones, drought and flood; the latter include anti-social activities, traffic, noise, water and air pollution, and land degradation (Figure 16.1; Williams, 1979). Figure 16.1 illustrates the relationships between the types of hazards in terms of their effects (short- and long term) and predictability (low to high). Single-factor hazards (for example, earthquakes) are intense geophysical hazards that are extremely difficult, if not impossible, to predict. While earthquakes may have relatively long-lasting economic effects, they usually threaten human life only for a matter of seconds. Volcanic eruptions threaten over a longer period, as do cyclones, and both are entirely geophysical and difficult to predict. The effects of drought and, to a greater extent, flood can be exacerbated by human intervention, often self-inflicted by carrying out commercial activity and inhabiting drought and flood-prone lands. While these are geophysical phenomena, their impacts can be diminished by not building on river floodplains. Also, human activities such as the clear-

ing of forests can indirectly increase the flood hazard within drainage basins. Wise management of catchments can therefore reduce the incidence of river flooding.

Land degradation lies farther up the curve of hazards (Figure 16.1) because it is a hazard with medium-term effects and is reasonably easily predicted. Ease of prediction has increased as more has become known about the interactions of flora, fauna, soils, agricultural and forest practices, and climate. To a large degree, land degradation is human-induced and therefore reversible if irreparable damage has not occurred. Pollution of water and air may occur from a point source or diffuse sources. In both instances these hazards are the product of poor environmental management and can, therefore, be treated if there is sufficient will and available resources to tackle problems. At the crest of the hazard curve are traffic hazards and anti-social activities, both multi-factorial, complex hazards with high predictability. In this Chapter, management strategies to overcome or reduce the worst effects of the hazards outlined on the curve, up to and including air and water pollution, are discussed in the context of Australian and New Zealand environments.

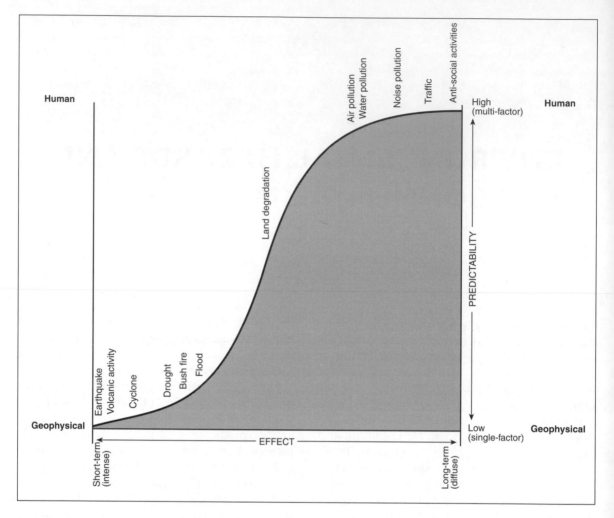

FIGURE 16.1 Natural and human-induced hazards (adapted from Williams, 1979)

GEOPHYSICAL HAZARDS

Earthquakes

Because Australia is not situated near plate boundaries, prediction of the potential location of earthquakes is difficult. However, areas of Australia considered at risk from earthquake damage include the western portion of Western Australia, including an area east of Perth centred on Meckering that was affected in 1968 by an earthquake of magnitude 6.8 on the Richter scale (Conacher and Murray, 1969). A belt running northwards within eastern South Australia to the border with the Northern Territory, southern and eastern Victoria and south-eastern New South Wales are areas at risk (Bryant, 1991). While these regions contain the state capital cities of Adelaide and Sydney, the most damaging earthquake to occur in Australia was at Newcastle, north of Sydney, in 1989. Here, 12 people lost their lives as buildings collapsed (Hunter, 1991), with A\$4 billion damage and insurance costs of approximately A\$1 billion.

New Zealand is much more prone to earthquake activity than Australia, hence its nickname the 'shaky isles', largely because of its position on plate boundaries, particularly the subducting section off eastern North Island. The eastern and southern parts of North Island and the northern part of South Island are the most affected. Although hundreds of earthquakes are recorded every year, many are unnoticed. The largest earthquake since European settlement occurred in 1855, magnitude 8.1 on the Richter scale and centred on Cook Strait (MfENZ, 1997).

Napier, on the east coast of North Island, suffered a damaging earthquake in 1931 that killed more than 250 people. Besides causing structural damage in towns and cities and causing loss of life, earthquakes are responsible for creating changes in land elevation and triggering landslides, particularly in mountain areas. It is estimated that the probability of a major earthquake occurring in the Wellington region within the next 100 years is between 40 and 85 per cent (Berryman, 1990). The city centre and transport arteries are therefore at risk, but building design to withstand shaking in earthquake-prone regions is now becoming more common.

Volcanoes

There are no active volcanoes in Australia, but in New Zealand there has been loss of life due to volcanic activity over the past 150 years. Since European settlement the largest eruption occurred in 1886 at Mount Tarawera, some 30 km to the south-east of Rotorua on North Island. Lake Tarawera was a major tourist attraction that had terraces and pools formed from thermally derived silica deposits over which Mount Tarawera towered. The appearance on the lake of a phantom Maori war canoe was said to foretell of an impending disaster. Four days later Mount Tarawera erupted. Within six hours over 1.4 million hectares of countryside had been covered in ash, with other parts affected by lava and mud (MfENZ, 1997).

A Maori village and the terraces and pools were destroyed, and 153 people were killed. The land around Mount Tarawera is now a Maori reserve.

More recently, Mount Ruapehu, in the Tongariro National Park, erupted in 1945, 1969, 1975, 1988, 1995 and 1996. The 1995 eruption produced spectacular emissions of ash and steam. It was, however, the deposition of lava during the period of the 1945–47 eruptions that indirectly caused great loss of life in 1953. Lava deposits from the eruptions of the 1940s at the Crater Lake on the northern side of Mount Ruapehu caused the lake level to rise to a point where the overflow burst in flood. The ensuing flood wave swept away a railway bridge moments before a train arrived and over 150 people were killed in the crash.

Atmospheric hazards

Hazards related to atmospheric conditions vary in intensity and scale of occurrence, from localised storms to regional-scale droughts. Tropical cyclones, tornadoes, thunderstorms, droughts and floods fall within this category, and all are of a largely geophysical nature, although floods may have a human-induced component (Figure 16.1).

Tropical cyclones are non-frontal cyclonic (low pressure) systems of tropical origin largely affecting Australia's northern, north-eastern and north-western coastal regions. Their main features and origins are described in detail by Oliver (1973) and Sturman and Tapper (1996). To summarise, they can produce winds greater than 300 km per hour and deliver over 3000 mm of rain. Their impact, therefore, includes destructive-force winds, severe flooding and storm surges on the coast (Chapman, 1994). It is estimated that between 1909 and 1980 there were ten cyclones per year in Australia on average (Bryant, 1991). Although not a large cyclone by world or even Australian standards, tropical cyclone Tracy was the most damaging. It brought destruction to the city of Darwin in the Northern Territory on Christmas Day 1974, with approximately 90 per cent of housing destroyed or damaged beyond habitation. Sixty-five people were killed (Chapman, 1994).

New Zealand is sometimes affected by tropical cyclones (once every five years: Sturman and Tapper, 1996). One of the most severe disasters to occur in New Zealand was due to the effects of cyclone Bola in 1988, which had an estimated 1 in 12 year recurrence interval. Rain, flooding and soil erosion caused almost $NZ120 million damage to property, and $NZ60 million of relief was provided (MfENZ, 1997). In an effort to reduce damage by tropical cyclones, forecasts of their likely movement are broadcast from Australian tropical cyclone warning centres and state emergency officers take action to minimise risks, particularly in urban areas (Chapman, 1994).

Tornadoes are rotating storms often associated with thunderstorms or bushfires (wild fires), and they can create severe damage where they make contact with the ground. While the most severe and damaging tornadoes occur in the United States, Australia and New Zealand witness their frequent occurrence, although they are less intense (Sturman and Tapper, 1996). In Australia, tornadoes are reported most frequently in the south-west of Western Australia and on the tablelands of eastern Queensland, while in New Zealand the Auckland and Bay of Plenty areas are the most affected.

Thunderstorms, although usually of only one to two hours duration, can also be damaging in the tropical and warm temperate regions of Australia and New Zealand. High wind shear, heavy precipitation (including hail) and lightning are typical. Chapman (1994) reports that the three-hour storm that affected Sydney on 21 January 1991 had wind gusts of up to 230 km per hour and hail up to 7 cm in diameter. Rainfalls exceeded the 100-year recurrence interval at some recording stations. A severe hailstorm struck Sydney in April 1999, causing widespread damage: 20 000 buildings and at least 30 000 vehicles were said to have been affected. More homes may have been damaged than during the Newcastle earthquake of 1989.

Drought, unlike other geophysical hazards, is generally defined by its effects rather than the occurrence of a well-defined short-term physical phenomenon. The temporary lack of water availability leads to a situation where demand is gradually not met, increasingly causing stress. In some lesser-developed parts of the world, notably Africa, the effects of drought often lead to starvation and widespread death. In Australia and New Zealand droughts generally have the effect of limiting agricultural production over wide areas, with attendant economic and social repercussions resulting from lack of domestic water and reduced incomes. Chapman (1994) emphasises distinctions between drought, aridity and desertification.

Above all, drought is a temporary phenomenon that can occur in any climate and is unpredictable in occurrence, frequency, duration and severity. On the other hand, aridity is a permanent condition (excepting global climatic change), where low precipitation, high evaporation and high solar energy inputs are normal. Desertification, by contrast, is a human-induced condition resulting from unwise land management practices leading to loss of vegetative cover, depletion of water supplies, loss of soil fertility and other types of environmental degradation.

As discussed in Chapter 4, the serious droughts of the years around the turn of the twentieth century, coupled with lack of knowledge of the carrying capacity and regenerative properties of the Australian semi-arid lands, led to severe land degradation (from which parts of the country have still to recover). There have, however, been other periods of drought in Australia (Figure 16.2), many of them clearly associated with the El Niño phenomenon (Chapter 3). Boughton (1993) states that between 1864 and 1983, there were 73 years when drought conditions prevailed, 61 per cent of the total, although it is rare for 'more than 20 per cent of Australia to be affected by drought at any one time' (Chapman, 1994: 125). New Zealand also suffers from extended periods of low rainfall,

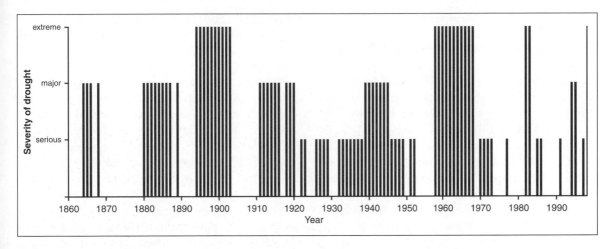

FIGURE 16.2 Years of significant drought in Australia (source: ABS, 1992b Commonwealth of Australia copyright reproduced by permission)

but they are less widespread and of shorter duration than in Australia. Hydroelectricity production (HEP), rural and urban water supplies and agricultural activities can be seriously reduced. The occurrence of drought is more common in areas of lower rainfall, such as the east of both islands. 'Over a dozen significant events have occurred between 1978 and 1994' (MfENZ, 1997: 7.19).

Often associated with drought are outbreaks of bushfire (or wildfire), although climate variability in Australia is an important factor, where during the wetter period there is a build-up of fuel on the forest and woodland floor. Loosely packed litter and dry grasses, the flammability of eucalypts and dry and windy weather conditions all contribute to the bushfire hazard.

Fire has been a management tool for both Aborigines and European settlers (Chapter 4), but the incidence of bushfires has increased since European settlement. The occurrence of remnant woodland and forest, and the large number of individual and small stands of trees within, and adjacent to, built-up areas are features of many Australian homesteads, villages, towns and cities. These conditions increase the potential for loss of property and life. In February 1967 devastating bushfires caused the loss of 62 lives and almost 1300

dwellings in bushland in the western suburbs of Hobart, Tasmania, during extremely hot weather following a moist spring that promoted vigorous vegetative growth (Solomon and Dell, 1967). This devastation was surpassed in the Ash Wednesday fires of 16 February 1983, at the end of the 1982–83 drought. In Victoria and South Australia over half a million hectares of country were burned and 76 people were killed (Chapman, 1994). During another hot, dry period in the summer of 1993–94, New South Wales was affected by fires that reached within 10 km of the Sydney central business district.

Using data from the Department of Bushfire Services of New South Wales, Chapman (1994: 30) showed that the most common cause of bushfires is 'miscellaneous, unknown' (27.5 per cent). Legal 'burning-off' to reduce the fuel load on the forest floor, optimally done during the coolest part of the year, accounted for 12.3 per cent of bushfires. Illegal burning-off accounted for 15.3 per cent, while lightning strikes and arson were responsible for a further 5.6 per cent and 8.4 per cent, respectively.

Whittow (1980) has recognised several hydrologic causes of floods, including rainstorm-river floods, snowmelt floods and coastal floods caused by temporarily elevated sea levels. In addition, the failure of dams and river levees, the natural rupture of lakes and land subsidence can also contribute to the hazard. Rainstorm-river floods are the most frequently occurring type in Australia and New Zealand. Their magnitude and frequency will depend on precipitation intensity and duration, the moisture status of the drainage basin, and the infiltration capacity of the soils and surfaces. Land use change within a catchment can influence the amount of water stored within it. If natural vegetation and soils are replaced by the relatively smooth and impervious surfaces of urban areas, runoff velocities and the volume and frequency of flows will increase. Therefore, urbanisation will very likely increase the flood hazard.

The impact of floods is related to the depth and duration of inundation and the velocity of the flood flows (Chapman, 1994). Damage to buildings, agricultural land, lines of communication and disruption to lives is said to cost Australia $A400 million per year on average (DESTA, 1996). In New Zealand, floods in the decade 1976–85 cost over $NZ800 million (1997 value), with a loss of 17 lives (Pearson, 1992). These impacts are, however, relatively small compared with the disruption and damage caused by floods in India and Bangladesh.

Geomorphologically, floodplains are part of the active fluvial system and they often have the best land for agriculture. While farmers are likely to be aware of the risk of flooding, it is urban dwellers who are often ignorant of the hazard and most liable to suffer. While many valleys of Australia and New Zealand are potentially flood-prone, it is those which are most closely settled that have attracted research attention and publicity. Although all capital cities and virtually all country towns in Australia are susceptible to flooding, storm-induced floods in Sydney and along the Illawarra coast to the south, seem to have increased in frequency (Bryant, 1991). One such storm occurred in February 1984, in an area of Wollongong approximately 100 km south of Sydney. The 24–hour rainfall was the greatest to occur in temperate Australia (796 mm, in an area with a mean annual

rainfall of 1600 mm). The flood caused a four-fold increase in stream channel capacity in the foothills of the Illawarra escarpment (Nanson and Hean, 1985).

In New Zealand, flooding can occur in all seasons and regions, although it is most frequent in the steep catchments of the west coast of South Island. The effects of increasing urbanisation on flood magnitude and frequency have been documented for a part of Auckland (Williams, 1976). Here, small floods increased in their frequency as urbanisation began, but large floods became more frequent only after 53 per cent of the drainage basin had become urbanised. Because there was an increase in the area of impervious surfaces, floods became more 'flashy' and had greater peak discharges. Artificial channels and sewers created more flow pathways, thus making the system generally more efficient for conveying storm runoff.

Response to geophysical hazards

It is important to recognise that disasters have both physical and social components, the latter including the vulnerability of the community. Chapman (1994) identifies several possible responses to geophysical hazards:

- avoid the hazard by relocating human activities
- modify the cause of the hazard
- modify the hazard environment
- modify the loss potential by accepting that the hazard will occur and cannot be controlled
- share the losses within the community
- do nothing.

In Australia, 'disaster management is moving away from the traditional reactive to a proactive framework' (Zerger, 1997: 469). For all geophysical hazards there have been, and will continue to be, emergency services organised by government and volunteer organisations. Enquiries set up after major disasters usually point to ways of improving the response. For example, there is a 'Newcastle Earthquake Database' on the World Wide Web which provides information on disaster management, earthquake engineering, economic impact, health issues, insurance and other services. Emergency Management Australia has been set up to reduce the impact of disasters in Australia and its region. At Macquarie University, in Sydney, the Natural Hazards Research Centre (NHRC) began operating in 1994, aiming to generate applied and strategic research and training for the insurance industry. In New Zealand there is the Institute of Geological and Nuclear Sciences Limited (IGNS), part of whose function it is to assess the risks and manage the impacts of earthquakes, volcanoes, landslides and tsunami.

Because New Zealand is in a highly active seismic zone, the IGNS provides a number of earthquake-related services, including:

- the assessment of faults, their type, rates of activity and history of movement
- the analysis of the effects of earthquakes on shaking intensity
- the mapping of earthquake zones
- the provision of advice on the design of earthquake-resistant structures
- an analysis of the vulnerability of roads, water supply and communication lines.

The NHRC at Macquarie University is carrying out research on loss estimation for potential earthquake damage in Australian capital cities. Detailed studies of losses in the 1989 Newcastle earthquake provide the basis for this work because this disaster was Australia's most expensive insurance loss. In addition, Standards Australia has released a new building standard (1998) for reducing earthquake damage to structures.

Volcanic hazard assessment services, relevant to the volcanic region of North Island, New Zealand, are available from the IGNS. While it is recognised that it is impossible to predict when the next eruption will occur, the institute advises on its likely effects. Using information on past eruptions, hazard maps, eruption scenarios and assessments on the vulnerability of lines of communication and water supplies can be provided.

Little can be done to alter the dynamics of tropical cyclones, and urbanisation continues to grow in northern Australia: 'an impediment to successful cyclone risk reduction in the Australian tropics remains not technical, but political' (Zerger, 1997: 470). Similarly, drought cannot be relieved by changing weather patterns, although cloud seeding to induce precipitation has been attempted without success. In general, mitigation of the effects of drought has included the provision of water by tanker, feed for animals, financial support and insurance. In urban areas restrictions on the use of water for non-essential purposes, such as the irrigation of parks and gardens, are common.

Rural Australia is, however, much more prone to economic hardship during extended periods of low rainfall. Farmers can receive advice on drought management from government departments and apply for financial assistance. This can be used for freight concessions to bring fodder, transport stock to regions unaffected by drought, and purchase seed and animals to restock after the drought (Chapman, 1994). Australian government and state ministers agreed to a national drought policy in 1992. The policy encourages primary producers to be self-reliant and maintain resources during drought, and to ensure rapid recovery. There are a number of financial support schemes available through the policy (DESTA, 1996).

Probably there is no more graphic image of hazards in Australia than bushfire. Because of its threat to life and property there is considerable debate on how the hazard can be reduced. Management of the fuel load on the forest floor is generally regarded as essential, but this can often lead to outbreaks of fire when it is done without due care. Control burning is controversial because its effects on the survival of species of plants and animals is sometimes unknown. In the event of serious fire, volunteer and state firefighters (Figure 16.3) are called in to try to control the fire by applying water, cutting fire-

FIGURE 16.3 A New South Wales State Forests fire-crew attends a bushfire (photo: J.P. Armstrong)

breaks and evacuating civilians. This can be extremely hazardous: the fire can jump forward to create new fronts, and shifts in wind strength and direction can redirect the fire-front and therefore endanger lives. The mobilisation of fire-crews in times of emergency utilises the Australian ethic of 'mateship': working together for a common cause. The summer of 1993–94 was one of severe fires in New South Wales. 'Volunteer fire-fighters were airlifted from all parts of Australia, and some interstate crews drove their fire-engines non-stop the thousand plus kilometres to fires in New South Wales. Some 2000 regular fire-fighters and 16 000 trained volunteers were co-ordinated in a paramilitary operation which was remarkably effective' (Chapman, 1994: 39).

Flood management has three components (Chapman, 1994):

- modification of the flood hazard by changing the behaviour of the flood
- minimising the risk of damage
- response measures.

Catchment management is employed in Australia and New Zealand to modify the flood hazard. The use of dams, contour banks and ponds to trap runoff, reafforestation to increase infiltration of rainfall and delay runoff, and the raising of river bank levees to increase river channel capacity are common measures. Risk can be minimised by restricting development on the floodplain, while response measures include flood warning systems and evacuation plans.

The New Zealand government passed a Soil Conservation and Rivers Control Act in 1941 which led to the establishment of catchment boards, responsible for flood control, mainly by building embankments (stopbanks) and drainage works. In Australia, the three components listed above have been employed in the Hunter Valley, some 200 km north of Sydney, New South Wales (Box 16.1).

MANAGING LAND AND WATER RESOURCES

The impact of human activity on land and water resources in Australia and New Zealand

---

Box 16.1 Flood mitigation in the Hunter Valley, New South Wales

Flooding has been a problem in the Hunter Valley since the advent of European settlement and a range of mitigation options have been employed since the middle of the nineteenth century.

The Hunter River basin has an area of 20 000 km² and contains valuable agricultural land, coal mines, commercial and industrial centres. During the early years of European settlement (from 1804 to the end of the nineteenth century), there was large-scale clearance of native timber from slopes and along streams. This, together with the development of pastures for cattle and sheep, led to greater rates of surface runoff and a perceived increase in flooding. During the period 1870 to 1895, there was a largely piecemeal construction of levees to protect agricultural land along the river. The levees, however, had the effect of raising water levels in unprotected areas. In addition, the deposition of sediment in stream channels reduced capacities and exacerbated flooding. In 1950 the Hunter Valley Conservation Trust was established because of widespread

concerns about the degradation of the river basin.

The largest flood on record occurred in February 1955 and there was considerable destruction of bridges, buildings and agricultural land, as well as loss of life and homes. In response to this event the Hunter Valley Flood Mitigation Act was passed by the New South Wales government in 1956. The Act enabled the integrated construction of levee and diversion banks, floodways, flood gates and drainage channels in the lower valley (Figure 16.4). Within the catchment, conservation works for water and soil continue to be carried out. In 1989, the Hunter Valley Conservation Trust became the Hunter Catchment Management Trust, under the New South Wales Catchment Management Act.

Responsibility for floodplain management rests with local councils who must prepare and implement management plans with the assistance of state government departments. As an aid to this process, flood inundation maps were prepared for some of the towns in the Hunter Valley.

---

was outlined in Chapter 4, where it was concluded that very little of either country was free from the marks of occupancy.

Prior to European settlement in New Zealand, Maoris recognised that scarce resources required protection. This was done by fortification, rules and customs to control over-exploitation. Because of low population densities throughout Australia before European invasion, there would probably have been little need for conservation strategies to be put in place by the Aborigines. Nevertheless, management of the environment was practised at least within the past 5000 years, a period of Aboriginal population expansion (Kohen, 1995) (Chapter 4). This expansion was probably very slow, and land management practices would have evolved gradually. However, with the advent of European settlement in New Zealand and Australia, things rapidly changed.

Three 'eras in evolving Australian environmental visions' have been recognised for the

period of European settlement (Frawley, 1994) which can equally be applied to New Zealand:

- Era 1: exploitative pioneering
- Era 2: national development and wise use of resources
- Era 3: modern environmentalism.

FIGURE 16.4 Flood mitigation levee protects the settlement of Lorn during a flood on the Hunter River, NSW (photo: R.J. Loughran)

From the time of European settlement until the end of the 19th century there was an almost immediate change from Aboriginal management to an exploitative attitude towards the environment (Era 1). Economic growth by closer settlement and increased agricultural production took place in both countries. This was coupled with a view that forest resources were inexhaustible. However, a Forests Act, aimed at slowing forest destruction, was introduced in New Zealand in 1874, but was repealed because it was regarded as contrary to economic development. There were, nevertheless, other acts brought in to prevent environmental damage and to control hazards and pests in New Zealand. This changing attitude heralded the beginning of Era 2.

The beginning of the twentieth century was a time of national development and the linking of environmental management with resource development (Era 2). In New Zealand it was recognised that further expansion of agricultural production by forest clearance was unwise, and more intensive use of land already in production was encouraged by government. This was paralleled in Australia through 'a public policy framework ... to apply scientific and economic principles to the efficient utilisation of resources' (Frawley, 1994: 66). For example, pastoralists in the Western Division of New South Wales were awarded longer leases in the hope that this increased security would promote better land management (Breckwoldt, 1988).

'Wise use' concepts are best illustrated in water management, where Australian states took on the role of developers and managers (Frawley, 1994). The use of water supply dams for irrigation was one way of expanding agricultural production into the Australian hinterland. Four large water projects founded on large dams were put in place after the Second World War, but they had their beginnings late in the nineteenth century or in the first-half of the twentieth century (Smith, 1998). These schemes included the Snowy Mountains scheme, the Ord River, HEP in Tasmania and the Burdekin River Dam in Queensland.

However, their economic worth and environmental viability have been questioned (see Chapter 8). Of these, 'the Ord Scheme is widely regarded as a colossal folly, and is used as an example of the worst excesses of the pork-barrelling of federal funds' (Smith, 1998: 170) (Box 16.2).

During Era 2 it was seen that catchment protection for soils was required, and that soil conservation to reduce sedimentation and to maintain soil productivity was necessary. A Soil Conservation Act came into force in New South Wales in 1938, having been preceded by many years of field demonstrations, overseas study tours and political lobbying. National forest reservations were proposed for timber production, but there was conflict because many still wished to clear additional land for agriculture. Wise conservation was, therefore, 'essentially utilitarian' (Frawley, 1994: 69).

It was at this time that ideas about preserving land and forests began to come to the fore. This led to the establishment of national parks for enjoyment and recreation, usually near centres of population. The Royal national park south of Sydney came into existence in 1879 as a public reserve, and was dedicated as a public park in 1886. This is the second oldest national park in the world after Yellowstone Park (1872) in the USA.

Modern environmentalism (Era 3) began in the 1960s. It was marked by increasing public concern about the state of the world environment and its perceived mismanagement. For the first time 'the environment' became a political issue. Community groups that were concerned about environmental issues, such as the logging of native forests and the loss of wilderness, lobbied governments and organised protest demonstrations. Government agencies, which were often seen as facilitators for the development of forestry, fishing, agriculture and minerals rather than as regulators, came into conflict with community conservation groups.

In New Zealand, for example, a Forest Service scheme to clearfell 340 000 hectares of beech and podocarp forest in South Island was

Box 16.2 The Ord River irrigation project

A major problem in tropical Australia has been how to intensify agricultural production. One solution, applied in a number of cases, has been to utilise irrigation via substantial investment in large irrigation projects designed to overcome the seasonal variability of rainfall. In drier areas water from these projects has been viewed as a basis for general economic exploitation or, as in the case of the Lower Burdekin in Queensland, it offers the potential for substantially increasing the intensity of agricultural production.

One of the biggest schemes is on the Ord River, near Wyndham in East Kimberley (WA). This is an area of 320 000 km² dominated by extensive pastoralism. In the late nineteenth century it was championed as a potential tropical paradise, but subsequent experience has revealed otherwise, through the effects of heavy monsoon rains followed by periods of prolonged drought. Pioneering settlement in the 1890s was associated with the Durack family who brought cattle to the area after a two-year cattle drive over 4800 km, and then operated a 2 830 000 ha estate. Many other cattle brought on long droves to the region by their accompanying would-be settlers, failed to survive Kimberley's harsh environment.

At Kununurra a township was created with federal funds as a 'safe' location for evacuees from Darwin after it had been bombed by the Japanese during the Second World War. The intention was to grow sugar in the vicinity, supported by irrigation water from the Ord River. A major project to this effect was started in 1959 with the intention of irrigating 12,000 ha. A second stage costing A$75 million was the creation of Lake Argyle in 1968, intended to be part of a federal cotton production scheme, with the government constructing a cotton gin and running a cotton stabilisation scheme. However,

between 1963 and 1973 there were only three years when the market value of the crop covered the growing costs. Major problems with disease and pests were experienced, and by 1975 all cotton growing in the area had ceased. Other problems that have affected all attempts to establish more intensive agricultural systems in the region are the high labour and food costs, and the high costs of importing the necessary inputs to such systems. Nevertheless, there have been more recent schemes aimed at expanding commercial cropping in the region. This has been linked to the Ord River HEP scheme, completed in 1996, which produces 30 MW of power, with transmission lines to the Argyle Diamond mine and to Kununurra.

The irrigation on the Ord is being expanded through the Ord Irrigation Co-operative formed in late 1995 which is operating and maintaining existing systems with over 100 farmers producing crops worth A$60 million per annum. An extra irrigated area of 64 000 ha is planned, with 50 000 ha for broadacre crops (especially sugar and cotton) and 14 000 ha for horticulture. This perseverance with cotton is reflected in similar schemes in other parts of tropical Australia, so that cotton is now the third most extensive field crop after wheat and sugarcane.

The other main crops grown in the country under irrigation are rice, grass, fruit, vegetables and ornamental plants. However, much of this is produced outside the tropics. For example, rice, the major irrigated crop, is grown mainly in the Murrumbidgee irrigation area in the Riverina of NSW. Other large irrigated areas are Sunraysia, along the Murray in the Victorian Mallee, the Goulburn Valley in northern Victoria, and the Burdekin irrigation area on the central coast of Queensland.

abandoned in 1975 because of public opposition (MfENZ, 1997). Proposals to explore for oil on the Great Barrier Reef in Queensland and mine mineral sands on the coast of New South Wales created conflict in Australia. The weight of these issues and the public support they received helped legitimise the environmental

movement and give it political influence. Consequently, government policy and methods of environmental management now include environmental protection, to the extent that they affect the every-day lives of Australians and New Zealanders. These include the regulation of vehicle emissions, the use of unleaded

Box 16.3 Landcare

In the 1980s, rural producer organisations recognised the need to combat land degradation, despite the presence of long-established government soil conservation departments in every state of Australia. While the government agencies were able to treat the symptoms of land degradation, they were unable to affect the cultural processes responsible for it (Baker, 1997). The National Farmers' Federation and the Australian Conservation Foundation proposed that Australia's pastoral and agricultural lands should be used within their capability by the year 2000, and that there should be sustainable land use from that time (Roberts, 1992). This proposal was adopted by the federal government, and it became a basis of the decade of Landcare under the Hawke Labor government. Funding was provided, for example, for tree planting, demonstration days, and controlling pests and weeds.

There are now approximately 4250 Landcare groups in Australia, and about one in three farmers are members, although the percentage varies from state to state. The highest membership is in Western Australia where, in 1992, 48 per cent of farmers belonged to a Landcare group. This probably reflects a well-funded publicity campaign in that state (Macgregor and Pilgrim, 1998).

There is, however, debate about the purpose of Landcare: economic sustainability (maintaining production) or ecological sustainability (Baker, 1997)? There is concern that Landcare is not addressing biodiversity conservation, and landholders may be more concerned about economic impacts. However, such criticisms may be unjustified 'or reflect unrealistic expectations of voluntary groups' (Curtis, 1997: 152).

petrol, controls on noise and air pollution in built-up areas and the non-removal of native trees from suburban gardens. In Christchurch, New Zealand, the city has been declared a Clean Air Zone to control air pollution from open fires. Homeowners receive grants to convert open fireplaces into other forms of non-polluting heating appliances. There are also restrictions on the types of fuel that can be burnt (MfENZ, 1997).

One of the most significant movements for land management and conservation to emerge in Australia has been Landcare. Landcare involves landholder groups, individuals and government working together on common land management problems (Box 16.3) (Roberts, 1992; Conacher and Conacher, 1995). Previously, land management and soil conservation operated as a 'top-down' approach in which landholders were advised by professional soil conservationists working within government departments. The National Landcare Program is now incorporated within the Australian government's Natural Heritage Trust, described by the government as 'the largest environmental rescue effort ever to be undertaken by an Australian government . . . It

represents a new era in environmental responsibility' (Natural Heritage Trust, 1999: www.nht.gov.au).

Water management in Australia is largely in the hands of state government and local authorities, often with the support of community groups. There is also a major water management initiative for the Murray–Darling basin. The Australian government, the governments of the four states that have territory within the basin (Queensland, New South Wales, Victoria and South Australia), and the community contribute to a management strategy. This aims to maintain or improve water quality, control or reverse land degradation, protect and rehabilitate the natural environment and conserve cultural heritage (DESTA, 1996). On a smaller scale, 'rivercare' community groups have been established in urban and rural areas.

Within New Zealand, regional councils are responsible for ensuring the maintenance of water quality, the abstraction of water, the discharging of pollutants and the placing of structures in lakes, streams and on the coast.

Forest management is likewise carried out by each state in Australia. By the 1920s all

FIGURE 16.5 Karri forest at Beedelup, Western Australia (photo: R.J. Loughran)

states had departments to manage forest resources, and in 1930 a national advisory bureau was created. The humid, steep and mountainous lands of Australia have the greatest areas of native forest (Figures 3.6, 16.5). Although there are privately owned forests, the majority are state-owned and logging is carried out under licence. There are many small communities in eastern and Western Australia that depend on forestry and sawmilling for their existence, and a decline in the number of forests available for exploitation could threaten their livelihood (Aplin, 1998).

Conflict between conservationists, who wish to see logging curtailed for the preservation of forest habitats, and the logging industry has become common. It is an environment-versus-jobs issue. Methods of forest management have also been a matter for concern, particularly where the profit motive seems to override a consideration of conservation values: for example, clearfelling or selective logging, and harvesting efficiency against potential damage to the environment. Replanting, seeding and natural regeneration are all now employed after harvesting (Figure 16.6).

The commercial exploitation of native forests in Australia has been accompanied by damage to the remaining trees. This takes the form of dieback caused by root attack which is provoked

FIGURE 16.6 The aftermath of logging at Nannup, Western Australia. Trash has been left to promote regeneration (photo: R.J. Loughran)

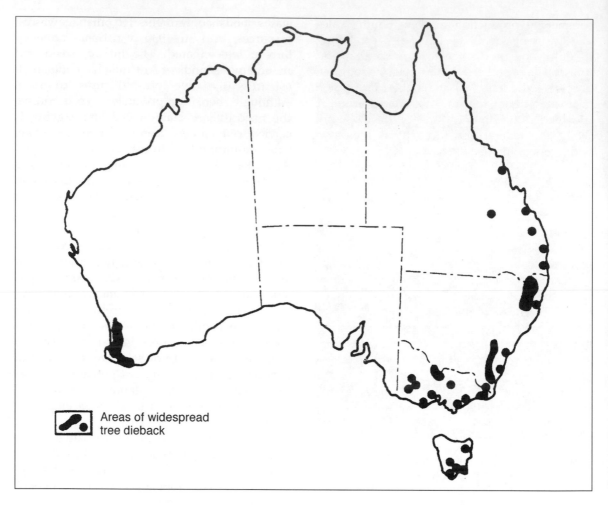

FIGURE 16.7 Areas affected by widespread dieback of trees (A.R.M. Young, 1996: 97 © Oxford University Press, reproduced by permission)

by the spread of pathogens encouraged by forest clearance activities. It has been especially common in one of the main timber species, jarrah. Where small stands have been left in the midst of grain or grazing pasture, dieback has been produced by the increased insect predation supported by the improved soils. Increased drought and salinity are other causes of dieback, which is now widespread in rural Australia (Figure 16.7), affecting the temperate eucalypt woodlands. Half of the forested and wooded lands in New South Wales are used for grazing by sheep and cattle and hence the problem of dieback is of major concern.

Since 1945 the conflict between development and conservation with respect to native forests has increased sharply. Conservation groups such as the Australian Conservation Federation and the Wilderness Society have been ranged against state forest authorities, the National Association of Forest Industries and Timber Town Associations. Major controversies have arisen over the exploitation of forests throughout the country, with the battles over Daintree, Lemonthyme and Terania Creek have become symbols of the ongoing struggle to halt forest destruction. From the 1960s rising Japanese demand for woodchip has encouraged commercial production in New South Wales and Tasmania, with the latter also expanding its pulpwood production for some new paper mills. The conflicts have been strong in Tasmania, the state with the highest proportion of its area under forest. The state government has tended to support forestry development whilst the federal government has identified with conservationists (see Chapter 8). Between 1975 and 1987 12 separate enquiries were held into forestry developments in Tasmania. In 1987 the Helsham Enquiry investigated whether the Lemonthyme and Southern forests qualified for world heritage listing. It ruled that they did qualify and subsequently, under pressure from electoral successes for the Green Party, the area under world heritage protection was expanded to twice the size recommended by the Enquiry. In the 1990s further challenges from conservationists have attacked the licences to export 2.8 million tonnes of woodchips from Tasmania held by the three principal operators (A.R.M. Young, 1996: 98–101).

After many decades of forest clearance throughout New Zealand, a State Forest Service was established in 1919. Its purpose was to preserve the remaining public indigenous forests and maintain steepland forest for erosion and flood control. Because the native forest trees take 60 to 120 years to reach maturity, the planting of faster-growing exotic species was necessary to meet New Zealand's needs. Between 1960 and 1990, timber production from indigenous trees declined from 50 per cent to only 5 per cent of the total, very largely due to pressure from environmental groups to preserve native forests that were under State Forest Service control (see Chapter 14). Attention was also turned on privately owned indigenous forests that were being exploited for logs and woodchip, or cleared for agriculture.

In 1992 a New Zealand Forest Accord was signed between the major forest companies and several environmental groups, 'with the industry representatives agreeing not to replace indigenous forest with exotic plantations and the environmentalists agreeing to support exotic forests as a renewable, environmentally friendly, alternative to the logging of native forests' (MfENZ, 1997: 8.31) (Box 16.4).

Box 16.4 Protection of native timber in New Zealand

New Zealand has 6.2 million ha of indigenous forest, of which 1.3 million ha (21 per cent) is in private hands, half belonging to Maori. However, by 1995 the area under plantations of exotic timber had risen to 1.6 million ha (nearly 10 per cent of New Zealand's land area). Concerns over the continued expansion of exotic species and destruction of native forest led to the 1992 Forest accord between environmental groups, government and the New Zealand Forest Owners Association (representing 90 per cent of the country's plantation forests). As part of the Accord, the environmentalist signatories acknowledged the role of plantation forests as a suitable means of sustainable production for wood products. Those areas of native forest on private land not covered by the accord are partly excluded in the 1993 Forest Amendment Act, which prohibits the export of woodchip or logs from native forests, with some exceptions, notably certain categories of Maori land. This has caused some controversy, for example over the rimu forest in Waitutu Forest on the borders of the Fiordland National Park, where Maori sold 20-year cutting rights to a commercial operator.

Wall and Cocklin (in Le Heron and Pawson, 1996) cite another example: that of the East Coast Forestry Project (ECFP) in North Island, in which the Ngati Porou have supported afforestation with exotic species under a government-funded scheme begun in 1992 to promote soil conservation whilst generating commercial returns and employment. The intention is to plant 200 000 ha of forest over a 28-year period through a Maori-owned commercial venture, Ngati Porou Forest Ltd. There have been conflicts with environmental groups and some mismatch between the need for the project to generate infrastructure and employment in the north of the region and the initial concentration of plantings in the south. Hence, whilst the Ngati Porou are the intended beneficiaries of the project, constraints are restricting this.

The Australian coastal zone has also been severely degraded (Chapter 4), and is threatened by a rise in sea level created by climate change. However, management of the coastal and fishing zone in Australia is fragmented and there is no real holistic policy (Aplin, 1998). Harvey *et al.* (1999) put forward methods for improving assessment of the vulnerability of Australia's coastal zone, suggesting there should be a move away from over-emphasis on sea-level rise and climate change. They focus instead on the importance of biophysical and human-induced coastal changes.

The largest protected marine park in the world, the Great Barrier Reef Marine Park off the coast of Queensland (Chapter 2), was placed on the world heritage list in 1981 (see below). Access for visitors and development is planned and controlled to maintain environmental protection.

Climate change and air pollution

While climate change has always been a feature of the global environment, there is strong evidence of global warming since at least the 1950s. It is recognised by many scientists that the addition of carbon dioxide, nitrous oxide and methane (greenhouse gases) to the atmosphere will produce warming, the consequences of which have been outlined for Australia by Henderson-Sellers and Blong (1989). They predict temperature rise, rising sea level, a southerly shift of Australia's rainbelts, more intense rainfalls, a more southerly penetration of cyclones, a reduction in snow cover in the Australian highlands and an increased incidence of El Niño.

These types of climate change could lead to a greater frequency of erosive events on the coast and inland, more floods, heatwaves, droughts and bushfires: 'The time to plan is now' (Henderson-Sellers and Blong, 1989, 194). Australia's government has not responded to this exhortation. At the Kyoto Climate Change Convention in December 1997, Australia was one of only three nations to be allowed to increase its greenhouse gas emissions.

However, flexibility within the Kyoto protocol allows governments a range of options when addressing climate change, one of which is the utilisation of carbon sinks. This includes the planting of forests to offset emissions of carbon dioxide, as is proposed in a scheme by NSW State Forests. New Zealand is committed to reducing net emissions of carbon dioxide to their 1990 levels by 2000, and then stabilising them.

BIODIVERSITY, RESERVES, NATIONAL PARKS AND WORLD HERITAGE

In one form or another, the preservation of landscapes has been practised for over 100 years in Australia and New Zealand. The criteria for

FIGURE 16.8 National parks in New Zealand

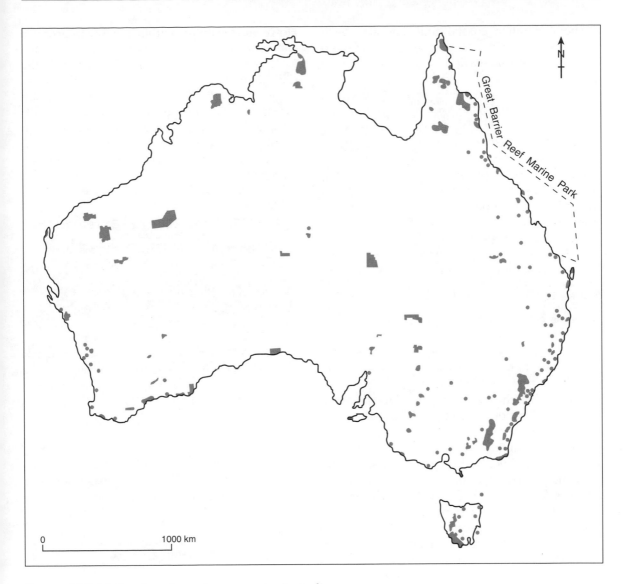

FIGURE 16.9 National parks and reserves in Australia

selecting areas as reserves and national parks have usually included proximity to centres of population (for recreation), scenic grandeur and economic worthlessness (Kirkpatrick, 1994). However, where there is economic value, the conservation interests of the national parks may be compromised. For example, in Kakadu National Park in the Northern Territory, the Australian government believes uranium

mining at Jabiluka and world heritage values can exist side by side (see Chapter 5).

Protected areas are not necessarily representative of habitats where biodiversity exists. In New Zealand, most of the protected land is in the mountains of both islands, and little of the coast is included (Figure 16.8).

There are approximately 3000 reserves in Australia covering 4 per cent (over 3 million

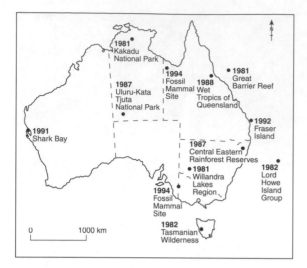

FIGURE 16.10 World heritage areas in Australia

TABLE 16.1 World heritage sites in Australia and New Zealand

Australia
The Great Barrier Reef, Queensland
Willandra Lakes Region, NSW
Kakadu National Park, Northern Territory
Tasmanian Wilderness World Heritage Area
Lord Howe Island, NSW
Central Eastern Rainforest Reserves, NSW and Queensland
Uluru-Kata Tjuta National Park, Northern Territory
Wet Tropics World Heritage Area, Queensland
Shark Bay, Western Australia
Fraser Island World Heritage Area, Queensland
Australian Fossil Mammal Sites, Queensland and South Australia
Sub-Antarctic islands of Heard, McDonald and Macquarie Islands

New Zealand
Tongariro National Park, North Island
Te Wahipounamu, south-west South Island

Sources: DESTA (1996); MfENZ (1997).

hectares) of the continent (Figure 16.9). Over 93 per cent of them are less than 10 000 ha, which may mean that they are not ecologically viable systems (Kirkpatrick, 1994). In New Zealand there are 13 national parks and 19 forest parks, covering an area of 4.2 million ha, or 16 per cent of the total land area. In addition there are 11 UNESCO world heritage properties in Australia and two natural sites in New Zealand (Table 16.1; Figure 16.10). Inclusion in the world heritage list signifies that a property has outstanding universal value.

In Australia and New Zealand, conservation authorities conduct a balancing act between protection and providing public access. Spectacular landscapes and historic sites are national assets and tourist attractions. The director-general of the New Zealand Department of Conservation (DoC) has written that 'DoC is not there to "lock up" land, but to open it up for public enjoyment while protecting its values' (Department of Conservation, 1997). Regulations and facilities for ecotourists are provided in such a way as to minimise damage to the environment (Figure 16.11). Increasing tourist demand is putting pressure on some popular national parks, however, and a system of permits may have to be introduced

to control walkers on the Cradle Mountain–Lake St Clair trail in the mountains of northern Tasmania, for example.

CONCLUSION

Conflicts between those who wish for increased economic development, perhaps at the expense of the environment, and environmentalists, have become a feature of the 1990s. Compromise is required on both sides but considerable progress has been made on a range of issues. The *Australia: state of the environment* report, (DESTA, 1996) lists eight 'commendable actions', including:

- the listing of world heritage properties
- the creation of structural solutions to complex management problems (e.g. the

Murray–Darling Basin Commission and the Great Barrier Reef Marine Park Authority)
- the limitation of water diversion from the Murray–Darling system
- the introduction of new fisheries acts for ecologically sustainable development
- the Landcare programme
- the phasing-out of ozone-depleting substances
- an increase in the use of renewable energy sources, such as solar power
- recycling of waste.

There are, however, a number of 'poor responses', including:

- an inability to manage the environment on a national scale because of interstate disagreements and a lack of common standards
- a lack of measures to combat threats to biodiversity
- government departments that promote economic development with little regard to environmental costs
- land clearing being allowed in some states and restricted in others
- poor urban and transport planning
- a shortfall in greenhouse gas emission reduction targets.

In New Zealand 'a lot has been done to recognise . . . environmental problems' (MfENZ, 1997: 10.22):

- The introduction of a Resource Management Act (RMA) in 1991 brings together laws governing land, air and water resources based on the notion of ecological sustainability.
- There is greater awareness of cultural heritage and the need to protect indigenous biodiversity.
- A large area of New Zealand is now permanently protected.
- Point-source pollution is being tackled.

The RMA, incorporating the principle of sustainability, reinforces New Zealand's 'clean

FIGURE 16.11 A board-walk in a fragile environment: Ulva Island, Stewart Island, New Zealand (photo: A.I. Loughran)

and green' image. The Act mandates environmental assessment of all resource consent applications, though concerns have been voiced regarding the extent to which the Act will protect from urban development the relatively small proportion of the country that is categorised as class 1 agricultural land (Grundy and Gleeson, 1996). Moreover, there has been much less concern than in the European Union regarding the adverse effects of industrial-style farming methods upon the environment. Perhaps the greatest concern has been voiced over the long term sustainability of pastoral use of the South Island 'High Country'.

Here half of the area is affected by the terms of lease of pastoral land under the 1948 Land Act which provides exclusive occupancy for purposes of pasturage, though other uses may be permitted upon application. However, there are strong pressures to convert the pastoral leases to freehold and to encourage land use diversification. The latter may include extensions of high visual impact forestry and reductions in stocking ratios on remaining rangeland, thereby permitting colonisation by adventive woody species (Swaffield in Le Heron and Pawson: 1996).

Other challenges for the future identified by the MfENZ include:

• the re-establishment of indigenous biodiversity
• the reduction of non-point (diffuse) pollution
• the introduction of sustainable land use.

With reference to the latter point, sustainability in relation to economic development has been under discussion for more than a decade. In both Australia and New Zealand the concept of ecologically sustainable development (ESD) has been adopted in legislation and policy, e.g. the RMA. However, political expediency often overrules environmental concerns, such as the increasing use and export of coal for power generation in Australia and overseas, despite the need to reduce greenhouse gas emissions.

FURTHER READING

A thorough description and review of Australia's environment and its (mis)management is provided by Aplin (1998). It includes discussion of philosophy and ethics, social aspects, and biophysical and physical sciences in relation to the Australian environment. Cocks (1992) is a stimulating review of the state of the environment in Australia, with consideration of how the environment could and should be better managed. Henderson-Sellers and Blong (1989) is a small book written for the (scientific) general public. It deals with how the greenhouse effect could affect the lives of Australians. See also MfENZ (1997), Smith (1998) and DESTA (1996).

17

THE WIDER CONTEXT AND THE FUTURE

It would not be an exaggeration to say that there has been a transformation post-1945 in Australasia's 'place in the world'. Throughout the nineteenth century and until the end of the Second World War, both Australia and New Zealand were identified very clearly as part of the British Empire – parts of the world coloured red on the atlases emanating from the UK. The population of both countries was derived largely from British and Irish stock, and their legal systems, official language, sporting and economic interests reflected their strong colonial ties. Those ties were symbolised in the Imperial Preference policy pursued by the UK and in Australasia's unwavering support for the UK's position in the two world wars. Indeed, the sacrifice made by the armed forces of Australia and New Zealand (the Anzacs) at Gallipoli in 1916 served to reinforce commitment to the UK and to the perceived common interests of the British Empire. Only since 1945 has there been a significant reorientation away from this focus upon a distant colonial power, and a search for a new identity.

This search has often proved difficult and there continue to be strong associations with the UK that have rendered the assumption of a new role and identity for Australasia a tortuous and contested process. This Chapter discusses critical elements in the creation of new 'identities' for Australia and New Zealand within the context of new geopolitical relations and the emergence of multilateral agreements in the Asia–Pacific region. It examines these evolving geopolitical relations, from the impact of military alliances to new trading links and the symbolic importance of the growing development of republicanism.

CHANGING INTERNATIONAL RELATIONS

For Australia the Second World War itself brought the recognition that the UK could not be relied upon to provide secure defence against enemy attack. As described below, this prompted moves to bring Australia and New Zealand under the American defensive umbrella and subsequently yielded greater trade with the USA. The military alliance with the United States led to both Australia and New Zealand sharing an involvement in the Vietnam War, further separating them from policies and interests pursued by the UK. More far-reaching than this in its effects, though, was the UK's entry to what was then (in 1973) termed the European Economic Community

(EEC), later renamed the European Union (EU).

Whilst this could not sever longstanding cultural and family ties between the UK and Australasia, it did hasten changing economic relationships and heightened a new awareness in Australia and New Zealand of the need to re-evaluate themselves in a regional context. Subsequently that context has been both a relatively local one, through closer economic links between Australia and New Zealand themselves, and much broader through growing trade with south-east Asia and involvement in geopolitical and economic initiatives encompassing the broad Asia–Pacific region.

From Gallipoli to Vietnam

For the first half of the twentieth century Australia's concern over the proximity of south-east Asia and its 'teeming masses' played a major role in its international relations, especially in the defence sphere and in the 'White Australia' policy. One outcome of the latter was the profound isolation of the country from its Asian neighbours, in favour of maintenance of various links to the UK and the Northern Hemisphere. Another manifestation was Australia's long term defence policy in which it gave strong support to Britain in both world wars. This support was effectively assured by the British admiralty's creation of a Pacific Fleet in 1913, including an Australian unit.

Post-1918 relations with the UK deteriorated, with several trade-related arguments, not to mention the infamous 'Bodyline' dispute during the cricket test series of 1932/3, and the rate of emigration from Britain fell. Nevertheless, the operation of Imperial Preference assured Australian farmers of a good market in the UK, and solidarity with the British cause was shown when the UK declared war on Germany in 1939 and Australia immediately followed suit.

Australia's strategic thinking post-1918 is indicated by the desire for some form of

'buffer' from the threat of invasion from the north. The buffer consisted of German New Guinea, including the northern Solomons and Nauru, awarded to Australia as a League of Nations mandate by way of German war reparations (Figure 17.1). This was a mini-imperialism, with Australian laws applying in these territories, including the Immigration Restriction Act which effectively prevented non-white immigration to territories in which the indigenous populations were all non-white.

Similarly, under Prime Minister Seddon in the 1890s, New Zealand had asserted a form of imperialism directed at obtaining control of several Pacific islands, notably Fiji, Samoa and Tonga. On a visit to the United States, Seddon even boldly expressed to President McKinlay a New Zealand interest in Hawaii. However, his grandiose schemes were soon rebuffed: the United States annexed Hawaii; the Governor of Fiji warned Fijians that New Zealand was after their land, and it remained a Crown colony; Samoa was abandoned by the British to the Germans and the Americans. Seddon's only 'success' was the annexation of the Cook Islands and, later, Niue. His foreign adventures, designed to win votes, achieved their aim of keeping him in power but without much enthusiasm from parliament for an imperial role other than that of being an independent country 'under Britain's wing'.

During the Second World War Australasia's substantial military co-operation with the United States pointed the way towards changed international relations post-war. Britain's inability to protect Singapore from the Japanese invasion, and its strategic emphasis upon the European and North African theatres of war, persuaded Australian Prime Minister John Curtin that Australia's principal military alliance should be with the United States, and that this would be the cornerstone of Australian defence in the south-west Pacific. Furthermore, Britain was forced to abandon Imperial Preference following its Lend–Lease Agreement with the United States, and the Attlee government, though divided on the issue, declined the opportunity to join the new

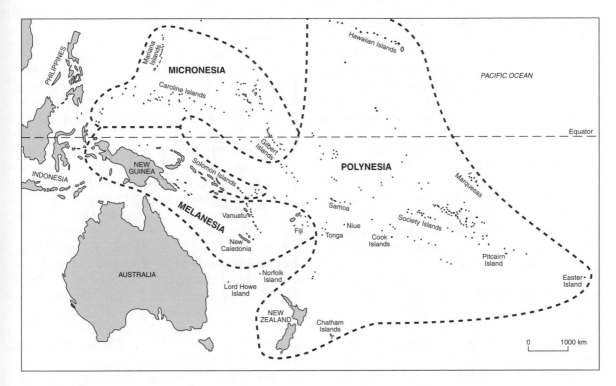

FIGURE 17.1 Melanesia, Micronesia and Polynesia

defence pact, ANZUS, agreed in 1951 between Australia, New Zealand and the United States. This agreement with the US helped tie Australia and New Zealand to the conflicts in Korea (1950–3) and Vietnam (1965–72).

In effect, US imperialism started to replace the British version, with the Americans seeking to use Australia as their base for a higher-profile presence in the South Pacific and building upon their links with Australia developed during the Second World War. Thus ANZUS functioned as a vehicle for protecting America's own Asia–Pacific interests, and on several occasions the US sided against Australian interests in the region, exemplified in American support for Indonesian-inspired insurrections in northern Borneo in 1965 and for Indonesia's claim to the western half of New Guinea, which has since become the Indonesian province of West Irian. However, American investment in Australia also increased, outstripping that from Britain by the late 1960s, as the US supplied

capital and technology, economic management and sophisticated producer and consumer goods. There were also several notable joint ventures, prefaced by General Motors combining with Holden in the 1930s.

For New Zealand the relationship with the UK remained of paramount importance until the 1950s. Only with the conclusion of the ANZUS and South-East Asia Treaty Organisation (SEATO) agreements, in 1951 and 1954 respectively, did defence links with other countries start to assume greater importance and New Zealand's geopolitical relationships begin to acquire a different form. In 1955 New Zealand withdrew its military support from the Commonwealth forces in the Middle East to focus attention upon the military action in Malaya as part of the ANZAM agreement, which had been signed in 1949 between the UK, Australia and New Zealand with respect to planning control of the defences within Australasia and Malaya. This reorientation was accompanied by the opening of New Zealand embassies and diplomatic offices throughout south-east Asia, marking the beginning of a wide-ranging set of relationships with Asian countries, including closer economic ties that ultimately proved vital to the New Zealand economy in the face of loss of easy access to its traditional UK market.

Membership of SEATO (which included the US, the UK, Australia, New Zealand, France, Thailand, the Philippines and Pakistan) and ANZUS led to New Zealand sending troops to conflicts in Thailand, Brunei, Borneo and, more significantly, to the Vietnam conflict in 1965. However, their role was not as prominent (or controversial) as that of the Australian forces in the same conflict, with never more than 550 New Zealand troops on the ground. This was part of the policy of Forward Defence which led to New Zealand being 'defended' in the Mekong Delta (Vietnam), a location as close to Western Europe as it is to New Zealand.

Australia first sent troops to Vietnam in April 1965. A total of 46 852 Australians served there, 37 per cent of them conscripted under the controversial scheme introduced by Prime Minister Menzies whereby all 20-year-old males registered for a ballot to determine whether they were called up for two years' military service. Domestic opposition to the war and to Australian involvement developed strongly in 1968 following the Vietcong's Tet Offensive.

By August 1969 public opinion polls revealed a majority of Australians opposed to continuing the conflict. Public protests grew, and an estimated one million took part in the last and biggest 'moratorium' (co-ordinated national demonstrations) in June 1971. By this time the size of the Australian military presence in Vietnam had been reduced, and when the Whitlam government came to office in December 1972 only a few 'advisers' remained. However, the legacy of divided opinion over the war was so great that it was not until 1987 that a 'welcome home' march was held in Sydney. The ultimate legacy of the war has been, first, the entry of refugees from Indo-China and, subsequently, large-scale

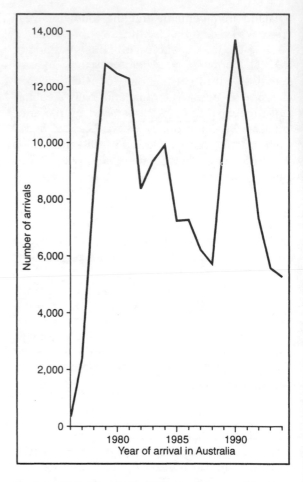

FIGURE 17.2 Number of Vietnam-born arrivals in Australia 1976–94 (source: ABS)

immigration from the region (see Chapter 9): 189 509 migrants between 1975 and 1995 (Burnley, 1994; Dunn, 1993; 1998; Viviani, 1996: 159) (Figure 17.2).

Whilst New Zealand and Australian troops were engaged in the Vietnam conflict, the UK withdrew its troops from Asia, confirming the significance of the ANZUS agreement to Australasian defence policy. Although the Labour opposition in New Zealand had not voted against the decision to send troops to Vietnam, by the time it was elected to government in 1972 it was following a policy opposing New Zealand's role in the conflict. By this time there had already been a reduced New

Zealand involvement, and the following year there was a complete withdrawal of troops.

Despite the unsuccessful intervention of ANZUS forces in Vietnam, the alliance remained the cornerstone of the military defence of both Australia and New Zealand until its existence was threatened by the anti-nuclear weapons policy of the New Zealand Labour Government, elected in 1984 (Box 17.1). The outcome of New Zealand enforcing a 'non-nuclear' policy has been to suspend American co-operation in the defence of New Zealand, though it might be argued that, with the ending of the 'Cold War' (and the removal of the 'Soviet threat'), such defence was less relevant anyway. However, close military co-operation has continued on a bilateral basis between Australia and the US, and between Australia and New Zealand.

THE ASIA–PACIFIC ECONOMIC CO-OPERATION (APEC) GROUP

The Asia–Pacific Economic Co-operation Group (APEC) was launched in November 1989 after a meeting of national leaders from the Asia–Pacific region in Canberra. The initiative for the meeting came from the drive for greater regional economic co-operation by Bob

---

Box 17.1 Nuclear-free New Zealand

New Zealand's stand against France's testing of nuclear weapons in French Polynesia from the 1970s has earned respect from its smaller neighbours and clearly established its role as a regional leader. Its stance, opposing the French and declaring itself to be 'nuclear-free' under the Lange government in 1984, has affected its relations with the Americans and has called into question the very existence of the ANZUS military alliance (Pugh, 1990). Numerous local authorities had declared themselves to be 'nuclear-free' before the newly-elected Lange government carried out its election promise of making the whole country a nuclear-free zone (Lange, 1990). In practical terms this meant an opposition to vessels that were nuclear-powered or were carrying nuclear weapons, immediately bringing the policy into conflict with the American navy's policy of neither confirming nor denying whether their vessels were carrying nuclear weapons. So when the USS Buchanan followed this dictum in 1985, she was not permitted to use any port in New Zealand. The ensuing diplomatic incident had substantial ramefications not only to the functioning of ANZUS but also to the broader canvas of New Zealand–US relations (Landais-Stamp and Rogers, 1989; Pugh, 1989).

Subsequently the Americans have excluded New Zealand from ANZUS exercises, thereby losing training capabilities for New Zealand's armed forces, loss of intelligence-sharing and access to US weaponry. The New Zealand military have also been denied access to command and staff colleges operated by the Americans. One result is that New Zealand is now heavily dependent on Australia for its security, as exemplified by the Australia–New Zealand frigate project.

There has been a loss of high-level access to the US government, with no New Zealand prime minister having a direct meeting with a US president between 1984 and early 1995, when there was a resumption of senior-level official contacts. Although trade between the two countries has risen considerably in volume since 1985, there have been periodic complaints by New Zealand that their goods have not received fair treatment, whilst objections have been lodged against the US Export Enhancement Programme as an example of unfair trading.

Some steps towards healing the rift have been taken in recent years, under the Clinton administration's policy of closer political co-operation, whilst retaining the fissure in the security relationship, similar to the US dual-track policy towards France (McCormick, 1995). New Zealand supported the American stance in the Gulf War of 1991, the development of APEC (see below), and various US-led UN resolutions, notably in 1993–4 when New Zealand had a seat on the UN Security Council.

Hawke, the Australian Prime Minister, and his counterparts in Japan and the US (Forbes, 1988). APEC has been viewed by the Australians as a vital adjunct to their regional and sub-regional trade programmes which have emphasised links throughout the Asia–Pacific region. However, it has also functioned as a forum through which the US has sought to promote greater freedom of trade, and as part of Japan's continued economic growth.

Initially, APEC had 12 member nations (Table 17.1). Subsequently, in 1991, membership was extended to the People's Republic of China, Chinese Taipei (Taiwan) and Hong Kong. Two years later Papua New Guinea and Mexico joined; more recently Chile, Vietnam, Russia and Peru have been added. Therefore member states have a population of two billion, a GDP of over US$13 trillion, 45 per cent of world trade and take 70 per cent of Australasia's exports. With the conclusion of the Uruguay Round of the General Agreement on Tariffs and Trade (GATT), APEC's initial goals have been extended under American and Australian pressure to direct APEC towards becoming an economic community more akin to the European Union (Table 17.1).

This has met with some resistance from those APEC members who are also part of the Association of South-East Asian Nations (ASEAN): currently consisting of Brunei, Indonesia, Laos, Malaysia, Mayanmar (Burma), the Philippines, Singapore, Thailand and Vietnam. ASEAN favours an APEC that remains as a loose consultative forum. Its members fear that a more integrated form of APEC could undermine their individual sovereignty and allow the more powerful economies, notably the US and Japan, to dominate. Furthermore, ASEAN itself has been moving towards formation of its own free trade area which the Indonesians and Malaysians, in particular, do not wish to see jeopardised by a stronger APEC.

Nevertheless, there have been moves towards developing closer economic integration within APEC, through moves to broaden the group's agenda from matters such as the

TABLE 17.1 APEC: membership, goals and activities

Membership
1989: Australia, Brunei, Canada, Indonesia, Japan, South Korea, Malaysia, New Zealand, the Philippines, Singapore, Thailand, United States
1991: People's Republic of China, Chinese Taipei (Taiwan), Hong Kong
1993: Mexico, Papua New Guinea
1994: Chile
1997: Peru
1998: Russia, Vietnam

Goals
- to help strengthen the multilateral free trade system (and support a successful conclusion to the Uruguay Round of the GATT)
- to provide an opportunity to assess prospects for and obstacles to increased regional trade and investment flows
- to identify a range of practical, common economic interests shared by APEC members.

Activities
- trade and investment liberalisation
  15 key sectors: environmental goods and services; fish and fish products; toys; forest products; gems and jewellery; oilseeds and oilseed products; chemicals; energy; food; rubber; fertilisers; automotive products; medical equipment and instruments; civil aircraft; telecommunications equipment
- business facilitation
- economic and technical co-operation (eco-tech)
  six priority areas: developing human resources; establishing stable capital markets; building economic infrastructure; harnessing technologies of the future; promoting environmentally sound growth; strengthening small and medium-sized enterprises.

improvement of air links to a more extensive programme of trade liberalisation. This growing focus on trade reflects the wishes of some APEC members to see the development of a Pacific–America Free Trade Area (PAFTA) to accompany the creation of the North

American Free Trade Area (NAFTA) between Canada, Mexico and the USA (de Oliver, 1993). For example, at the APEC meeting in Indonesia in 1994, it was agreed (in the Bogor Declaration) that by 2000 the members would have begun to reduce trade barriers. Also, the five most developed nations (Australia, Canada, Japan, New Zealand and the US) consented to abolish all restrictions on imports from the rest of the membership by 2010. Other members will have until 2020 to follow suit.

Following the Osaka Action Agenda, agreed at the APEC Heads of State meeting in 1995 and supplemented by the 1996 Manila Action Plan, APEC has pursued its goals through activities in three areas: trade and investment liberalisation, business facilitation, and economic and technical co-operation (see Table 17.1). With respect to the first of these, members take action on a unilateral voluntary basis through annual individual action plans aimed at meeting the elements contained in the Bogor Declaration, and specifically towards developing free trade in 15 key industrial and service sectors.

Business facilitation has involved measures such as simplifying and harmonising customs procedures, mutual recognition of testing authorities for meeting industrial product standards, promoting investment by strengthening protection of intellectual property rights, and easing restrictions on regional travel by business people. Economic and technical co-operation has involved various capacity-building activities conducted by APEC bodies, especially aimed at enhancing members' ability to benefit from the liberalising agenda and reducing disparities within the diverse APEC region.

Increasingly therefore, APEC is intended to promote economic growth rather than be the focus of debate about security issues or politics (Rudner, 1995), but it is unclear just what its precise role will be. One suggestion is that it will take the form of 'open regionalism' whereby barriers to trade and investment within the membership group will be reduced through phased cuts of tariffs and quotas (Elek, 1992a; 1992b). This would also involve making access to APEC economies easier and cheaper for non-members. This places an onus upon trade reforms within APEC consistent with World Trade Organisation (WTO) moves towards trade liberalisation.

In a broader context, the positions adopted by governments within the APEC meetings have reflected different views of the benefits and dangers posed by greater ease of international trade (Robinson, 1995b). Some of the poorer countries have been wary of exposing domestic industry to greater foreign competition, whilst the richer countries have been striving to generate increased opportunities for their exporters. The close link between the rich governments and commercial interests is clear here, and may be regarded as another manifestation of globalisation in which business seeks to transcend the limitations imposed by international boundaries and by restrictions on both trade and movements of capital.

This global dimension to economic activity is blurring distinctive national identities and rendering outcomes of national trading policies less definitive. Therefore, one of the problems is how to resolve the 'open regionalism' of APEC with the 'closed regionalism' of NAFTA, ASEAN and Closer Economic Relations (CER) between Australia and New Zealand (see below) (Gallant and Stubbs, 1997).

From the late 1980s at least two-thirds of all exports from APEC countries have been destined for other APEC members, with approximately 17 per cent destined for the European Union (Forbes, 1997: 15). Both Australia and New Zealand have experienced a very fast growth in trade with APEC members, reflecting strong government policies to increase economic linkages with east and south-east Asian countries during the last decade (Cleary and Bedford, 1993).

For example, the New Zealand government is pursuing an Asia 2000 initiative to attract investment from south-east Asia and to tap new technologies, market knowledge and management skills from Asian partners (Beal, 1995). It is the intention that this initiative will lead to closer trading links with ASEAN

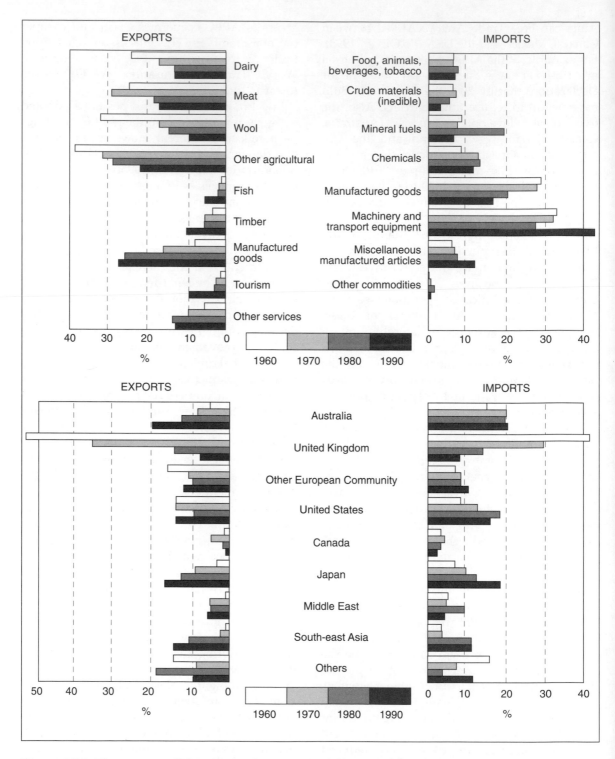

Figure 17.3 The pattern of New Zealand's imports and exports 1960–90 by sector and destination (source: SNZ)

TABLE 17.2 Overseas trade in merchandise, 1996/7

(a) Australia

Country	Exports A$000m	%	Imports A$000m	%	Sector	Exports %	Imports %
Japan	16.4	21.6	10.8	13.9	Agriculture	11	1
USA	4.6	6.1	17.6	22.6	Fish	1	–
NZ	5.6	7.4	3.6	4.6	Mining	22	4
South Korea	6.6	8.7	2.3	3.0	Food	14	4
China	3.8	5.0	4.0	5.1	Textiles	4	7
Taiwan	3.4	4.5	2.6	3.3	Wood	1	3
Canada	1.3	1.7	1.6	2.1	Print/paper	1	2
APEC	57.6	75.8	51.9	66.7	Oil/coal/chemicals	8	17
UK	2.8	3.7	4.9	6.3	Metal products	22	8
All EU	8.5	11.2	19.4	24.9	Machinery	15	53
Others	9.0	13.0	6.5	8.4	Non-metal products	1	1
Total	76.0	100.0	77.8	100.0		100	100

(b) New Zealand

Country	Exports NZ$000m	%	Imports NZ$000m	%	Sector	Exports %	Imports %
Japan	3.1	14.8	2.8	13.1	Dairy produce	20	*
USA	2.1	10.0	3.6	16.9	Meat	13	*
Australia	4.3	20.5	5.1	23.9	Wool/skins	8	*
South Korea	1.0	4.8	0.4	1.9	Other agricultural	6.5	*
China	0.6	2.9	0.9	4.2	Fish	5	*
Taiwan	0.6	2.9	0.6	2.8	Timber/paper/pulp	12.5	3
Canada	0.3	1.4	0.4	1.9	Mech. appliances	4	15
APEC	14.5	69.0	15.4	72.3	Aluminium	4	–
UK	1.4	6.7	1.1	5.2	Electrical machinery	3	10
All EU	3.4	16.2	4.2	19.7	Mineral fuels	3	7
Others	3.1	14.8	1.7	8.0	Others	21	65
Total	21.0	100.0	21.3	100.0		100	100

*included under 'Others'
Source: Yearbooks.

countries, and there was an input of NZ$1.7 million to the Asia 2000 Foundation by the New Zealand government in 1997. A similar amount was raised from corporate sources and offshore donors, to contribute to grants under business, education, and media and culture programmes. The share of both New Zealand's exports to and imports from south-east Asia has more than trebled since 1970, and by 1997 New Zealand's share of trade (exports and imports) with ASEAN represented 7.3 per cent of all the country's trade. Australia has established an ASEAN–Australia Economic Cooperation Programme to foster joint ventures, and, as a dialogue partner of ASEAN, Australia may seek to establish further special access to any newly established ASEAN common market.

Box 17.2 Japanese direct investment in Australia

As indicated in Chapter 15, a prominent feature of Australian economic development since the 1960s has been the growth in foreign-direct investment (FDI). Indeed, the stock of FDI in both Australia and New Zealand is over 20 per cent, generally higher than that in many developed countries. Traditionally the source of this investment was the UK, with the United States playing a greater role from the 1950s. However, FDI from Japan and, more recently, other east Asian countries, has grown substantially from the 1960s. In particular, Australia was a prominent target of Japanese FDI between 1980 and 1989, linked to the Japanese 'bubble economy' of increased manufacturing output and a near doubling in the value of the yen (Rimmer, 1993; 1997a). There was direct investment in the real-estate industry by life insurance companies for asset management, for example the EIE Development Co. which invested in Bond University on the Gold Coast, resorts in Queensland (Figure 17.4) and offices in Sydney. Subsequently, the 'bursting' of the bubble economy reduced Japanese FDI, producing a decline in investment in finance and insurance, services and real estate. There was a relative retreat from North America, Australia and Singapore in favour of China, Hong Kong and Malaysia.

One of the major foreign investors in Australia was the Japanese construction company Kumagai Gumi (KG) Co. Ltd. Between 1983 and 1986 this firm captured 41 contracts in Australia worth A$3.3 billion, thereby becoming Australia's largest foreign investor in new property development. The company's sales in 1985 were almost five times greater than those of the largest Australian contractor. The entry of KG to the Australian market rapidly expanded the number of projects undertaken there by Japanese international construction contractors. Previously there had been 11 major projects undertaken by such contractors, including the construction of port facilities at Port Hedland for export of iron ore to Japan and the Toyota Motor Corporation's plant in Melbourne (Figure 17.5(a)). No Japanese construction companies were awarded contracts in Australia between 1972 and the appearance of KG a decade later, initially as a provider of finance to projects where one of the principal partners had left, e.g. the Wandering Star tourist resort in northern New South Wales, and luxury quay apartments in Sydney. Subsequently, KG became associated with a broad spectrum of high-profile projects as a developer (Rimmer, 1988), with local contractors performing the actual construction. The focus was upon tourism and out-of-town retailing projects, especially in Sydney, Brisbane and Canberra (Figure 17.5(b)). A similar pattern has been followed by other Japanese companies entering the Australian market, with a marked concentration upon Sydney (Figure 17.5(c)). These firms have benefited from the move from public to private money for funding major developments, but have helped render Australian real estate more vulnerable to investment decisions made in Japan. Indeed, there have been periodic attempts to restrict an 'open door' for foreign investment when Australian concerns have grown over its effects on domestic enterprises.

With the demise of the bubble economy, FDI from other sources has become more prominent, e.g. South Korea (Waitt, 1994). Australia has also gained investment from a new source: economic refugees. For example, as the return of Hong Kong to Chinese administration approached, there was a strong increase in migration from Hong Kong to Australia, with a surge in the emigration of business people bringing substantial funds with them. Funds transferred rose to over A$1000 million per annum for the first time in 1987/8, having been only A$42.8 million in 1982/3 (Nash, 1993). Between 60 and 90 per cent of these funds have been invested in real estate.

As a result of the Asian financial crisis in 1998, cross-border mergers and acquisitions in Australia fell from US$11 billion the previous year to US$7.4 billion. However, the overall level of mergers and acquisitions in the country rose by 16.3 per cent to A$51.2 billion in more than 1000 deals, of which A$23.6 billion involved foreign acquisitions, half of which went to US firms and one-sixth to UK firms. This reflected the weakness of the Australian dollar, the bottoming of the commodity cycle and global perceptions of Australia as a vital commodity supplier (primarily to Asia). Resources companies accounted for about 10 per cent of Australian takeover activity compared with 1 per cent worldwide (e.g. Billiton of the UK, which purchased QNI, a major nickel producer). However, insurance and property-related transactions were more important, and the biggest takeover bid in the country's history was launched in the form of a A$3.3 billion bid by AMP, Australia's largest insurance company and financial services group, for the general insurer GIO Holdings Australia.

FIGURE 17.4 Japanese-funded tourist development at Sanctuary Cove on Queensland's Gold Coast (photo: G.M. Robinson)

Changing trading patterns for both Australia and New Zealand testify to the growing importance of trade with members of APEC. It is the size and wealth of these markets that has made them so significant, but with an accentuated effect upon Australasia because of the growth in trade flows within the Asia–Pacific region and the emergence of new regional links through investment, tourism and financial flows (Box 17.2). As shown in Figure 17.3, for New Zealand the diversification of trading partners and the move away from the reliance on the UK has been a feature of the last four decades, whilst a diversification in the content of exports has also occurred during this time (as discussed in Chapter 7) (Robinson, 1993a). The UK, which took over half of New Zealand's exports in 1960, accounted for less than 7 per cent by value in 1997, though it is still the second most important overseas market for meat and dairy produce, and the third most important for wool and fruit. In effect, Australia, Japan and the United States have replaced the UK as New Zealand's main trading partners. Four markets (the EU, Australia, Japan and the United States) now account for 60 per cent of New Zealand's export earnings and 70 per cent of its imports.

There have been considerable successes in overseas campaigns by New Zealand-based firms seeking new markets. For example, sales of dairy produce to Russia have nearly quintupled in value in the 1990s, so that Russia is now the fourth-largest market. Sales of forest products to Japan and South Korea have risen by 70 per cent in the 1990s, and China has become the leading purchaser of New Zealand wool. Against this, the volatility of the Middle Eastern market for meat has seen a sharp decline in purchases by Iran (the third most important market in the mid-1980s) compared with a three-fold increase in sales to Saudi Arabia in the 1990s. Export of halal slaughtered meat to Malaysia and Indonesia has increased strongly in recent years.

The growing importance of the Asia–Pacific arena to the Australian economy can be seen in the growing volume of trade with Japan and the so-called 'Tigers' of Taiwan, South Korea, Hong Kong and Singapore. Together these markets account for one-third of both the country's exports and imports. Sales of iron ore

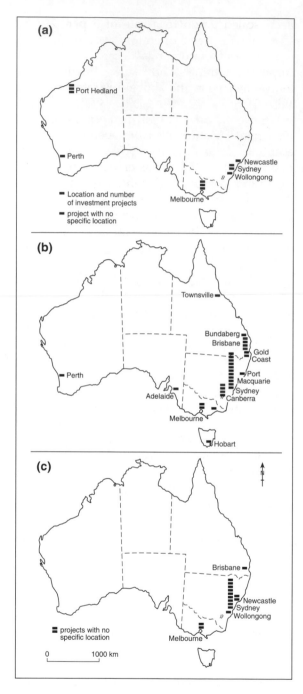

FIGURE 17.5 (a) Projects undertaken by Japanese contractors prior to 1982; (b) contracts undertaken by Kumagai Gumi or its subsidiary in Australia, 1983–7; (c) projects undertaken in Australia by Japanese construction firms other than Kumagai Gumi, 1983–7 (Rimmer 1988)

and coal have made Japan the number one market, followed by the ASEAN countries and the EU (Robinson, 1996a). The EU is the largest supplier of goods, followed by the US and Japan. However, Australia has a substantial trade deficit with the US and the EU (Table 17.2).

Future prospects for increased trade in the Asia–Pacific region depend upon the continuation of trade liberalisation measures within APEC. Liberalisation will probably enhance the farm commodity exports of the US and efficient producers of particular commodities (Ufkes, 1993). The latter may include certain types of production in Australasia, notably dairy produce and a range of meat products.

However, before full advantage can be taken of the highly efficient production and marketing systems in Australasia, liberalisation will have to extend to reduced discrimination against New Zealand produce as currently still practised in the EU and the United States. In the past this discrimination has mainly taken the form of quotas, but these may be outlawed by the WTO. This holds the prospect of increased exports of agro-commodities from Australasia to the markets of North America and Europe. The recent dispute with the EU regarding its tariffs on New Zealand-made spreadable butter illustrate the continuing difficulties faced in overcoming obstacles to freer trade. Similarly, Californian kiwifruit growers are now protected by US anti-trust laws which place special duties on New Zealand kiwifruit exports (Hoadley, 1997).

CLOSER ECONOMIC RELATIONS (CER)

From the early 1980s successive governments in Australia and New Zealand have sought to secure new markets for their goods through bilateral agreements with trading partners. Of these the most wide-ranging has been between the two countries themselves in the form of the

TABLE 17.3 The objectives of the Closer Economic Relations Agreement (CER)

- to strengthen the broader relationship between Australia and New Zealand;
- to develop closer economic relations between the member states through a mutually beneficial expansion of free trade between Australia and New Zealand;
- to eliminate barriers to trade between Australia and New Zealand under an agreed timetable and with a minimum of disruption; and
- to develop trade between Australia and New Zealand under conditions of fair competition.

Source: Article 1, CER Agreement.

Closer Economic Relations (CER) Agreement, concluded in 1983 and building upon the earlier New Zealand–Australia Free Trade Agreement (NZAFTA) which was signed in 1966 (Holmes, 1989). CER had the express intent of removing all tariffs and quantitative restrictions on the passage of goods between the two countries by 1995 (see Table 17.3).

In 1988 a review of CER brought forward to 1990 the date for full free trade in goods. It also widened the agreement to include services, business law and industry assistance. Trans-Tasman anti-dumping provisions were also removed. In 1989 the Trade in Services Protocol came into force, providing free trade in services based on the concept of national treatment, with some exceptions proscribed by each government. These included certain aspects of telecommunications, broadcasting, maritime transport, civil aviation, postal services and insurance services. In part, this reflected New Zealand's greater pursuit of deregulation in these areas.

An agreement on aviation was reached for the phasing-in of a single aviation market. The intended target date for the latter was 1 November 1994, but this was delayed by the Australians who have also maintained higher protective tariffs for longer, especially on textiles, clothing, footwear and automobiles. Nevertheless, most of its external tariffs had

been scaled down to a rate of 5 per cent by 1996, the base level operating in New Zealand, where only automobiles, car components, carpets, clothing and some textiles and footwear have tariffs above this rate (Sandrey, 1997: 54–5).

For New Zealand, CER has represented an important constituent of the opening-up of its economy following the post-1984 economic reforms (BIE, 1995). It has enabled the country to expand its exports of manufactured goods to Australia (especially timber products), exploiting the 'window of opportunity' offered by CER as part of the adjustment to greater international competition. Sandrey (1997: 55) sees CER as a critical part of the move by both Australia and New Zealand to more open economies, smoothing the path for a ready acceptance and support for the initial APEC declaration (see above). The benefits of the agreement may seem more obvious for New Zealand because of its potential to expand the country's 'domestic' market from its own population base of 3.7 million consumers to a combined size of 22 million, a similar proportional increase in market size to that enjoyed by the UK entering the EEC in 1973, but smaller than the increase (a factor of 13) for Canada on joining NAFTA.

One clear impact of CER has been the growing volume of trans-Tasman trade, in both absolute and relative terms, and a shift in New Zealand's favour of the ratio of trade between the two countries. Australia's share of New Zealand's imports has changed little since 1966, but New Zealand's percentage share of the Australian market has more than doubled. Exports to Australia (by value) from New Zealand increased by 19 per cent per annum in the 1980s compared with 14 per cent per annum in exports to other countries. In 1990 Australia accounted for 18.3 per cent of New Zealand's exports whilst supplying just over one-fifth of its imports. By 1997 Australia was receiving 20.3 per cent of New Zealand's total exports and supplying 23.8 per cent of total imports – an overall trade in excess of NZ$9 billion, though the balance has tipped strongly in Australia's favour in recent years. Australia's main exports to New Zealand are fuels, vehicles and machinery/mechanical appliances – the latter is the leading item exported from New Zealand to Australia.

Mutual investment between the two countries has also been stimulated by CER as have trans-Tasman joint venture agreements (Le Heron and Pawson, 1996: 43–51). The way in which the two countries now represent a single market is best illustrated in the food and beverage retailing sector. For example, Lion Nathan now controls over half of the beer market in New Zealand and nearly half of that in Australia, whilst there are strong interconnections between the major supermarket operators.

CER has stimulated cross-Tasman takeovers in both countries, e.g. New Zealand's Whitcoulls' purchasing Angus & Robertson, Australia's largest bookstore chain, and National Australia Bank taking over the Bank of New Zealand. Elsewhere, firms are increasingly treating the two countries as one market. An example of this is in the whitegoods industry where the growth of export markets helped change the industry's character in New Zealand, with one company, Fisher and Paykel, using Fordist mass-production techniques and product agreements with foreign companies, Hoover, Kelvinator and Bendix, to expand production. Australia was the main export market in the 1970s and further impetus to this export trade was given by CER and the 1980s' economic reforms. In 1991 Fisher & Paykel started a production line for refrigerators in Queensland and has targeted niche markets in Asia (e.g. deep-freezes to Japan) and Europe. The firm now has 40 per cent of the Australian whitegoods market, half supplied from Brisbane and half from New Zealand, and has offices in Hong Kong, Singapore and Taiwan (Latham, in Le Heron and Pawson, 1996: 173–5).

In addition to closer bilateral economic links, the two countries have maintained their close defence collaboration and have pursued similar political strategies towards the wider Asia–Pacific region, as discussed above. Their common stand against French nuclear testing in the South Pacific in the mid-1990s was another illustration of how common interests have been pursued, and this is echoed in some of their ongoing relations with the smaller South Pacific island nations (Box 17.3).

A REPUBLICAN FUTURE?

This book opened with a discussion of Australasia's changing 'place in the world'. Its content has given substance to the nature of that change, highlighting economic and social transformations that have made both countries more aware of their existence within the broad Asia–Pacific region and raised difficult questions about Australian and New Zealand identity (see Theophanous, 1995). They have become more cosmopolitan and outward-looking countries, developing a variety of new links to their region whilst retaining certain ties to the UK based on history and culture. There remains one critical UK tie which some maintain renders them less than wholehearted regional participants, and affects the way in which they are perceived by the rest of the world. That is the monarchy, and the fact that the head of state is Queen Elizabeth II, Queen of the UK as well as being Queen of Australia and Queen of New Zealand. There also remain a number of UK acts of parliament which are in force as part of the law of Australia and New Zealand. Both countries have a Governor-General, appointed by the Queen, whose assent is required before bills approved by the respective national parliaments become law.

Only in the last 10 to 15 years has this monarchical role been questioned, with the ensuing debate taking a markedly more robust form in Australia. Here argument about the country renouncing its formal ties to the UK and becoming a republic has gathered pace during the 1990s, led initially by the then Prime Minister, Paul Keating, who championed the cause of republicanism, with a desire

Box 17.3 International relations with small South Pacific countries

As the country with the largest population of Polynesians, New Zealand has enjoyed a 'special' relationship with many of the small Polynesian island communities. Its relationships with the Cook Islands and Niue have remained particularly strong through 'colonial' ties, although the nature of the formal links has changed. Following annexation, Cook Island citizens also had citizenship of New Zealand. In 1965 they were given self-government, assuming a dominion status within the area within which New Zealand currency is the legal form of exchange. Similarly, in 1974 Niue also became self-governing whilst retaining an even closer association with New Zealand than that enjoyed by the Cook Islands.

One of the recurrent themes of New Zealand's post-war foreign policy has been the development of closer ties with Pacific island countries, starting with its role as a member of the South Pacific Commission in 1947. This was essentially aimed at promoting research in the South Pacific, but other developments have been associated with various forms of aid and trade. Trade with the small South Pacific nations has increased, but still accounts for no more than 3 per cent of New Zealand's total exports. Imports from the region have been encouraged since 1981 through the South Pacific Regional Trade and Economic Co-operation Agreement (SPARTECA), under which both New Zealand and Australia provide, on a non-reciprocal basis, duty-free and unrestricted access into their markets for most of the products exported by the Pacific Forum countries (see below) (Robinson, 1996a). SPARTECA also includes provisions relating to dumped and subsidised goods, and special treatment for the smallest island countries. It encourages enhancement of the export capacities

of the island countries and requires Australia and New Zealand to give special consideration to island products offered for export. Although the agreement has been significant for some of the small island nations, it has little overall effect on either the imports or exports of their two large neighbours. Indeed, despite SPARTECA, both countries enjoy positive trading balances with most of their South Pacific neighbours and have some significant investments there, for example Australian mining interests in Papua New Guinea. The chief limitations of SPARTECA are the very small size of the economies represented (Sutherland, 1982) and their excessive reliance on primary production.

The South Pacific Forum was established in 1971, bringing together the heads of government of the independent and self-governing countries of the South Pacific: Australia, New Zealand, the Cook Islands, Fiji, Nauru, Tonga, Western Samoa, Niue, Papua New Guinea, Kiribati, Tuvalu, Vanuatu, Solomon Islands, Marshall Islands and the Federated States of Micronesia and Palau. It was formed to tackle common issues from a regional perspective, thereby enhancing the collective regional voice of the South Pacific. In the mid-1990s the Forum was particularly vocal in condemnation of French testing of nuclear weapons in French Polynesia, and its pressure contributed to the French agreeing on a cessation of testing (Smith, 1996). Environmental issues have been a recurrent feature of the Forum's annual meetings, with a Fisheries Agency created to assist members in management and conservation of the region's marine resources. A South Pacific Regional Environment Programme focuses on protection and management of environmental resources.

to see Australia entering the twenty-first century as a republic (Figure 17.6). This has had substantial support as well as some strong opposition. However, in New Zealand there has been little pro-active lobbying for the republican cause despite an expression of support being voiced by then Prime Minister James Bolger in the early 1990s (Figure 17.7). In some respects the clamour of the debate in Australia compared with the response in New

Zealand represents a critical differentiation between the two countries and the way in which their inhabitants regard themselves. A more prosaic explanation for the difference is that New Zealand has just had a major constitutional reform, the adoption of the mixed member proportional voting system (Lange, 1997) and is still digesting its effects, whereas Australia has had no such change. Moreover, in Australia, there is a legacy to be exploited

FIGURE 17.6 Canberra from Red Hill with the Parliament Building in centre; the capital of a new republic? (photo: Tourist Board)

from the Governor-General's dismissal of the Whitlam government in 1975.

The republican debate in Australia has advanced rapidly through the 1990s, with fluctuating responses to opinion pollsters' questions on whether the public was in favour of changing to a republic. However, official polls have taken the place of the unofficial litmus test, setting the country on course for a full-scale referendum on the subject. In December 1997 a postal ballot of registered voters returned 46 members (60.5 per cent) supporting the republican cause out of 76 persons elected to a Constitutional Convention meeting in February 1998. However, the low turn-out of 57 per cent raised questions regarding the country's overall views on republicanism.

The low turnout may have reflected apathy towards the Convention or to uncertainty

regarding a vote in favour of republicanism, given a continuing split amongst the pro-republican lobby. On one side is the Australian Republican Movement (ARM), favouring a 'minimalist' republic with a president to be appointed by a joint sitting of the federal parliament. The other side favours the election of a president by popular vote. The 76 elected delegates joined 40 politicians from federal, state and territory parliaments, and 36 delegates appointed by the Prime Minister. A straw poll of this second bloc of 76 delegates revealed 28 republicans, 17 monarchists and 31 uncommitted. The promise from the Prime Minister was that if the Convention reached consensus on a preferred model for a republic, a national referendum would be granted on that model before the end of 2000. Alternatively, Australians could be asked to vote on

FIGURE 17.7 Will the statue of Richard Seddon, outside 'the Beehive' (parliament's executive offices) in Wellington, see the creation of a new republic in the twenty-first century? (photo: G.M. Robinson)

preferred options through a plebiscite. A decision would then have to be made by the federal government on whether to put this option to another referendum.

The Constitutional Convention in February 1998 voted in favour of a parliamentary appointment system for a new head of state. By a majority of 133 to 17, with two abstentions, the Convention voted for the resolution 'that this Convention recommends to the Prime Minister and Parliament that the republican model, and other related changes to the Constitution, supported by this Convention, be put to the people in a constitutional referendum'. The model referred to is for a republic with a president nominated by the public, selected by the Prime Minister and ratified by a two-thirds majority of the federal parliament

(Howell, 1998). The referendum to be held before the end of 1999 will have to decide between the existing constitutional monarchy, with the Queen as the legal head of state, and the new system.

Therefore there is the prospect of a fundamental change in the status of Australia, which republican supporters feel would confirm the country's status within its region rather than as an adjunct or distant appendage to a colonial power. Yet, it is far from clear that there is overwhelming public support for a republic. A split between the young and those of non-British stock – republican – and older generations with memories of the Second World War or having emigrated from the UK – monarchists – is a too simplistic portrayal of the divisions. There are various complicating factors, not least of which is the fact that the current Prime Minister, John Howard, supports maintenance of the status quo, whilst uncertainty relating to who would be the first president of the republic may also have a negative effect on the republican vote.

At the turn of the century, Australia and New Zealand will be in the media spotlight: the 'republic' referendum in Australia; New Zealand hosting the 1999 APEC summit; the Olympic Games in Sydney; the America's Cup yacht races in Auckland; the Commonwealth Games in Melbourne in 2006. Media attention will no doubt emphasise the strong sporting traditions of the two countries, allied to a distinctiveness that transcends their well-recognised flora and fauna and the origins of the modern states as part of the British Empire. It remains to be seen whether Australia will become a republic in the early years of the new century, but many predict that this constitutional change will not be long in arriving and that it will be followed eventually by a similar change in New Zealand. As this Chapter has stressed, these prospective changes will reinforce the nature of economic development, in which both countries will be tied increasingly to the Asia–Pacific region. This growing involvement with trade and investment in and from this region has recently been problematic as a result of the recent economic downturn in south-east and east Asia. Australia appears to have weathered this set-back better because of its strong resource base, whilst New Zealand's reliance on the agro-commodities sector, forest products and the Asian export market has been a weakness. However, the longer-term prospect of greater involvement in the region is bright, given its market size and potential. Moreover, the role of Australasia as an international tourist destination may help revive economic growth quite rapidly if the world tourist industry continues to expand early in the new century.

It can be observed that the twentieth century has brought about vast economic and social changes to both countries, precipitating fundamental reassessments of their relationships with the rest of the world. In a relatively short period Australia has become a vibrant multicultural nation, helping to develop closer ties throughout the Asia–Pacific region, whilst gradually coming to terms with a long and unfortunate domestic history of poor relations between the majority and the indigenous peoples.

Multiculturalism has been less marked in New Zealand where more emphasis has been placed on resolving conflicts arising from the long neglect of the Treaty of Waitangi. Nevertheless, the focus on the Asia–Pacific region is also apparent here, especially with a more diverse population through multiethnic immigration during the last two decades. The economic and social geographies of Australia and New Zealand have changed tremendously in the twentieth century, with an accelerating rate of change from the early 1980s. Whilst this change may not continue so rapidly early in the twenty-first century, there is no doubt that the evolving economies and societies, as well as their interactions with the physical environment, will continue to present fascinating challenges for geographers to study.

FURTHER READING

Two useful overviews of the changing 'place in the world' of Australia and New Zealand are, respectively, Chalkley and Winchester (1991) and Forer (1995). Contrasting views of New Zealand's 'place' and changing identity are provided by Brown (1997), Crosbie (1989) and Harland (1992a; 1992b). Once more, Le Heron and Pawson (1996) have much to say on this subject. The ANZUS crisis is discussed by Landais-Stamp and Rogers (1989), Lange (1990) and Pugh (1989). The aftermath of the break-down in New Zealand–US relations is discussed by McCormick (1995), and the ending of French nuclear testing in French Polynesia by Smith (1996). The development of APEC is analysed by Forbes (1997) and Rudner (1995). An outline of changing trading relations is given by Robinson (1993a; 1995b; 1996a). Different aspects of the 'Asianisation' of both countries are given in Beal (1995), Dunn (1993; 1998) and Viviani (1996). Sandrey (1997) discusses CER. Japanese FDI in Australia is analysed perceptively in several studies by Rimmer (1992; 1993; 1994; 1997). A South Korean example is given by Waitt (1994). Howell (1998) provides an assessment of the republican debate in Australia.

REFERENCES

Adler, S. and Brenner, J. 1992, 'Gender and space: Lesbians and gay men in the city', *International Journal of Urban and Regional Research*, 16(1), 24–34.

Alexander, I. (1986). 'Land use and planning in Australian cities: capital takes all', pp. 113–130 in McLoughlin and Huxley (eds) (1986).

Aldrich, R. (ed.) 1994, *Gay perspectives II: More essays in Australian gay culture*, (Department of Economic History with the Australian Centre for Gay and Lesbian Research, University of Sydney, Sydney).

Allan, R.J., Lindesay, J.A. and Parker, D.E. (eds) 1996, *El Niño south oscillation and climatic variability* (Commonwealth Scientific and Industrial Research Organisation Publishing, Collingwood).

Alston, M. (ed.) 1991, *Family farming – Australia and New Zealand* (Centre for Rural Social Research, School of Humanities and Social Sciences, Charles Stuart University-Riverina, Wagga Wagga).

Anderson, A.J. 1989, *Prodigious birds: Moas and moa-hunting in prehistoric New Zealand* (Cambridge University Press, Cambridge).

Anderson, A.J. 1991, 'The chronology of colonization in New Zealand', *Antiquity*, 65, 767–95.

Anderson, A.J. and McGlone, M. 1992, 'Living on the edge (prehistoric land and people in New Zealand)', in Dodson (ed.) (1992).

Anderson, C. 1985, 'Queensland Aboriginal peoples today', *Queensland Geographical Journal*, 4th Series, 1, 296–320.

Anderson, K. 1993, 'Otherness, culture and capital: Chinatown's transformation under Australian multiculturalism', pp. 68–89 in Clark, Forbes, and Francis (eds) (1993).

Anderson, K. 1999, 'Reflections on Redfern', in Stratford (ed.) (1999).

Anderson, K. and Gale, F. (eds) 1992, *Inventing places: studies in cultural geography* (Longman Cheshire, Melbourne).

Anderson, K. and Gale, F. (eds) 1999, *Cultural geographies* (Addison Wesley Longman, Harlow).

Anderson, P., Bhatia, K. and Cunningham, J. 1994, *Mortality of indigenous Australians* (AGPS, Canberra).

Andrews, J. 1966, 'The emergence of the wheat belt in south eastern Australia to 1930', pp. 5–24 in Andrews, J. (ed.), *Frontiers and men: a volume in memory of Griffith Taylor* (F.W. Cheshire, Melbourne).

Aplin, G.J. 1996, *Global environmental crises: an Australian perspective* (Oxford University Press, Melbourne).

Aplin, G.J. 1998, *Australians and their environment* (Oxford University Press Australia, South Melbourne).

Archer, J. 1996, *The great Australian dream: the history of the Australian house* (HarperCollins, Sydney).

Archie, C. 1995, *Maori sovereignty: the Pakeha perspective* (Hodder Moa Beckett, Auckland).

Argy, F. 1998, *Australia at the crossroads: Radical free market or a progressive liberalism?* (Allen and Unwin, Sydney).

Armstrong, H. and Gross, R. 1995, *Tri-continental: the rise and fall of a merchant bank* (Melbourne University Press, Melbourne).

Atkinson, G. 1991, 'Soil survey and mapping', pp. 89–111 in Charman and Murphy (eds) (1991).

Attwood, B. (ed.) 1996, *In the age of Mabo: history, Aborigines and Australia* (Allen and Unwin, St Leonards, NSW).

Auckland City Council 1997, *Auckland city's people report* (ACC, Auckland). Internet site: http://www.akcity.govt.nz/about/intro/au cklands_people/data4.htm

Auckland City Council 1998, *Waterfront 2000* (ACC, Auckland). Internet site: http:// www.akcity.govt.nz/council/projects/water front2000/projects1.htm

Australian Bureau of Statistics (ABS) 1992a, *Social indicators 1992* (ABS, Canberra).

ABS 1992b, *Australia's environment, issues and facts* (ABS, Canberra).

ABS 1994, *Australian social trends 1994* (ABS, Canberra).

ABS 1996, *Australian social trends 1996* (ABS, Canberra).

ABS 1997a, *1996 census of population and housing: selected social and housing characteristics* (ABS, Canberra).

ABS 1997b, *Australian social trends 1997* (ABS, Canberra).

ABS 1998a, *Australian social trends 1998* (ABS, Canberra).

ABS 1998b, *Migration: Australia 1996–1997* (ABS, Canberra).

ABS 1998c, *Sydney: a social atlas* (AGPS, Canberra).

ABS 1998d, *Year book Australia* (ABS, Canberra).

ABS and Australian Institute of Health and Welfare (ABS/AIHW) 1997, *The health and welfare of Australia's Aboriginal and Torres Strait Islander peoples* (ABS, Canberra).

Australian Government, 1994a, *Working nation: policies and programs* (AGPS, Canberra).

Australian Government, 1994b, *Working nation: the White Paper on employment and growth* (AGPS, Canberra).

Australian Housing and Urban Research Institute (AHURI), 1995, *The internationalisation of the Far North Queensland regional economy: final report to the Far North Queensland Region Economic Development Strategy* (AHURI, Brisbane).

Badcock, B. 1995, 'Towards more equitable cities: a receding prospect?', pp. 196–217 in Troy (ed.) (1995).

Badcock, B. 1997, 'Recently observed polarising tendencies and Australian cities', *Australian Geographical Studies*, 35(3), 243–59.

Baker, R. 1992, 'Gough Whitlam time: land rights in the Borroloola area of Australia's Northern Territory', *Applied Geography*, 12, 162–75.

Baker, R. 1997, 'Landcare: policy, practices and partnerships', *Australian Geographical Studies* 35, 61–73.

Ballance, P.F. and Williams, P.W. 1992, 'The geomorphology of Auckland and Northland', pp. 210–32 in Soons and Selby (eds) (1992).

Bamber, G.J. and Lansbury, R.D. 1997, 'Employment relations in the Australian automotive industry: a question of survival', pp. 81–101 in Kitay and Lansbury, (eds) (1997).

Banham, M. 1998, 'The challenge of urban Maori: reconciling conceptions of indigeneity and social change', *Asia Pacific Viewpoint*, 39, 303–14.

Barlow, B.A. 1994, 'Phytogeography of the Australian region', pp. 3–15 in Groves (ed.) (1994).

Barnard, A. (ed.) 1962, *The simple fleece* (Melbourne University Press, Melbourne).

Barnes, T.T. and Hayter, R. (eds) 1998, *Troubles in the rainforest: British Colombia's Forest economy in transition* (Western Geographical Press, Victoria, BC), Canadian Western Geographical Sries, 33.

Barnet, R. and Cavanagh, J. 1995, *Global dreams: imperial corporations and the New World Order* (Simon and Schuster, New York).

Barr, N. and Cary, J. 1992, *The Australian search for sustainable land use: greening a brown land* (Macmillan, Melbourne).

Barry, R.G. 1969, 'The world hydrological cycle', pp. 11–29 in Chorley (ed.) (1969).

Bartlett, C.A. and Ghosha, S. 1989, *Managing across borders: the transnational solution* (Harvard Business School Press, Boston).

Basher, L.R., Matthews, K.M. and Zhi, L. 1995, 'Surface erosion assessment in the

Canterbury downlands using 137^{C}s distribution', *Australian Journal of Soil Research* , 33, 787–803.

Baum, S. 1997, 'Sydney: social polarization and global city status', *Urban Futures*, 22, 21–36.

Bauer, F.H. 1984, 'US investment in the Northern Territory pastoral industry', *Working Papers in Economic History*, no. 25 (School of Social Studies, Australian National University, Canberra).

Beal, T.M. 1995, 'New Zealand – coming to terms with Asia', *British Review of New Zealand Studies*, 8, 59–84.

Bedford, R. 1996, 'International migration, 1995', *New Zealand Journal of Geography*, October, 1–13.

Bedford, R., Goodwin, J., Ho, E. and Lidgard, J. 1997, 'Migration in New Zealand 1986–1996: a regional perspective', *New Zealand Journal of Geography*, October, 16–31.

Beer, A. 1998, 'Immigration and slow-growth economies: the experience of South Australia and Tasmania', *Australian Geographer*, 29, 223–40.

Begg, M. 1997, 'Sharemilking: one way to get started', pp. 199–202 in Stokes and Begg (eds) (1997).

Begg, M. and Begg, D. 1997, 'Dairying: a vertically integrated agro-industrial system', pp. 159–76 in Stokes and Begg (eds) (1997).

Belich, J. 1986, *The New Zealand Wars and the Victorian interpretation of racial conflict: the Maori, the British and the New Zealand Wars* (McGill-Queen's University Press, Montreal and Kingston).

Belich, J. 1996, *Making peoples: a history of the New Zealanders* (Allen Lane/The Penguin Press, Auckland).

Bell, A., 1988/89, 'Trees, water and salt – a fine balance', *Ecos*, 58, 2–8.

Bell, D.J. 1991, 'Insignificant others: lesbian and gay geographies', *Area*, 23(4), 323–29.

Bell, S. 1993, *Australian manufacturing and the state: the politics of industry policy in the post-war era* (Cambridge University Press, Cambridge).

Bennett, B. 1998, 'Dealing with dryland salinity', *Ecos*, 96, 9–20.

Berry, M. and Huxley, M. 1992, 'Big build:

property capital, the state and urban change in Australia', *International Journal of Urban and Regional Research*, 16, 35–59.

Berryman, K. 1990, 'Late Quaternary movement on the Wellington Fault in the Upper Hutt area, New Zealand', *New Zealand Journal of Geology and Geophysics*, 33, 257–70.

Betts, K. 1991, 'Australia's distorted immigration policy', pp. 149–77 in Goodman, O'Hearn and Wallace-Crabbe (eds) (1991).

Binney, J. 1995, *Redemption songs: a life of Te Kooti Arikirangi Te Turuki* (Auckland University Press/Bridget Williams Books, Auckland).

Birrell, B. 1996, 'Our nation: the vision and practice of multiculturalism under labor', *People and Place*, 4(1).

Birrell, B., Maher, C. and Rapson, V. 1997, 'Welfare dependence in Australia', *People and Place*, 5(2), 68–77.

Bishop, P. 1996, 'Off road: four wheel drive and the sense of place', *Environment and Planning D: Society and Space*, 14, 257–21.

Blainey, G. 1966, *The tyranny of distance* (Melbourne University Press, Melbourne).

Blainey, G. 1984, *All For Australia* (Methuen Haynes, Sydney).

Bliss, E. (cd.) 1997, 'Islands: economy, society and environment', *Proceedings of the Joint 2nd Conference of the Institute of Australian Geographers and New Zealand Geographical Society, Hobart*.

Blunden, G., Moran, W. and Bradley, A. 1997, '"Archaic" relations of production in modern agricultural systems: the example of sharemilking in New Zealand', *Environment and Planning A*, 29, 1759–76.

Bolam, A. 1994, 'Small urban Australia: central in the national economy but still peripheral in policy', *Australian Geographer*, 25, 1225–32.

Bonnano, A. 1993, 'The agro-food sector and the transnational state: the case of the EC', *Political Geography*, 12, 341–60.

Boston, J., Martin, J., Pallot, J. and Walsh, P. (eds) 1991, *Reshaping the state. New Zealand's bureaucratic revolution* (Oxford University Press, Auckland).

Boughton, W.C. 1993, 'Atmospheric processes and runoff', In McTainsh, G.H. and Boughton, W.C. (eds) (1993).

Bouquet, M. and Winter, M. (eds) 1987, *Who from their labours rests?* (Avebury, Aldershot).

Bowen, B. and Gourlay, P. 1993, *The economics of coal export controls* (Australian Bureau of Agricultural and Resource Economics, Canberra).

Bowler, J.M. 1986, 'Quaternary landform evolution', pp. 117–47 in Jeans (ed.) (1986).

Breckwoldt, R. 1988, *The dirt doctors* (Soil Conservation Service of New South Wales, Sydney).

Bridgman, H.A., Warner, R.F. and Dodson, J. 1995, *Urban biophysical environments* (Oxford University Press, Melbourne).

Britton, S., Le Heron, R.B. and Pawson, E.J. (eds) 1992, *Changing places in New Zealand: a geography of restructuring* (New Zealand Geographical Society, Christchurch).

Brookfield, F.M. 1995, 'The Treaty of Waitangi, the Constitution and the future', *British Review of New Zealand Studies*, 8, 3–20.

Brooking, T. 1997, *Lands for the people? The Highland Clearances and the colonisation of New Zealand: a biography of John McKenzie* (University of Otago Press, Dunedin).

Brooking, T. and Rabel, R. 1995, 'Neither British nor Polynesian: a brief history of New Zealand's other immigrants', pp. 23–49 in Greif (ed.) (1995).

Brown, R. 1997, *New Zealand identity in a transnational age* (Kakapo Books, London).

Bryant, E.A. 1991, *Natural hazards* (Cambridge University Press, Cambridge).

Bunker, R. 1987, 'Urban consolidation and Australian cities', pp. 61–78 in Hamnett and Bunker (eds) (1987).

Bunker, R. 1995, 'State planning operation', pp. 142–63 in Troy (ed.) (1995).

Burch, D., Lawrence, G., Rickson, R.E. and Goss, J. (eds) 1998, 'Australasia's food and farming in a globalised economy: recent developments and future prospects', *Monash Publications in Geography and Environmental Science*, no. 50.

Burch, D. and Pritchard, W.N. 1996, 'The uneasy transition to globalization: restruc-

turing of the Australian tomato processing industry', pp. 107–26 in Burch *et al.* (eds) (1996).

Burch, D., Rickson, R.E. and Lawrence, G. (eds) 1996, *Globalization and agrifood restructuring: perspectives from the Australasia region* (Avebury, Aldershot).

Burch, D., Goss, J. and Lawrence, G. (eds) 1999, *Restructuring global and regional agricultures: transformations in Australian agri-food economies and spaces* (Ashgate, Aldershot).

Bureau of Industry Economics (BIE) 1994, *Regional development: patterns and policy implications* (AGPS, Canberra).

BIE, 1995, *Impact of the CER Trade Agreement: lessons for regional economic co-operation* (BIE Report no. 95/17, Canberra).

Bureau of Meteorology, Australia, 1989, *Climate of Australia* (AGPS, Canberra).

Bureau of Tourism Research (BTR) 1995, *Impact – a monthly facts sheet on the economic impact of tourism and broad tourism trends* (AGPS, Canberra).

Burnley, I.H. 1976, *The social environment* (Longman Cheshire, Sydney).

Burnley, I.H. 1989, 'Settlement dimensions of the Vietnam-born population in metropolitan Sydney', *Australian Geographical Studies*, 27, 129–54.

Burnley, I.H. 1994, 'Immigration, ancestry and residence in Sydney', *Australian Geographical Studies*, 32, 69–89.

Burnley, I.H. 1998, 'Immigrant city, global city? Advantage and disadvantage among communities from Asia in Sydney', *Australian Geographical Studies*, 29, 49–70.

Burnley, I.H. and Forrest, J. (eds) 1985, *Living in cities* (Allen and Unwin, Sydney).

Burnley, I.H., Murphy, P.A. and Fagan, R.H. 1997a, *Immigration and Australian cities* (The Federation Press, Sydney).

Burnley, I.H., Murphy, P.A. and Jenner, A. 1997b, 'Selecting suburbia: residential relocation to outer Sydney', *Urban Studies*, 34(7), 1109–27.

Burnley, I.H., Pryor, R.J. and Rowland, D.T. (eds) 1980, *Mobility and community change in Australia* (Queensland University Press, St Lucia).

Butlin, N.G. 1962, 'Distribution of sheep population: preliminary statistical picture, 1860–1957', pp. 262–6 in Barnard (ed.) (1962).

Campbell, D.I. and Murray, D.L. 1990, 'Water balance of snow tussock grassland in New Zealand', *Journal of Hydrology*, 118, 229–45.

Campbell, H. 1996, 'Organic agriculture in New Zealand: corporate greening, transnational corporations and sustainable agriculture', pp. 153–72 in Burch, Rickson and Lawrence (eds)

Cannon, M. 1987, *The exploration of Australia* (Readers Digest, Surry Hills, NSW).

Cant, G. 1990, 'Waitangi: Treaty and Tribunal', *New Zealand Journal of Geography*, 85, 7–12.

Cant, G., Overton, J. and Pawson, E.J. (eds) 1993, 'Indigenous land rights in Commonwealth countries: dispossession, negotiation and community action, *Proceedings of a Commonwealth Geographical Bureau Workshop, Feb., 1992*, Department of Geography, Christchurch.

Capling, A. and Galligan, B. 1992, *Beyond the protective state: the political economy of Australia's manufacturing industry policy* (Cambridge University Press, Melbourne).

Carnahan, J.A. 1986, 'Vegetation', pp. 260–82 in Jeans (ed.) (1986).

Castles, F.G. (ed.) 1991, *Australia compared: people, policies and politics.* (Allen and Unwin, Sydney).

Castles, F.G., Gerritson, R. and Vowles, J. 1996, *The great experiment* (Allen and Unwin, Auckland).

Castles, S. 1992, 'The "new" migration and Australian immigration policy', pp. 45–72 in Inglis, *et al.* (eds) (1992).

Castles, S. 1997, 'Multicultural citizenship: a response to the dilemma of globalisation and national identity', *Journal of Intercultural Studies*, 18(1), 5–22.

Castles, S., Kalantzis, M., Cope, B. and Morrissey, M. 1992, *Mistaken identity: multiculturalism and the demise of nationalism in Australia* (Pluto Press, Sydney).

Catley, R. 1996, *Globalising Australian capitalism* (Cambidge University Press, Cambridge).

Caulfield, J. 1992, 'Planning policy options for Brisbane's growth', *Power and policy in Brisbane* (Centre for Australian Public Sector Management, Griffith University, Nathan).

Caulfield, J. and Minnery, J. 1994, 'Planning as legitimation: a study of the Brisbane Strategy Plan', *International Journal of Urban and Regional Research*, 18, 673–89.

Ceplecha, V.J., 1971, 'The distribution of the main components of the water balance in Australia', *Australian Geographer*, 11, 455–62.

Chalkley, B.S. and Winchester, H.P.M. 1991, 'Australia in transition', *Geography*, 76, 97–108.

Chamberlin, B. 1996, *Farming and subsidies: debunking the myths* (Euroa Farms Ltd, Pukekohe).

Chapman, D.M. 1994, *Natural hazards* (Oxford University Press, Melbourne).

Charman, P.E.V and Murphy, B.W. (eds) 1991, *Soils – their properties and management* (Sydney University Press and Oxford University Press, Melbourne).

Cheyne, C., O'Brien, M. and Belgrave, M. 1997, *Social policy in Aoteoroa/New Zealand: a critical introduction* (Oxford University Press, Auckland).

Chisholm, M. and Smith, D.M. (eds) 1990, *Shared space, divided space: essays on conflict and territorial organisation* (Unwin Hyman, London).

Chorley, R.J. (ed.) 1969, *Water, earth and man* (Methuen, London).

Clark, A.H. 1949, *The invasion of New Zealand by people, plants and animals* (Rutgers University Press, New Brunswick, NJ).

Clark, G.L., Forbes, D. and Francis, R. (eds) 1993, *Multiculturalism, difference and postmodernism.* (Longman Cheshire, Melbourne).

Clark, M. 1981, *A short history of Australia* (South Melbourne), 2nd edition.

Clarke, F.G. 1992, *Australia: a concise political and social History* (Harcourt Brace Jovanovich, Sydney).

Cleary, M. and Bedford, R.D. 1993, 'Globalisation and the new regionalism: some implications for New Zealand', *New Zealand Journal of Geography*, 88, 19–22.

Cliff, A.D., Gould, P.R., Hoare, A.G. and Thrift, N.J. (eds) 1995, *Diffusing geography: essays for Peter Haggett* (Blackwell, Oxford).

Cloke, P.J. 1989, 'State deregulation and New Zealand's agricultural sector', *Sociologia Ruralis*, 29, 34–48.

Cloke, P.J., Le Heron, R.B. and Roche, M. 1990, 'Towards a geography of political economy perspective on rural change: the example of New Zealand', *Geografisker Annaler*, 72B, 13–25.

Cocks, K.D. 1992, *Use with care – managing Australia's natural resources in the twenty-first century* (New South Wales University Press, Kensington, NSW).

Coles, R. (ed.) 1997, *The end of public housing: a discussion forum organised by the Urban Research Program*, 25 October 1996 (Urban Research Program, Australian National University, Canberra).

Colhoun, E.A., Hannan, D. and Kiernan, K. 1996, 'Late Wisconsin glaciation of Tasmania', *Papers and Proceedings of the Royal Society of Tasmania*, 130, 33–45.

Colley, P. 1997, 'Investment practices in coal: the practice and profit of quasi-integration in the Australia–Japan coal trade', *Energy Policy*, 25, 1013–25.

Colley, P. 1998, 'Trading practices in the coal market: application of the theory of bilateral monopoly to the Australian-Japan coal trade', *Resources Policy*, 24, 59–75.

Collins, J. 1988, *Migrants in a distant land: Australia's post-war immigration.* (Pluto Press, Sydney and London).

Colls, K. and Whitaker, R. 1990, *The Australian weather book* (Child and Associates, Frenchs Forest, NSW).

Conacher, A.J. and Conacher, J. 1995, *Rural land degradation in Australia* (Oxford University Press, Melbourne).

Conacher, A.J. and Murray, I.D. 1969, 'The Meckering earthquake, Western Australia, 14 October 1968', *Australian Geographer*, 11, 179–84.

Connell, J. and Howitt, R. (eds) 1991, *Mining and indigenous peoples in Australasia* (Sydney University Press, Sydney).

Cook, L. 1997, 'New Zealand's current and future population dynamics', *Population Conference, Wellington 12–14 November*, Wellington.

Cooke, R.U. and Doornkamp, J.C. 1990, *Geomorphology in environmental management – a new introduction* (Clarendon Press, Oxford), 2nd edition.

Coombes, B. 1997, 'Rurality, culture and local economic development', unpublished Ph.D. thesis, Department of Geography, University of Otago.

Coombes, B. and Campbell, H. 1996, 'Pluriactivity in (and beyond?) a regulationist crisis', *New Zealand Geographer*, 52(2), 11–17.

Corbett, D. 1987, *The geology and scenery of South Australia* (Wakefield Press, Netley, South Australia).

Cornforth, I. 1998, *Practical soil management* (Lincoln University Press, Canterbury, and Whitireia Publishing with Daphne Brasell Associates Ltd, Wellington).

Courtenay, P.P. 1982, *Northern Australia: patterns and problems of tropical development in an advanced country* (Longman Cheshire, Melbourne).

Cox, E. 1995, *A truly civil society* (ABC Books, Sydney).

Crozier, M.J., Gage, M., Pettinga, J.R., Selby, M.J. and Wasson, R.J. 1992, 'The stability of hillslopes', pp. 63–90 in Soons and Selby (eds) (1992).

Crabb, P. 1991, 'Paying for water: there are no free drinks!', *Australian Geographer*, 22, 126–8.

Craik, J. 1987, 'A crown of thorns in paradise: tourism on Queensland's Great Barrier Reef', pp. 135–58 in Bouquet and Winter (eds) (1987).

Craik, J. 1991, *Resorting to tourism: cultural policies for tourist development in Australia* (Allen and Unwin, Sydney).

Crang, M. 1996, 'Watching the city: video surveillance and resistance', *Environment and Planning A*, 28(12), 2099–104.

Crawford, P. 1993, *Nomads of the wind* (BBC Books, London).

Crawshaw, M. 1998, 'Hero worship', *Metro (Auckland)*, no. 200 (Feb.), 30–5.

Crosbie, A.J. 1989, 'New Zealand: remote on the periphery', *British Review of New Zealand Studies*, 2, 38–50.

Crough, G. and Weelwright, T. 1982, *Australia: a client state* (Penguin, Ringwood, Vic.).

Cruickshank, M. 1998, 'The power of language is Moana's political weapon of choice', *Sunday Star Times*, p. 3.

Cumberland, K.B. 1949, 'Aotearoa Maori: New Zealand about 1780', *Geographical Review*, 39, 401–24.

Cumberland, K.B. 1962, 'Moas and men: New Zealand about AD 1250', *Geographical Review*, 52, 151–73.

Cumberland, K.B. 1981, *Landmarks* (Readers Digest, Surry Hills, NSW).

Cunningham, C., Jones, M. and Taylor, N. 1994, 'The child-friendly neighbourhood: some questions and tentative answers from Australian research', *International Play Journal*, 2, 79–95.

Curtis, A.L. 1997, 'Landcare, stewardship and biodiversity conservation', In Klomp and Lunt (eds) (1997).

Dale, A. 1992, 'Aboriginal councils and natural resource use planning: participation by bargaining and negotiation', *Australian Geographical Studies*, 30, 9–26.

Daly, M.T., Stimson, R.J. and Jenkins, O. 1996, 'Tourism and foreign investment in Australia: trends, prospects and policy implications', *Australian Geographical Studies*, 34, 169–84.

Dalziel, P. 1994, 'A decade of radical economic reform in New Zealand', *British Review of New Zealand Studies*, 7, 49–72.

Dalziel, P. and Lattimore, R. 1996, *The New Zealand macroeconomy: a briefing on the reforms* (Oxford University Press, Melbourne).

Dalziel, R. 1992, 'The politics of settlement', pp. 87–111 in Rice (ed.) (1992).

Damer, D. 1995, 'The making of urban New Zealand', *Journal of Urban History*, 22(1), 6–39.

Dargavel, J. 1991, 'Governments and the environment: managing Tasmania's forest sector', pp. 128–45 in Rich and Linge (eds) (1991).

Dargavel, J. 1995, *Fashioning Australia's forests* (Oxford University Press, Melbourne).

Dargavel, J. and Tucker, R. (eds) 1991, *Changing Pacific forests: historical perspectives* on the forest economy of the Pacific Basin (Forest History Centre, Durham, NC).

Daumanting, I. 1997, *Australia's federation, 1901* (National Australia Day Council, Canberra).

Davey, J.A. 1993, *From Birth to Death III* (Institute of Policy Studies, Wellington).

Davey, J.A. 1994, 'Ethnic differences and the policy response in New Zealand', *British Review of New Zealand Studies*, 7, 93–112.

Davidson, B.R. 1965, *The northern myth: a study of the physical and economic limits to agricultural and pastoral development in tropical Australia* (Melbourne University Press, Carlton).

Davidson, C. 1995, 'Employment in New Zealand after the "revolution": the outcome of restructuring', *British Review of New Zealand Studies*, 8, 99–116.

Davidson, J.M. 1984, *The prehistory of New Zealand* (Longman Paul, Auckland).

Davies, J.L. 1986, 'The coast', pp. 203–22 in Jeans (ed.) (1986).

Davis, B.W. 1980, 'The struggle for south-west Tasmania', pp. 152–69 in Scott (ed.) (1980).

Davis, B.W. 1986, 'Tasmania: the political economy of a peripheral state', pp. 209–225 in Head (ed.) (1986).

Davis, M. 1992, *City of quartz: excavating the future in Los Angeles* (Vintage Books, New York).

Davis, M. 1997, *Gangland: cultural elites and the new generationalism* (Allen and Unwin, Sydney).

Davis, P., McLeod, K., Ransom, M. and Ongley, P. 1997, *The New Zealand socioeconomic index of occupational status.* (Statistics New Zealand, Wellington).

Davis, S.L. and Prescott, J.R.V. 1992, *Aboriginal frontiers and boundaries in Australia* (Melbourne University Press, Carlton, Vic).

Davison, G., Dingle, T. and O'Hanlon, S. (eds) 1995, *The cream brick frontier: histories of Australian suburbia* (Monash Publications in History, no. 19, Monash University, Victoria).

Dawson, J., 1988, *Forest, vine and snow tussocks – the story of New Zealand plants* (Victoria University Press, Wellington).

Department of Conservation (DoC) 1997, *Conservation action: Department of Conservation achievements and plans, 1996/1997 – 1997/1998* (DoC, Wellington, NZ.

Department of the Environment, Sport and Territories, Australia (DESTA) 1996, *Australia: state of the environment* (CSIRO Publishing, Collingwood, Melbourne).

Department of Immigration and Multicultural Affairs (DIMA) 1998a, *Over fifty years of post-war migration* (DIMA, Canberra).

Department of Immigration and Multicultural Affairs (DIMA) 1998b, *Recent developments in the immigration and multicultural affairs portfolio* (DIMA, Canberra).

Dillon, M. 1991, 'Interpreting Argyle: Aborigines and diamond mining in north-west Australia', pp. 139–52 in Connell and Howitt (eds) (1991).

Dixon, R.A. and Dillon, M.C. (eds) 1990, *Aborigines and diamond mining: the politics of resource development in the east Kimberley* (University of Western Australia Press, Perth).

Dodson, J. (ed.) 1992, *The native lands: prehistory and environmental change in Australia and the Southwest Pacific* (Longman Cheshire, Melbourne).

Dodson, J. 1996, 'Human rights and the extinguishment of native title', *Australian Aboriginal Studies*, 2, 12–23.

Doherty, L. 1998, 'Drinking problem', *The Sydney Morning Herald*, 8 August, p. 35.

Donahue, R.L., Miller, R.W. and Shickluna, J.C. 1983, *Soils – an introduction to soils and plant growth* (Prentice-Hall, Englewood Cliffs, NJ), 5th edition.

Donaldson, M. 1982, 'The state, the steel industry, and BHP', *Australian and New Zealand Journal of Sociology*, 18, 339–63.

Douglas, E.M.K. 1984, 'Land and the Maori identity in contemporary New Zealand', *Plural Societies*, 125, 33–51.

Douglas, R. 1980, *There's got to be a better way!* (Fourth Estate, Wellington).

Dovers, S. (ed.) 1994, *Australian environmental history: essays and cases* (Oxford University Press, Melbourne).

Drakakis-Smith, D. and Hirst, J. 1981, 'The spatial delineation of Aboriginal Australia', *Australian Geographer*, 15, 52–4.

Duffy, M. 1994, 'Suburbia's new cathedrals: Australia's shopping malls', *Independent Monthly* (Mar.), 28–33.

Duncan, C. 1987, 'Mineral resources and mining industries', pp. 320–52 in Jeans (ed.) (1987).

Duncan, I., Bollard, A., Buchan, D. and Rivers, M.J. 1994, 'Corporatisation and privatisation – welfare effects', Working Paper, New Zealand Institute of Economic Research, No. 94/19.

Duncan, M.J. 1992, 'Flow regimes of New Zealand rivers', pp. 13–27 in Mosley (ed.) (1992).

Dunn, K.M. 1993, 'The Vietnamese concentration in Cabramatta: site of avoidance and deprivation or island of adjustment and protection?', *Australian Geographical Studies*, 31(2), 228–45.

Dunn, K.M. 1998, 'Rethinking ethnic concentration: the case of Cabramatta, Sydney', *Urban Studies*, 35(3), 503–27.

Dunn, K.M., McGuirk, P.M. and Winchester, H.M. 1995, 'Place making: the social construction of Newcastle', *Australian Geographical Studies*, 33, 149–66.

Durie, M. 1998, *Te mana te kawanatanga: the politics of Maori self-determination* (Oxford University Press, Auckland).

Dyster, B. and Meredith, D. 1990, *Australia in the international economy in the 20th century* (Cambridge University Press, Cambridge).

Easton, B. (ed.) 1989, *The making of Rogernomics* (Auckland University Press, Auckland).

Easton, B. 1994, 'Economic rationalism and the New Zealand welfare state', *British Review of New Zealand Studies*, 7, 25–32.

Edgington, D.W. 1982, 'Changing patterns of CBD office activity in Melbourne', *Australian Geographer*, 15, 231–42.

Edgington, D.W. 1988, 'Suburban economic development in Melbourne', *Australian Planner*, 26, 7–14.

Edgington, D.W. 1989, 'The consequences of economic restructuring for Melbourne's metropolitan policy', *Urban Policy and Research*, 7, 51–9.

Elek, A. 1992a, 'Pacific economic co-operation: policy choices for the 1990s', *Asian–Pacific Economic Literature*, 6(1), 1–15.

Elek, A. 1992b, 'Trade policy options for the Asia–Pacific region in the 1990s: the potential of open regionalism', *American Economic Review*, 82(2), 74–8.

Ell, S. 1995, *'There she blows': sealing and whaling days in New Zealand* (Bush Press, Auckland).

Engwicht, D. 1992, *Towards an eco-city: calming the traffic* (Envirobook, Sydney).

Epps, R. 1995, 'The sustainability of Australian agricultural production systems: a realistic objective or simply a desirable aim?', *Australian Geographer*, 226, 173–9.

Erskine, W.D. 1996, 'Rapid response and recovery of a sand-bed stream to a catastrophic flood', *Zeitschrift für Geomorphologie*, 40, 359–83.

Erskine, W.D. and Warner, R.F. 1988, 'Geomorphic effects of alternating flood- and drought-dominated regimes on NSW coastal rivers', pp. 223–44 in Warner (ed.) (1988).

Eyles, G.O., 1983, 'The distribution and severity of present erosion in New Zealand', *New Zealand Geographer*, 39(1), 12–28.

Eyles, R.J. and McConchie, J.A. 1992, 'Wellington', pp. 382–406, in Soons, J.M. and Selby, M.J. (eds) *Landforms of New Zealand* (Longman Paul, Auckland).

Fagan, R.H. 1984, 'Corporate structure and regional uneven development in Australia', pp. 91–123 in Taylor (ed.) (1984).

Fagan, R.H. 1986, 'Australia's BHP Ltd – an emerging transnational resources corporation', *Raw Materials Report*, 4, 46–55.

Fagan, R.H. 1987, 'Local enterprise initiatives: long-term strategy for localities or "flavour of the month"?', *Australian Geographer*, 18, 51–6.

Fagan, R.H. 1990a, 'Elders IXL Ltd: finance, capital and the geography of corporate restructuring', *Environment and Planning A*, 22, 647–66.

Fagan, R.H. 1990b, 'The restructuring of Elders IXL Ltd: finance and the global shift', *Australian Geographer*, 21, 90–2.

Fagan, R.H. 1991, 'Industrial policy and the macro-economic environment', *Australian Geographer*, 22, 102–5.

Fagan, R.H. 1994, 'Working nation and the outer suburbs: the example of Western Sydney', *Australian Geographer*, 25, 121–5.

Fagan, R.H. 1995, 'Economy, culture and environment: perspectives on the Australian food industry', *Australian Geographer*, 26, 1–10.

Fagan, R.H. 1996, 'Exploring economic integrations with Asia: the Australian food industry and global change, pp. 205–25 in Robinson, R. (ed.) *Pathways to Asia: the politics of engagement* (Allen and Unwin, Sydney).

Fagan, R.H. 1997a, 'Local food/global food: globalization and local restructuring', pp. 197–208 in Lee and Wills (eds) (1997).

Fagan, R.H. 1997b, 'Global-local relations and public policy: Australian industry and the Pacific Rim', pp. 129–50 in Rimmer (ed.) (1997b).

Fagan, R.H. and Le Heron, R.B. 1994, 'Reinterpreting the geography of accumulation: the global shift and local restructuring', *Environment and Planning D: Society and Space*, 12, 265–85.

Fagan, R.H. and O'Neill, P.M. 1994, 'Introduction: industry, employment and regions: geographical perspectives on *Working Nation*', *Australian Geographer*, 25, 101–3.

Fagan, R.H. and Rich, D.C. 1990, 'Industrial restructuring in the Australian food industry: corporate strategy and the global economy', pp. 175–94 in Wilde, P.D. and Hayter, R. (eds) *Industrial transformation and challenge in Australia and Canada* (Carleton University Press, Ottawa).

Fagan, R.H. and Webber, M.J. 1994, *Global restructuring: the Australian experience* (Oxford University Press, Oxford).

Fahey, B.D. and Rowe, L.K. 1992, 'Land-use impacts', pp. 265–84 in Mosley (ed.) (1992).

Fairweather, J. 1982, 'Agrarian restructuring in New Zealand', Research Reports, Agri-business and Economics Research Unit, Lincoln University, no. 213.

Farmer, R.S.J. 1996 'Economic deregulation and changes in New Zealand's immigration policy: 1986 to 1991', *People and Place*, 4(3), 55–63.

Farmer, R.S.J. 1997, 'New Zealand's targeted immigration policy, 1991–1996', *People and Place*, 5(1), 1–15.

Faulkner, H.W. and Walmsley, D.J. 1998, 'Globalisation and the pattern of inbound tourism in Australia', *Australian Geographer*, 29, 91–106.

Ferguson, J. and Simpson, R. 1995, *The Australian rural labour market* (National Farmers' Federation, Canberra).

Fincher, R.M. 1997, 'Gender, age and ethnicity in immigration for an Australian nation', *Environment and Planning A*, 29, 217–36.

Fincher, R.M. 1998, 'Population questions for Australian cities: reforming our narratives', *Australian Geographer*, 29, 31–48.

Finkelstein, J. 1973, 'Survey of NZ tank evaporation', *Journal of Hydrology (NZ)*, 12, 119–31.

Finlayson, B.L. and McMahon, T.A. 1988, 'Australia v. the world: a comparative analysis of streamflow characteristics', pp. 17–40 in Warner (ed.) (1988).

Fischer, K.F. 1984, *Canberra: myths and models.* (Institute of Asian Affairs, Hamburg).

Fitzgerald, S. 1997, *Is Australia an Asian country?* (Allen and Unwin, St Leonards, NSW).

Fitzharris, B.B. and Endlicher, W. 1996, 'Climatic conditions for wine grape growing', *New Zealand Geographer*, 52, 1–11.

Fitzharris, B.B., Mansergh, G.D. and Soons, J.M. 1992, 'Basins and lowlands of the South Island', pp. 407–23 in Soons and Selby (eds) (1992).

Flannery, T. 1994, *The future eaters* (Reed Books, Chatswood, NSW).

Forbes, D. 1988, 'Aid, trade and the "new realism": Australia's links with east and south-east Asia', *Australian Geographer*, 19, 182–94.

Forbes, D. 1997, 'Regional integration, internationalisation and the new geographies of the Pacific Rim', pp. 13–28 in Watters, McGee and Sullivan (eds) (1997).

Forer, P. 1995, Off the sheep's back? Two decades of change in New Zealand', pp. 241–61 in Cliff *et al.* (eds) (1995).

Forster, C. 1995, *Australian cities: continuity and change* (Oxford University Press, Melbourne).

Fougere, G. 1994, 'Health', pp. 146–60 in Spoonley, Pearson and Shirley (eds) (1994).

Frawley, K.J. 1994, 'Evolving visions: environmental management and nature conservation in Australia', pp. 55–78 in Dovers (ed) (1994).

Freestone, R. 1993, 'Heritage, urban planning and the postmodern city', *Australian Geographer*, 24, 17–24.

French, Justice R.S. 1996, 'The Wentworth Lecture. Native title: the beginning or the end of justice?', *Australian Aboriginal Studies*, 2, 2–11.

Friedmann, H. and McMichael, P. 1989, 'Agriculture and the state system: the rise and fall of national agricultures, 1870 to the present', *Sociologia Ruralis*, 29, 93–117.

Fuchs, R.J., Chamies, J. and Lo, F. (eds) 1994, *Mega-city growth and the future.* (United Nations University Press, New York).

Gale, F. 1972, *Urban Aborigines* (Australian National University Press, Canberra).

Gale, F. 1987, 'Aborigines and Europeans', pp. 127–43 in Jeans (ed.) (1987).

Gale, F. 1990, 'Aboriginal Australia: survival by separation', pp. 217–34 in Chisholm and Smith (eds) (1990).

Gale, S.J. 1992, 'Long-term landscape evolution in Australia', *Earth Surface Processes and Landforms*, 17, 323–43.

Gallant, N. and Stubbs, R. 1997, 'APEC's dilemmas: institution building around the Pacific Rim', *Pacific Affairs*, 70, 203–18.

Galloway, R.W. 1963, 'Glaciation in the Snowy Mountains: a re-appraisal', *Proceedings of the Linnean Society of New South Wales*, 88, 180–98.

Gardner, H. 1989, 'The parliament of the suburbs: the Melbourne and Metropolitan Board of Works', *Canberra Series in Adminstrative Studies*, 9 (Canberra College of Advanced Education, Canberra).

Garnham, B. 1997, 'Tourism in change', *New Zealand Journal of Geography*, April, 5–11.

Gentilli, J, 1986, 'Climate', pp. 14–48 in Jeans (ed.) (1986).

Gertler, M.S. 1997, 'Globality and locality: the future of "geography" and the nation-state', pp. 12–33 in Rimmer (ed.) (1997b).

Gleeson, B., Hay, C. and Law, R. 1998, 'The geography of mental health in Dunedin, New Zealand', *Health and Place*, 4(1), 1–14.

Go, F.M. and Jenkins, C.L. (eds) 1997, *Tourism and economic development in Asia and Australasia* (Cassell, London and Washington).

Goodman, D., O'Hearn, D. J. and Wallace-Crabbe, C. (eds) 1991, *Multicultural Australia: the challenges of change* (Scribe, Melbourne).

Goodman, D. and Redclift, M. 1991, *Refashioning nature?* (Blackwell, Oxford).

Goodman, J. and Pauly, L. 1993, 'The obsolescence of capital controls?', *World Politics*, October, 50–83.

Gould, J.D. 1982, *The rake's progress: The New Zealand economy since 1945* (Oxford University Press, Auckland).

Graetz, B. and McAllister, I. 1994, *Dimensions of Australian society* (MacMillan Education, Melbourne), 2nd edition.

Gray, A. 1989, 'Aboriginal migration to the cities', *Journal of the Australian Population Association*, 6, 122–44.

Gray, I., Lawrence, G. and Dunn, T. 1993, *Coping with change: Australian farmers in the 1990s* (Centre for Rural Social Research, Wagga Wagga).

Graycar, A. and Jamrozik, A. 1993, *How Australians live: social policy in theory and practice* (Macmillan Education Australia, Melbourne), 2nd edition.

Green, D.G. 1996, *From welfare state to civil society: towards welfare that works in New Zealand* (New Zealand Business Roundtable, Wellington).

Greif, S.W. (ed.) 1995, *Immigration and national identity in New Zealand: one people, two peoples, many peoples* (Dunmore Press, Palmerston North).

Grey, A. 1994, *Aotearoa and New Zealand: a historical geography* (Canterbury University Press, Christchurch).

Griffin, A. and Darcy, S. 1997, 'Australia: consequences of the newly adopted pro-Asia orientation', pp. 67–90 in Go and Jenkins (eds) (1997).

Grosvenor, S., Le Heron, R.B. and Roche, M.M. 1995, 'Sustainability, corporate growers, regionalisation and Pacific–Asia links in the Tasmanian and Hawkes Bay apple industries', *Australian Geographer*, 26, 163–72.

Grosvenor, S. and Wood, L.J. 1996, 'Recent changes in the Tasmanian dairy industry', *New Zealand Geographer*, 52, 56–64.

Groves, R.H. (ed.) 1994, *Australian vegetation* (Cambridge University Press, Cambridge), 2nd edition.

Grundy, K.J. and Gleeson, B.J. 1996, 'Sustainable management and the market: the politics of planning reform in New Zealand', *Land Use Policy*, 13, 197–211.

Gunder, M. 1996, 'Urban policy formulation under efficiency: the case of Auckland City Council's Britomart Development', *Urban Policy and Research*, 14(3), 199–214.

Gunder, M. 1998, 'The free lunch public transport centre: a New Zealand case study on how to acquire 2900 car parking spaces and $0.4 billion in public debt', *World Transport Policy and Practice*, 4(3), 8–15.

Hairsine, P., Murphy, B., Packer, I. and Rosewell, C. 1993, 'Profile of erosion from a major storm in the south-east cropping zone', *Australian Journal of Soil and Water Conservation*, 6(4), 50–5.

Hall, C.M. 1995, *Introduction to tourism in Australia: impacts, planning and development* (Longman, Melbourne), 2nd edition.

Hamnett, C. 1991, 'The blind men and the elephant: the explanation of gentrification', *Transactions of the Institute of British Geographers*, 16, 173–89.

Hamnett, S. and Bunker, R. (eds) 1987, *Urban Australia: planning issues and policies* (Mansell Publishing, London).

Hansen, J. 1994, 'There goes the neighbourhood', *Metro (Auckland)*, No. 162(Dec.), 112–23.

Hansen, J. 1995, 'A world of their own', *Metro (Auckland)*, No. 168(June), 70–7.

Harbridge, R. (ed.) 1993, *Employment contracts: New Zealand experiences* (Victoria University Press, Wellington).

Harbridge, R. and Hince, K. 1994, 'Employment relations in New Zealand: a review of bargaining and worker representation in 1993/4', *British Review of New Zealand Studies*, 7, 113–27.

Harland, B. 1992a, 'New Zealand and the emerging tripolar world', *British Review of New Zealand Studies*, 5, 37–50.

Harland, B. 1992b, *On our own: New Zealand in the emerging tripolar world* (Institute of Policy Studies, Victoria University of Wellington).

Harris, C.R. 1986, Native vegetation; pp 29–54, in Nance and Speight (eds) (1986).

Harris, R. and Leiper, N. 1995, *Sustainable tourism: an Australian experience* (Butterworth-Heinemann, Sydney).

Harvey, D. 1989, *The condition of postmodernity: an enquiry into the origins of cultural change* (Blackwell, Oxford).

Harvey, N., Clouston, E. and Carvalho, P. 1999, 'Improving coastal vulnerability assessment methodologies for integrated coastal zone management: an approach from South Australia', *Australian Geographical Studies*, 37, 50–69.

Haughton, G. 1988, 'Community and industrial restructuring in the Illawarra', *Environment and Planning A*, 21, 233–47.

Haughton, G. 1990, 'Manufacturing recession? BHP and the recession in Wollongong', *International Journal of Urban and Regional Research*, 14, 70–88.

Haworth, R.J. and Ollier, C.D. 1992, 'Continental rifting and drainage reversal: the Clarence River of eastern Australia', *Earth Surface Processes and Landforms*, 17, 387–97.

Hayward, D. 1996, 'The reluctant landlords: a history of public housing in Australia', *Urban Policy and Research*, 14(1), 5–35.

Head, B. (ed.) 1986, *The politics of development in Australia* (Allen and Unwin, Sydney).

Head, L. 1989, 'Pre-historic Aboriginal impacts on Australian vegetation: an assessment of the evidence', *Australian Geographer*, 20, 167–72.

Heal, A. 1994, 'Villa file', *Metro (Auckland)*, No. 161(Nov.), 80–9.

Heal, A. 1997, 'Road rage', *Metro (Auckland)*, No. 191(May), 58–71.

Healy, J. 1992, 'Central volcanic region', pp. 256–86 in Soons and Selby (eds) (1992).

Healy, T.R. and Kirk, R.M. 1992, 'Coasts', pp. 161–86 in Soons and Selby (eds) (1992).

Heathcote, R.L. 1965, *Back of Bourke: a study of land appraisal and settlement in semi-arid Australia* (Melbourne University Press, Carlton).

Heathcote, R.L. (ed.) 1988, *The Australian experience* (Longman Cheshire, Melbourne).

Heathcote, R.L. 1994, *Australia* (Longman, Harlow), 2nd edition.

Heathcote, R.L. and Mabbutt, J.A. (eds) 1988, *Land, water and people: geographical essays in Australian resource management* (Allen and Unwin, Sydney).

Heathcote, R.L. and Thom, B.G. (eds) 1979, *Natural hazards in Australia* (Australian Academy of Science, Canberra).

Heerdegen, R.G. and Shepherd, M.J. 1992, 'Manawatu landforms – product of tectonism, climate change and process', pp. 308–33 in Soons and Selby (eds) (1992).

Henderson, I. 1997, 'Economic policy: successes and failures of reform', pp. 113–23 in Singleton (ed.) (1977).

Henderson-Sellers, A. and Blong, R.J. 1989, *The greenhouse effect: living in a warmer Australia* (New South Wales University Press, Kensington).

Hensher, D., King, J. and Oum, T. (eds) 1996, *World transport research*, Proceedings of the Seventh World Conference on Transport Research, Volume 3: *Transport policy* (World Conference on Transport Research, Sydney).

Hewitt, A.E. 1992a, 'Soil classification in New Zealand: legacy and lessons', *Australian Journal of Soil Research*, 30, 843–54.

Hewitt, A.E. 1992b, *New Zealand soil classification* (Department of Science and Industrial Research, Lower Hutt) Land Resources Scientific Report, No. 19.

Hicks, D.M., Hill, J. and Shankar, U. 1996, 'Variation of suspended sediment yields around New Zealand: the relative importance of rainfall and geology'. *International*

Association of Hydrological Sciences Publications, no. 236 (Proceedings of the Exeter Symposium, Erosion and Sediment Yield: Global and Regional Perspectives), pp. 149–56.

Hillier, J. 1996, 'The gaze in the city: video surveillance in Perth', *Australian Geographical Studies*, 34(1), 95–105.

Hillier, J. and McManus, P. 1993, 'Pull up the drawbridge: fortress mentality in the suburbs', in Watson and Gibson (eds) (1993).

Hillman, M., Adams, J. and Whitelegg, J. 1990, *One false move: a study of children's independent mobility* (Policy Studies Institute, London).

Hills, B. 1998, 'Fortress Sydney: It's a place called home', *Sydney Morning Herald*, 4 April, p. 9.

Hoadley, S. 1997, *The US–New Zealand kiwifruit dispute* (Australia–New Zealand Studies Center, Pennsylvania State University, University Park, PA).

Hodge, S. 1996, 'Disadvantage and "otherness" in Western Sydney', *Australian Geographical Studies*, 34(1), 32–44.

Holland, P. and Johnston, W.B. (eds), 1987, *Southern approaches: geography in New Zealand* (New Zealand Geographical Society, Christchurch).

Holmes, F. (ed.) 1989, *Stepping stones to freer trade?* (Institute of Policy Studies, Wellington).

Holmes, J.H. 1994, 'Coast versus inland? Two different Queenslands', *Australian Geographical Studies*, 32, 167–82.

Hood, D.E. 1987, 'Employment changes in Tasmanian manufacturing, 1980–1985: an enterprise-based approach', *Australian Geographer*, 18, 149–60.

Hopkins, A. 1991, 'Recent trends in New Zealand's livestock industry', *Geography*, 76, 269–71.

Horvath, R. and Engels, B. 1985, 'The residential restructuring of inner Sydney', pp. 143–59 in Burnley and Forrest (eds) (1985).

Howell, P. 1998, 'An Australian Convention conceived in controversy', *The Round Table*, 347, 343–56.

Howitt, R. 1991a, 'Aborigines and gold mining in central Australia', pp. 119–38 in Connell and Howitt (eds) (1991).

Howitt, R. 1991b, 'Aborigines and restructuring in the mining sector: vested and representative interests', *Australian Geographer*, 22, 117–19.

Howitt, R. 1992, 'Weipa: industrialisation and indigenous rights in a remote Australian mining area', *Geography*, 77, 223–35.

Howitt, R., Connell, J. and Hirsch, P. (eds) 1996, *Resources, nations and indigenous peoples: case studies from Australasia, Melanesia and South-East Asia* (Oxford University Press, Melbourne).

Howitt, R. and Jackson, S. 1998, 'Some things do change: indigenous rights, geographers and geography in Australia', *Australian Geographer*, 29, 155–74.

Hucker, B. 1995, 'Britomart: a bottomless pit', *Planning Quarterly*, 119, 11–12.

Hughes, P.J. and Sullivan, M.E. 1981, 'Aboriginal burning and late Holocene geomorphic events', *Search*, 12, 277–8.

Hughes, R. 1987, *The fatal shore: a history of the transportation of convicts to Australia 1787–1868* (Collins Harvill, London).

Hughson, J. 1992, 'Australian soccer: ethnic or Aussie? The search for an image', *Current Affairs Bulletin*, 68(10), 12–16.

Hughson, J. 1997, 'Football, folk dancing and fascism: diversity and difference in multicultural Australia', *Australian and New Zealand Journal of Sociology*, 33(2), 167–86.

Hugo, G. 1995, 'The turnaround in Australia: some first observations from the 1991 Census', *Australian Geographer*, 25, 1–17.

Hugo, G. 1996, 'Counterurbanisation', pp. 126–46 in Newton and Bell (eds) (1996b).

Hungerford, L. 1996, 'Australian sugar in the global economy: recent trends, emerging problems', pp. 12–18 in Burch, Rickson and Lawrence (eds) (1996).

Hunter, B. and Gregory, R.G. 1997, 'Recently observed polarising tendencies and Australian cities', *Australian Geographical Studies*, 35(3), 243–59.

Hunter, C. 1991, *Earthquake tremors felt in the Hunter Valley since white settlement* (Hunter House Publications, Newcastle, NSW).

Huxley, M. 1999, 'Reconfiguring local governance and the regulation of space in Victoria, Australia', unpublished paper presented to the RGS-IBG Annual Conference, University of Leicester, 5 January.

Hyde, N. 1996, 'Harbouring doubts', *Metro (Auckland)*. No. 177(Mar.), 50–5.

Independent Commission on International Development Issues 1980, *North–South: a programme for survival* (Pan Books, London and Sydney).

Industry Commission, 1993, *Impediments to regional adjustment* (AGPS, Canberra).

Inglis, C., Gunasekaran, S., Sullivan, G., and Chung-Tong, W. (eds) 1992, *Asians in Australia: the dynamics of migration and settlement* (Allen and Unwin, Sydney).

Ip, M. 1995, 'Chinese New Zealanders: old settlers and new immigrants', pp. 161–99 in Greif (ed.) (1995).

Irwin, G. 1992, *The prehistoric exploration and colonisation of the Pacific* (Cambridge University Press, Cambridge).

Isbell, R.F. 1992, 'A brief history of national soil classification in Australia since the 1920s', *Australian Journal of Soil Research*, 30, 825–42.

Jackson, M. 1992, 'Justice: unitary or separate', pp. 171–5 in Novitz and Willmott (eds) (1992).

Jackson, R.T. 1987, 'Commuter mining and the Kidston gold mine: goodbye to mining towns?', *Geography*, 72, 162–5.

Jacobs, J.M. 1988, 'Politics and the cultural landscape: the case of Aboriginal land rights', *Australian Geographical Studies*, 26, 249–63.

Jacobs, J.M. 1996, *Edge of Empire: postcolonialism and the city* (Routledge, London and New York).

Jacobs, J.M. 1998, 'Resisting reconciliation: the secret geographies of (post)colonial Australia', pp. 203–16 in Pile and Keith (eds) (1998).

Jakubowicz, A. 1984, 'State and ethnicity: multiculturalism as ideology', pp. 14–28 in Jupp (ed.) (1984).

James, C. 1992, *New territory: the transformation of New Zealand 1984–92* (Bridget Williams Books, Wellington).

Jamrozick, A., Boland, C. and Urquart, R. 1995, *Social change and cultural transformation in Australia* (Cambridge University Press, Melbourne).

Jeans, D.N. (ed.) 1986, *The natural environment* (Sydney University Press, Sydney).

Jeans, D.N. (ed.) 1987, *Australia: a geography*, Volume 1: *The natural environment* (Sydney University Press, Sydney).

Jennings, J.N. 1967, 'Two maps of rainfall intensity in Australia', *Australian Geographer*, 10, 256–62.

Jennings, J.N. and Mabbutt, J.A. 1986, 'Physiographic outlines and regions', pp. 80–96 in Jeans (ed.) (1986).

Jensen, A. 1986, 'Inland waters', pp. 148–81 in Nance and Speight (eds) (1986).

Johnson, L. and Wright, S. 1994, '(White) papering over the regional problem: restructuring in Geelong and government response', *Australian Geographer*, 25, 115–21.

Johnson, R.W. (ed.) 1976, *Volcanism in Australasia* (Elsevier, Amsterdam).

Johnston, L. 1997, 'Queen(s') Street or Ponsonby Poofters? Embodied HERO Parade sites', *New Zealand Geographer*, 53(2), 29–33.

Johnston, R.J., Gregory, D. and Smith, D. M. 1995, *The dictionary of human geography* (Blackwell, Oxford), 3rd edition.

Jones, F.L. 1996a, 'Ethnic enclaves: a transitory phenomenon', *People and Place*, 4(2), 32–3.

Jones, F.L. 1996b, 'National identity and social values', *People and Place*, 4(4), 17–26.

Jones, F.L. 1997, 'Ethnic diversity and national identity', *Australian and New Zealand Journal of Sociology*, 33(3), 285–305.

Jones, H. 1992, 'The new global context of international migration policy options for Australia in the 1990s', *Area*, 24, 359–66.

Jones, H. 1998, 'The Asianisation of Australia's immigration programme; diversity, preferences and citizenship', *Australian Studies*, 13, 91–115.

Jones, M. 1986, *A sunny place for shady people: the real Gold Coast story* (Allen and Unwin, Sydney).

Jordens, A. 1995, *Redefining Australians:*

immigration, citizenship and national identity
(Hale and Iremonger, Sydney).

Journeaux, P. 1996, 'Trends in New Zealand agriculture', *New Zealand Journal of Geography*, 91, 1–9.

Jupp, J. (ed.) 1984, *Ethnic politics in Australia*. (Allen and Unwin, Sydney).

Jupp, J. 1991, 'Managing ethnic diversity: how does Australia compare?', pp. 39–54 in Castles (ed.) (1991).

Jupp, J. 1995, 'From "White Australia" to "Part of Asia": recent shifts in Australian immigration policy towards the region', *International Migration Review*, 29, 207–28.

Jupp, J. 1997, 'Immigration and national identity: multiculturalism', pp. 132–44 in Stoker (ed.) (1997).

Kamp, P.J.J. 1992a, 'Tectonic architecture of New Zealand', pp. 1–30 in Soons and Selby (eds) (1992).

Kamp, P.J.J. 1992b, 'Landforms of Hawke's Bay and their origin: a plate tectonic interpretation', pp. 344–66 in Soons and Selby (eds) (1992).

Kamp, P.J.J. 1992c, 'Landforms of Wairarapa: a geological perspective', pp. 367–81 in Soons and Selby (eds) (1992).

Kawharu, I.H. 1968, 'Urban immigrants and Tangata Whenua', pp. 174–86 in Schwimmer, E. (ed.) (1968).

Kawharu, I.H. (ed.) 1989, *Waitangi: Maori and Pakeha perspectives on the Treaty of Waitangi* (Oxford University Press, Auckland).

Kawharu, I.H. 1992, 'The Treaty of Waitangi: a Maori point of view', *British Review of New Zealand Studies*, 5, 23–37.

Kearns, R.A. 1991, 'The place of health in the health of place: the case of the Hokianga Special Medical Area', *Social Science and Medicine*, 33, 519–30.

Kearns, R.A. and Joseph, A.E. 1997, 'Restructuring health and rural communities in New Zealand', *Progress in Human Geography*, 21, 18–32.

Kearns, R.A. and Smith, C.J. 1994, 'The residential mobility experiences of marginalised groups', *Tijdschrift voor Economische en Sociale Geografie*, 85(2), 114–29.

Kearns, R.A., Smith, C.J. and Abott, M.W. 1991, 'Another day in paradise? Life on the margins in New Zealand', *Social Science and Medicine*, 33(4), 369–79.

Kelliher, F.M. and Scotter, D.R. 1992, 'Evaporation, soil and water', pp. 135–45 in Mosley (ed.) (1992).

Kellow, A. 1986, 'Electricity planning in Tasmania and New Zealand: political processes and the technological imperative', *Australian Journal of Public Administration*, 45, 2–17.

Kellow, A. 1989, 'The dispute over the Franklin River and south-west wilderness area in Tasmania, Australia', *Natural Resources Journal*, 29, 129–46.

Kellow, A. 1996, *Transforming power: the politics of electricity planning* (Cambridge University Press, Cambridge).

Kelsey, J. 1995, *Economic fundamentalism*. (Auckland University Press, Auckland).

Kelsey, J. 1997, *The New Zealand experiment: a world model for structural adjustmnent?* (Auckland University Press/Bridget Williams Books, Auckland).

Kenworthy, J. 1995, 'Automobile dependence in Bangkok: an international comparison with implications for planning policies', *World Transport Policy and Practice*, 1(3), 31–41.

Kirkpatrick, J. 1994, *A continent transformed* (Oxford University Press, Melbourne).

Kissling, C. 1998, 'Beyond the Australasian single aviation market', *Australian Geographical Studies*, 36, 170–6.

Kissling, C. and Douglass, M. 1993, 'Transportation', pp. 33–56 in Memon and Perkins (eds) (1993).

Kitay, J. and Lansbury, R.D. (eds) 1997, *Changing employment relations in Australia* (Oxford University Press, Melbourne).

Klomp, N. and Lunt, I. (eds) 1997, *Frontiers in ecology: building the links* (Elsevier Science, Oxford).

Knopp, L. 1990, 'Some theoretical implications of gay involvment in an urban land market', *Political Geography Quarterly*, 9(4), 337–52.

Kohen, J.L. 1995, *Aboriginal environmental*

impacts (University of New South Wales Press, Sydney).

Lack, J. and Templeton, J. 1995, *Bold experiment: a documentary history of Australian immigration since 1945* (Oxford University Press, Melbourne).

Landais-Stamp, P. and Rogers, P. 1989, *Rocking the boat: New Zealand, the United States and the nuclear-free zone controversy in the 1980s* (Berg, Oxford).

Lane, M.B. 1997, 'Aboriginal participation in environmental planning', *Australian Geographical Studies*, 35, 308–23.

Langdale, J.V. 1991, 'Restructuring telecommunications', *Australian Geographer*, 22, 124–6.

Lange, D. 1990, *Nuclear free – the New Zealand way* (Penguin Books, Auckland).

Larritt, C.S. 1995, 'Taking part in Mutawintji: Aboriginal involvement in Mootwingee National Park', *Australian Geographical Studies*, 33, 242–56.

Lauria, M.A.K.L. 1985, 'Toward an analysis of the role of gay communities in the urban renaissance', *Urban Geography*, 6(2), 152–69.

Laurie, J. 1996, 'South Auckland success story – How Bizarre', *The Australian*, 11 July, p. 32.

Law, R. 1997, 'Masculinity, place and beer advertising in New Zealand: the Southern Man campaign', *New Zealand Geographer*, 53(2), 22–2.

Lawrence, G. 1996, 'Contemporary agri-food restructuring: Australia and New Zealand', pp. 45–72 in Burch, Rickson and Lawrence (eds) (1996).

Le Heron, R.B. 1979, 'A round of growth: the verdict on New Zealand's programme of regional development', *New Zealand Geographer*, 35, 71–9.

Le Heron, R.B. 1987a, 'Processing industries – converting farm products', pp. 74–80 in Saunders (ed.) (1987).

Le Heron, R.B. 1987b, 'Rethinking regional development', pp. 261–82 in Holland and Johnston (eds) (1987).

Le Heron, R.B. 1988, 'State, economy and crisis in New Zealand in the 1980s: implications for land-based production of a new mode of regulation', *Applied Geography*, 8, 273–90.

Le Heron, R.B. 1989, 'A political economy perspective on the expansion of New Zealand livestock farming, 1960–1984', *Journal of Rural Studies*, 5, 17–43.

Le Heron, R.B. 1990, 'Reorganisation of the New Zealand meat freezing industry: political dilemmas and social impacts', pp. 108–127 in Rich and Linge (eds) (1991).

Le Heron, R.B. 1991a, 'New Zealand agriculture and changes in the agriculture–finance relation during the 1980s', *Environment and Planning A*, 23, 1653–70.

Le Heron, R.B. 1991b, 'New Zealand's export meat freezing industry: political dilemmas and spatial impacts', pp. 108–27 in Rich and Linge (eds) (1991).

Le Heron, R.B. 1992, 'Meat freezing industry restructuring', pp. 107–14 in Britton, Le Heron and Pawson (eds) (1992).

Le Heron, R.B. 1993, *Globalized agriculture: political choice* (Pergamon Press, Oxford).

Le Heron, R.B. and Pawson, E.J. (eds) 1996, *Changing places: New Zealand in the nineties* (Longman Paul, Auckland).

Le Heron, R.B. and Roche, M.M. 1995, 'A "fresh" place in food's space', *Area*, 27, 23–33.

Le Heron, R.B., Roche, M.M. and Johnston, T. 1994, 'Pluriactivity: an exploration of issues with reference to New Zealand's livestock and fruit agro-commodity systems', *Geoforum*, 25, 155–72.

Lee, R. and Wills, J. (eds) 1997, *Geographies of economies* (Edward Arnold, London).

Leeper, G.W. (ed.) 1970, *The Australian environment* (Commonwealth Scientific and Industrial Research Organisation and Melbourne University Press, Melbourne), 4th edition.

Lees, L. and Berg, L.D. 1995, 'Ponga, glass and concrete: a vision for urban socio-cultural geography in Aoteoroa/New Zealand', *New Zealand Geographer*, 51(2), 32–41.

Legat, N. 1995a, 'City limits', *Metro (Auckland)*, No. 171 (Sept.), p. 65.

Legat, N. 1995b, 'Cross purposes', *Metro (Auckland)*, No. 167(May), pp. 80–91.

Levitus, R. 1991, 'The boundaries of Gagudju Association membership: anthropology, law and public policy', pp. 153–68 in Connell and Howitt (eds) (1991).

Lindsey, D.G. and Kearns, R.A. 1994, 'The writing's on the wall: graffiti, territory and urban space in Auckland', *New Zealand Geographer*, 50(2), 7–13.

Linge, G.R.J. 1979, *Industrial awakening: a geography of Australian manufacturing 1788 to 1890* (Australian National University Press, Canberra).

Linge, G.R.J. 1991a, 'If we are to prosper, the economy must become more productive', *Australian Geographer*, 22, 105–8.

Linge, G.R.J. 1991b, 'Just-in-time in Australia', *Australian Geographer*, 22, 67–74.

Lloyd, P.J. (ed.) 1984, *Mineral economics in Australia* (George Allen and Unwin, Sydney).

Lockie, S. 1998, 'Landcare and the state: action at a distance', pp. 15–28 in Burch, Lawrence, Rickson and Goss (eds) (1998).

Logan, T. 1986 'A critical examination of Melbourne's District Centre policy', *Urban Policy and Research*, 4(2), 2–14.

Logan, W.S. 1985, *The gentrification of inner Melbourne* (University of Queensland Press, St Lucia).

Loughran, R.J., Elliott, G.L. and McFarlane, D.J. (in preparation) *A national reconnaissance survey of soil erosion in Australia using the caesium-137 method* (University of Newcastle, Australia).

Lowe, D.J. and Green, J.D., 1992, 'Lakes', pp. 107–43 in Soons and Selby (eds) (1992).

McAneney, K.J., Judd, M.J. and Weeda, W.C., 1982, 'Loss in monthly pasture production resulting from dryland conditions in the Waikato', *New Zealand Journal of Agricultural Research*, 25, 151–6.

McCall, G. and Connell, J. (eds) 1993, *A world perspective on Pacific Islander migration: Australia, New Zealand and the USA* (Pacific Studies Monograph, 6, Centre for South Pacific Studies, University of New South Wales, Sydney).

McColl, F.D. 1984, 'Foreign investment in Australian mining', pp. 121–56 in Lloyd (ed.) (1984).

McCormick, J.M. 1995, 'Healing the American

rift with New Zealand', *Pacific Affairs*, 68, 392–410.

MacGregor, C.J. and Pilgrim, A. 1998, 'Is Landcare funding hitting the target?', *Natural Resource Management*, 1, 4–8.

McKenna, M. 1996, *The captive republic* (Cambridge University Press, Cambridge).

McKenna, M. 1998, 'Growing apples and the "growing pains" of restructuring for New Zealand's pipfruit industry', *New Zealand Geographer*, 54, 37–45.

McKenzie, K. 1998, 'Don't fence me in', *Sydney Morning Herald*, 26 February, p. 6.

McKinnon, M. (ed.) 1997, *New Zealand historical atlas* (Department of Internal Affairs and David Bateman Ltd, Auckland).

Mackintosh, W.A. 1964, *The economic background of dominion–provincial relations* (University of Toronto Press, Toronto).

McLaren, R.G. and Cameron, K.C. 1996, *Soil science – sustainable production and environmental protection* (Oxford University Press, Auckland).

McLoughlin, D. 1989, 'Gone south: the new migrants', *North and South*, December, 52–6.

McLoughlin, D. 1997, 'NZ government rethink on immigration', *Business Times (South East Asia)*, 17 October.

McLoughlin, J.B. 1991, 'Urban consolidation and urban sprawl: a question of density', *Urban Policy and Research*, 9(3), 148–56.

McLoughlin, J.B. 1992, *Shaping Melbourne's future? Town planning, the state and civil society* (Cambridge University Press, Cambridge).

McLoughlin, J.B. and Huxley, M. (eds) 1986, *Urban planning in Australia: critical readings* (Longman Cheshire, Melbourne).

McTainsh, G.H. 1993, 'Soils', in McTainsh and Boughton (eds) (1993).

McTainsh, G.H. and Boughton, W.C. (eds) 1993, *Land degradation processes in Australia* (Longman Cheshire, Melbourne).

Mabbutt, J.A. 1986, 'Desert lands', pp. 180–202 in Jeans (ed.) (1986).

Maddock, R. and McLean, I.W. (eds) 1987, *The Australian economy in the long run* (Cambridge University Press, Melbourne).

Maher, C. 1997, 'Urban consolidation in the context of contemporary development trends', *Historic Environment*, 13(1), 35–46.

Maher, N. 1994, 'Minangkaban migration: developing an ethnic identity in a multicultural society', *Australian Geographical Studies*, 32, 58–68.

Mahony, D. 1978, *Beaches: learning to live with the sea* (Charden Publications, Summer Hill, NSW).

Malcolm, W., Sale, P. and Egan, A. 1996, *Agriculture in Australia: an introduction* (Oxford University Press, Melbourne).

Marceau, J. 1997, 'An uncertain world: global localisation and the emerging industrial territory of Australia', pp. 34–56 in Rimmer (ed.) (1997b).

Marcuse, P. 1996, *Is Australia different? Globalization and the new urban poverty* (Australian Housing and Urban Research Institute, Melbourne).

Marden, P. and Mercer, D. 1998, 'Locating strangers: multiculturalism, citizenship and nationhood in Australia', *Political Geography*, 17, 939–58.

Martin, G.W. (ed.), 1978, *The founding of Australia* (Sydney University Press, Sydney).

Martin, G.W. 1992, 'James Busby and the Treaty of Waitangi', *British Review of New Zealand Studies*, 5, 5–22.

Martin, G.W. 1997, *Edward Gibbon Wakefield: abductor and mystagogue* (Ann Barry, Edinburgh).

Martin, S. and McLeay, F. 1998, 'The diversity of farmers' risk management strategies in a deregulated New Zealand environment', *Journal of Agricultural Economics*, 49, 218–33.

Mascarenhas, R.C. 1991, 'State-owned enterprises', pp. 27–51 in Boston *et al.* (eds) (1991).

Massey, P. 1995, *New Zealand: market liberalisation in a developed economy* (Macmillan, Basingstoke).

Matheson, P. 1994, '150 years after. The influence of the Scottish disruption of 1843 on New Zealand', *British Review of New Zealand Studies*, 7, 7–4.

Mees, P. 1995, 'Urban transport policy practices in Australia', *World Transport Policy and Practice*, 1(1), 20–4.

Meinig, D. 1959, 'Colonization of wheat lands: some Australian comparisons', *Australian Geographer*, 7, 145–56.

Meinig, D. 1962, *On the margins of the good earth: the South Australian wheat frontier, 1869–1884* (Rand McNally, Chicago).

Memon, P.A. 1993, *Keeping New Zealand green: recent environmental reforms* (University of Otago Press, Dunedin).

Memon, P.A. and Cullen, R.C. 1996, 'Rehabilitation of indigenous fisheries in New Zealand', pp. 252–64 in Howitt, Connell and Hirsch (eds) (1996).

Memon, P.A. and Gleeson, B.J. 1995, 'Towards a new planning paradigm? Reflections on New Zealand's Resource Management Act', *Environment and planning B: Planning and Design*, 22, 109–24.

Memon, P.A. and Perkins, H.C. (eds) 1993, *Environmental planning in New Zealand.* (Dunmore Press, Palmerston North).

Mercer, D.C. 1985, 'Australia's constitution, federalism and the "Tasmanian dam case"', *Political Geography Quarterly*, 4, 92–110.

Mercer, D.C. 1997, 'Aboriginal self-determination and indigenous land title in post-Mabo Australia', *Political Geography*, 16, 189–212.

Metcalfe, A.W. and Bern, J. 1994, 'Stories of crisis: restructuring Australian industry and rewriting the past', *International Journal of Urban and Regional Research*, 18, 658–72.

Ministry for the Environment, New Zealand (MfENZ) 1997, *The state of New Zealand's environment* (Ministry for the Environment and GP Publications, Wellington).

Mitchell, K. 1993, 'Multiculturalism, or the united colours of capitalism', *Antipode*, 25, 263–95.

Mitchell, T. 1996, *Popular music and local identity: rock, pop and rap in Europe and Oceania* (Leicester University Press, Leicester).

Moran, W. 1987, 'Marketing structures and rural land use change', *New Zealand Geographer*, 43, 164–9.

Moran, W. 1997, 'Farm size change in New

Zealand', *New Zealand Geographer*, 53(1), 3–13.

Moran, W., Blunden, G., Workman, M. and Bradley, A. 1996, 'Family farmers, real regulation, and the experience of food regimes', *Journal of Rural Studies*, 12, 245–58.

Moran, W. and Nason, S.J. 1981, 'Spatio-temporal localisation of New Zealand dairying', *Australian Geographical Studies*, 19, 47–66.

Morgan, G. 1993, 'Acts of enclosure: crime and defensible space in contemporary cities', in Watson and Gibson (eds) (1993).

Morgan, R.P.C. 1995, *Soil erosion and conservation* (Longman, Harlow), 2nd edition.

Moriarty, P. and Beed, C. 1992, 'The end of the road for cars', *Urban Futures*, 2(2), 10–14.

Morison, I. 1987, 'Whatever became of Canberra's Y-Plan?, *Australian Planner*, 25(4), 21–7.

Morison, I. 1995, 'Beyond the city-state: metropolitan Canberra', *Urban Policy and Research*, 13(2), 117–24.

Morrison, P.S. 1989, *Labour adjustments in metropolitan regions* (Institute of Policy Studies, Wellington).

Morrison, P.S. 1990, 'Migrants, manufacturing and metropolitan labour markets in Australia', *Australian Geographer*, 21, 151–63.

Morrison, P.S. 1995, 'The geography of rental housing and the restructuring of housing assistance in New Zealand', *Housing Studies*, 10(1), 39–56.

Morrison, P.S. 1997, 'Unemployment and non-employment: urban inequalities in New Zealand 1981–1991', *Institute of Australian Geographers/New Zealand Geographical Society Joint Conference*, Hobart, Institute of Australian Geographers/New Zealand Geographical Society.

Morrison, P.S. and McMurray, S. 1997, 'The inner city apartment versus the suburb', *Environment and Planning A*, 29, 913–28.

Mosely, P. 1995, *Ethnic involvement in Australian soccer: a history 1950–1990* (National Sports Research Centre, Australian Sports Commission, Canberra).

Mosley, M.P. (ed.) 1992, *Waters of New Zealand* (New Zealand Hydrological Society: Wellington North).

Mulvaney, D.J. and White, J.P. (eds) 1987, *Australians to 1788* (Sydney University Press, Sydney).

Mules, T. 1998, 'Decomposition of Australian tourist expenditure', *Tourism Management*, 19, 267–71.

Mulgan, R. 1989, *Maori, Pakeha and democracy* (Auckland University Press, Auckland).

Mullins, P. 1990, 'Tourist cities as new cities: Australia's Gold Coast and Sunshine Coast', *Australian Planner*, 28(3), 37–41.

Mullins, P. 1992, 'Cities for pleasure: the emergence of tourism urbanization in Australia', *Built Environment*, 18, 187–98.

Mullins, P. 1993, 'Class relations and tourism urbanisation', in Watson and Gibson (eds) (1993).

Mullins, P. 1994, 'Class relations and tourism urbanization: the regeneration of the petite bourgeoisie and the emergence of a new urban form', *International Journal of Urban and Regional Research*, 18, 591–608.

Multicultural Marketing News 1996, 'The battle to keep ethnicity in soccer', *Multicultural Marketing News*, 41(July–August), 7.

Mulvaney, D.J. and White, J.P. (eds) 1987, *Australians to 1788* (Sydney University Press, Sydney).

Murphy, C.L. 1992, *Soil landscapes of the Gosford–Lake Macquarie 1:100 000 sheet* (Department of Conservation and Land Management, Sydney).

Murphy, L. 1997, 'The New Zealand experience of public housing reform', pp. 1–16 in Coles, R. (ed.) (1997).

Murphy, P.A. 1995, 'Winners, losers and curate's eggs: urban and regional outcomes of Australian economic restructuring 1971–1991', *Geoforum*, 26, 337–49.

Murphy, P.A. and Burnley, I. 1996, 'Exurban migration', pp. 242–58 in Newton and Bell (eds) (1996b).

Murphy, P.A. and Watson, S. 1994, 'Social polarization and Australian cities', *International Journal of Urban and Regional Studies*, 18(4), 573–90.

Murphy, P.A. and Watson, S. 1997, *Surface city: Sydney at the millennium* (Pluto Press, Sydney).

Nairn, I.A., Hewson, C.A.Y., Latter, J.H. and Wood, C.P. 1976, 'Pyroclastic eruptions of Ngauruhoe volcano, central North Island, New Zealand, 1974 January and March', pp. 385–405 in Johnson, R.W. (ed) (1976).

Nance, C. and Speight, D.L. (eds) 1986, *A land transformed: environmental change in South Australia* (Longman Cheshire, Melbourne).

Nanson, G.C. and Erskine, W.D. 1988, 'Episodic changes of channels and flood-plains on coastal rivers in New South Wales', pp. 201–21 in Warner (ed.) (1988).

Nanson, G.C. and Hean, D. 1985, 'The West Dapto flood of February, 1984: rainfall characeristics and channel changes', *Australian Geographer*, 16, 249–58.

Nash, A. 1993, 'Hong Kong's business future: the impact of Canadian and Australian business migration programmes', pp. 309–40 in Yeung (ed.) (1993).

Neall, V.E. 1992, 'Landforms of Taranaki and the Wanganui Lowlands', pp. 287–307 in Soons and Selby (eds) (1992).

Newman, P. 1995, 'The end of the urban freeway', *World Transport Policy and Practice*, 1(1), 12–19.

Newman, P. and Kenworthy, J. 1989, *Cities and automobile dependence: an international source-book.* (Gower, Aldershot).

Newman, P. and Kenworthy, J. 1991, *Towards a more sustainable Canberra: an assessment of Canberra's transport, energy and land use* (Institute for Science and Technology Policy, Perth).

Newman, P. and Kenworthy, J. 1992, *Winning back the cities* (Australian Consumers' Association and Pluto Press, Sydney).

Newton, P.W. and Bell, M. 1996a, 'Mobility and change: Australia in the 1990s', pp. 1–17 in Newton and Bell (eds) (1996b).

Newton, P.W. and Bell, M. (eds) 1996b, *Population shift: mobility and change in Australia* (AGPS, Canberra).

New Zealand Dairy Board (NZDB) 1994, *Annual Report 1994* (NZDB, Wellington).

New Zealand Meteorological Service 1983, *Climate regions of New Zealand* (Miscellaneous Publication No. 174, Wellington).

Northcote, K.H. 1978, 'Soils and land use', *Atlas of Australian resources* 1 (Division of National Mapping, Canberra).

Northcote, K.H. 1979, *A factual key for the recognition of Australian soils* (Rellim Technical Publications, Glenside, South Australia), 4th edition.

Novitz, D. and Willmott, B. (eds) 1992, *New Zealand in crisis: a debate about today's critical issues* (GP Publications, Wellington).

O'Connor, K.B. 1989, 'Australian ports, metropolitan areas and trade-related services', *Australian Geographer*, 20, 167–72.

O'Connor, K.B. 1993, 'Creativity and metropolitan development: a study of media and advertising in Australia', *Australian Journal of Regional Studies*, 6, 1–14.

O'Connor, K.B. 1998, 'The international air linkages of Australian cities, 1985–1996', *Australian Geographical Studies*, 36, 143–55.

O'Connor, K.B., Stimson, R.J. and Taylor, S.P. 1998, 'Convergence and divergence in the Australian space economy', *Australian Geographical Studies*, 36, 205–22.

Office for the Co-ordinator General of Queensland (OCGQ) 1994, *Cairns International Airport economic impact study* (prepared by Ernst and Young for OCGQ, Brisbane).

Office of Multicultural Affairs (OMA) 1989, *National Agenda for a Multicultural Australia* (Australian Government Publishing Service, Canberra).

Ollier, C.D. 1982, 'The Great Escarpment of eastern Australia', *Journal of the Geological Society of Australia*, 29, 13–23.

Ollier, C.D. 1986, 'Early landform evolution', pp. 97–116 in Jeans (ed.) (1986).

Olive, L.J. and Rieger, W.A. 1986, 'Low Australian sediment yields – a question of inefficient sediment delivery?', *International Association of Hydrological Sciences Publications*, no. 159 (Proceedings of the Albuquerque Symposium, Drainage Basin Sediment), pp. 355–64.

Oliver, J. 1973, 'Australian weather example no. 1, tropical cyclone', *Australian Geographer*, 12, 257–63.

de Oliver, M. 1993, 'The hegemonic cycle and free trade: the United States and Mexico', *Political Geography*, 12, 457–72.

O'Loughlin, C.L. and Pearce, A.J. 1982, 'Erosion processes in the mountains', pp. 68–79 in Soons and Selby (eds) (1982).

O'Neill, P.M. 1991, 'Plants on stand-by: the textiles and clothing industry in non-metropolitan areas of New South Wales and Victoria', *Australian Geographer*, 22, 108–12.

O'Neill, P.M. 1994, '*Working Nation*, economic change and the regions', *Australian Geographer*, 25, 109–15.

O'Neill, P. and Fagan, R.H. 1994. 'Introduction: industry, employment and regions: geographical perspectives on *Working Nation*', *Australian Geographer*, 25, 101–3.

Opperman, M. 1993, 'Regional market segmentation analysis in Australia', *Journal of Travel and Tourism Marketing*, 2(4).

Opperman, M. 1994, 'Regional aspects of tourism in New Zealand', *Regional Studies*, 28, 155–67.

Orams, M. 1999, *Marine tourism: development, impacts and management* (Routledge, London and New York).

Orange, C. 1990, *An illustrated history of the Treaty of Waitangi* (Allen and Unwin, Wellington).

Orchard, L. 1995, 'National urban policy in the 1990s', pp. 65–86 in Troy (ed.) (1995).

Owens, J.M.R. 1990, 'The Treaty of Waitangi – event and symbol', *British Review of New Zealand Studies*, 3, 6–15.

Page, S.J. and Piotrowski, S. 1990, 'A critical evaluation of international tourism in New Zealand', *British Review of New Zealand Studies*, 3, 87–108.

Paris, C. 1992, 'Better late than never?: the "Building Better Cities Program"', *Brisbane City Council Chair in Urban Studies Colloquium Series*, no. 4 (Queensland University of Technology, Brisbane).

Paris, C. 1994, 'New patterns of urban and regional development in Australia: demographic restructuring and economic change', *International Journal of Urban and Regional Research*, 18, 555–72.

Parker, P. 1992, 'Information specialisation and trade: trading companies and the coal trade of Europe and Japan', *Pacific Economic Research Papers*, no. 207 (Australia–Japan Research Centre, Australian National University, Canberra).

Pasquarelli, J. 1998, *The Pauline Hanson story by the man who knows* (New Holland Publications, Frenchs Forest, NSW).

Paton, T.R., Humphreys, G.S. and Mitchell, P.B. 1995, *Soils – a new global view* (University College London Press, London).

Pawson, E.J. 1991, '1990 and the Treaty', *New Zealand Geographer*, 47, 32–5.

Pawson, E.J. 1992, 'Two New Zealands: Maori and European', pp. 15–35 in Anderson and Gale (eds) (1992).

Pawson, E.J. 1999, 'Postcolonial New Zealand? Or two New Zealands? Maori and Pakeha', in Anderson and Gale (eds) (1999).

Payne, M. 1993, *Tourism in the Pacific Rim: growth in a region of opportunity* (Financial Times Business Information, London).

Pearce, D.G. 1990, 'Tourism, the regions and restructuring in New Zealand', *Journal of Tourism Studies*, 1(2), 33–42.

Pearce, D.G. 1995, 'CER, trans-Tasman tourism and a single aviation market', *Tourism Management*, 16(2):111–20.

Pearce, D.G. and Simmons, D.G. 1997, 'New Zealand: tourism – the challenges of growth', pp. 197–221 in Go and Jenkins (eds) (1997).

Pearson, C.P. 1992, 'Analysis of floods and low flows', pp. 95–116 in Mosley (ed.) (1992).

Pearson, D. 1994, 'Self-determinism and indigenous peoples in comparative perspective: problems and possibilities', *Pacific Viewpoint*, 35, 29–42.

Pearson, D. and Ongley, P. 1996, 'Multiculturalism and biculturalism: the recent New Zealand experience in comparative perspective', *Journal of Intercultural Studies*, 17(1–2), 5–28.

Peck, A.J. 1993, 'Salinity', pp. 234–70 in McTainsh and Boughton (eds) (1993).

Peck, J.A. 1990, 'Circuits of capital and industrial restructuring: adjustments in the Australian clothing industry', *Australian Geographer*, 21, 33–52.

Peel, M. 1995, 'The urban debate: from "Los Angeles" to the urban village', pp. 39–64 in Troy (ed.) (1995).

Pellow, A. 1996, *Transforming power: the politics of electricity planning* (Cambridge University Press, Cambridge).

Perkins, H.C., Memon, P.A., Swaffield, S. and Gelfand, L. 1993, 'The urban environment', pp. 11–32 in Memon and Perkins (eds) (1993).

Perry, D. and Morad, M. 1997, 'Waitomo's lament: the quest for sustainable sheep and beef livestock farming in the Waikato region', pp. 190–8 in Stokes and Begg (eds) (1997).

Perry, M. 1991, 'Some implications of the internationalisation of commercial capital for New Zealand', *New Zealand Geographer*, 47, 26–31.

Pickard, J. 1993, 'Western New South Wales – increased rainfall, not a miracle, leads to recovery', *Australian Journal of Soil and Water Conservation*, 6(4), 4–9.

Pigram, J.J.J. 1986, *Issues in the management of Australia's water resources* (Longman Cheshire, Melbourne).

Pile, S. and Keith, M. (eds) 1998, *Geographies of resistance* (Routledge, London).

Pillans, R.B., Pullar, W.A., Selby, M.J. and Soons, J.M. 1992, 'The age and development of the New Zealand landscape', pp. 31–62 in Soons and Selby (eds) (1992).

Pomeroy, A. 1986, 'A sociological analysis of structural change in pastoral farming in New Zealand', unpublished Ph.D. thesis, Department of Sociology, University of Essex.

Pool, I. 1991, *Te iwi Maori: a New Zealand population, past, present and projected* (Auckland University Press, Auckland).

Porter, M. 1990, *The competitive advantage of nations* (Macmillan, London).

Powell, D. 1993, *Out west: perceptions of Sydney's western suburbs.* (Allen and Unwin, Sydney).

Powell, J.M. 1968, 'A pioneer sheep station: the Clyde Company in Western Victoria, 1836–1840, *Australian Geographical Studies*, 6, 59–66.

Powell, J.M. 1976, *Environmental management in Australia, 1788–1914* (Oxford University Press, Melbourne).

Powell, J.M. 1988, *An historical geography of modern Australia: the restive fringe* (Cambridge University Press, Cambridge).

Power, J., Wettenhall, R.L. and Halligan, J. (eds) 1981, *Local government systems of Australia* (AGPS, Canberra).

Pratt, J. 1994, 'Crime, deviance and punishment', pp. 213–37 in Spoonley, Pearson and Shirley (eds) (1994).

Pritchard, W.N. 1996, 'Restructuring in the Australian dairy industry: the reconstruction and reinvention of co-operatives', pp. 139–52 in Burch, Rickson and Lawrence (eds) (1996).

Pritchard, W.N. 1998, 'The emerging contours of the third food regime: evidence from Australian dairy and wheat sectors', *Economic Geography*, 74, 64–74.

Pugh, M.C. 1989, *The ANZUS crisis. Nuclear visiting and deterrence* (Cambridge University Press, Cambridge).

Pugh, M.C. 1990. 'Reflections on the ANZUS crisis', *British Review of New Zealand Studies*, 3, 27–33.

Pyne, S.J. 1991, *Burning bush. A fire history of Australia* (Henry Holt, New York).

Quinlan, H.G. 1998, 'Air services in Australia: growth and corporate change, 1921–1996', *Australian Geographical Studies*, 36, 156–69.

Rawling, J. 1987, 'Capital, the state and rural land holdings: the example of the pastoral industry in the Northern Territory', *Australian Geographer*, 18, 23–32.

Resource Assessment Commission (RAC) 1992, *A survey of Australia's forest resources* (AGPS, Canberra).

Reynolds, H. 1996, *Aboriginal sovereignty* (Allen and Unwin, Sydney).

Rice, G.W. (ed.) 1992, *The Oxford history of New Zealand* (Oxford University Press, Oxford), 2nd edition.

Rich, D.C. 1987, *The industrial geography of Australia* (Croom Helm, Beckenham).

Rich, D.C. and Linge, G.J.R. (eds) 1991, *The state and the spatial management of industrial change* (Routledge, London and New York).

Riddolls, P.M. 1987, *New Zealand geology: containing geological map of New Zealand 1:2 000 000* (Department of Scientific and Industrial Research Science Information Publishing, Wellington).

Riley, S.J. 1988, 'Secular change in the annual flows of streams in the NSW section of the Murray–Darling basin', pp. 245–266 in Warner (ed) (1988).

Rimmer, P.J. 1967, 'The search for spatial regularities in the development of Australian seaports 1861–1961/2', *Geografiska Annaler*, 49B, 42–54.

Rimmer, P.J. 1988, 'Japanese construction contractors and the Australian states: another round of interstate rivalry', *International Journal of Urban and Regional Research*, 12, 404–24.

Rimmer, P.J. 1992, 'Japan's "resort archipelago": creating regions of fun, pleasure, relaxation, and recreation', *Environment and Planning A*, 2, 1599–1625.

Rimmer, P.J. 1993, 'Japan's "bubble economy" and the Pacific: the case of the EIE group', *Pacific Viewpoint*, 34, 25–44.

Rimmer, P.J. 1994, 'Japanese investment in golf course development: Australia–Japan links', *International Journal of Urban and Regional Research*, 14, 234–55.

Rimmer, P.J. 1997a, 'Japan's foreign direct investment in the Pacific Rim, 1985–1993', pp. 113–32 in Watters, McGee and Sullivan (eds) (1977).

Rimmer, P.J. (ed.), 1997b, *Pacific Rim development: integration and globalisation in the Asia–Pacific economy* (Allen and Unwin, Sydney).

Rimmer, P.J. 1998, 'Ocean liner shipping services: corporate restructuring and port selection/competition', *Asia Pacific Viewpoint*, 39, 193–208.

Roberts, B.R. 1992, *Land care manual* (New South Wales University Press, Kensington, NSW).

Robinson, G.M. 1988, 'Spatial changes in New Zealand's food processing industry, 1973–84', *New Zealand Geographer*, 44, 69–79.

Robinson, G.M. 1993a, 'Trading strategies for New Zealand: the GATT, CER and trade liberalisation', *New Zealand Geographer*, 49, 13–22.

Robinson, G.M. 1993b, 'Brisbane: planning for Australia's largest metropolitan local authority', *Queensland Geographical Journal*, 4th series, 8, 55–74.

Robinson, G.M. 1993c, 'Local development initiatives in Australia', pp. 235–56 in Bruce, D. and Whitla, M. (eds), *Community-based approaches to rural development: principles and practice* (Rural and Small Town Research and Studies Programme, Mount Allison University, Sackville, New Brunswick).

Robinson, G.M. 1994a, 'Fast-track planning in Queensland: The Bjelke-Petersen era', *Progress in Rural Policy and Planning*, 4, 275–86.

Robinson, G.M. 1994b, 'Instant urban development: Brisbane's South Bank Parklands', *Geography*, 79, 269–71.

Robinson, G.M. 1994c, 'Politics, power and state control in Melbourne during the 1970s and 1980s', *Australian Studies*, 8, 61–88.

Robinson, G.M. 1995a, 'The deregulation and restructuring of Australia's cane sugar industry', *Australian Geographical Studies*, 33, 212–27.

Robinson, G.M. 1995b, 'New Zealand's trading policy in an age of globalisation: GATT, APEC and CER', *Pacific Viewpoint*, 36(2), 129–41.

Robinson, G.M. 1996a, 'Globalisation and trading strategies in the South Pacific', pp. 323–38 in Yeung (ed.) (1996).

Robinson, G.M. 1996b, 'Stud farming in the Hunter Valley: expansion and restructuring in the equine industry', *Australian Geographer*, 27(1), 133–48.

Robinson, G.M. 1996c, 'Globalisation and trading strategies in the South Pacific', *Australian Studies*, 10, 70–85.

Robinson, G.M. 1997a, 'Greening and globalizing: agriculture in "the new times"', pp. 41–53 in Ilbery, B.W., Chiotti, Q., and Rickard, T. (eds), *Agricultural restructuring and sustainability: a geographical perspective* (CAB International, Wallingford).

Robinson, G.M. 1997b, 'Farming without subsidies: lessons for Europe from New Zealand?', *British Review of New Zealand Studies*, 10, 89–104.

Robinson, M.E. 1976, *The New South Wales wheat frontier, 1851 to 1911* (Australian National University, Canberra).

Roche, M.M. 1990, *History of New Zealand forestry* (New Zealand Forestry Corporation and GP Books, Wellington).

Roche, M.M. 1991, 'Privatising the exotic forest estate: the New Zealand experience', pp. 139–54 in Dargavel and Tucker (eds) (1991).

Roche, M.M. 1996, 'Britain's farm to global seller: food regimes and New Zealand's changing links with the Commonwealth', pp. 339–60 in Yeung (ed.) (1996).

Roche, M.M., Johnston, T. and Le Heron, R.B. 1992, 'Farmers' interest groups and agricultural policy in New Zealand during the 1980s', *Environment and Planning A*, 24, 1749–67.

Roper, B. and Rudd, C. (eds) 1993, *State and economy in New Zealand* (Oxford University Press, Auckland).

Rowley, C.D. 1970, *The destruction of Aboriginal society* (Australian National University Press, Canberra).

Rudd, C. 1990, 'Politics and markets: the role of the state in the New Zealand economy', pp. 83–100 in Holland, M. and Boston, J. (eds), *The fourth Labour government. Politics and policy in New Zealand* (Oxford University Press, Auckland).

Rudner, M. 1995, 'APEC: the challenges of Asia Pacific economic co-operation', *Modern Asian Studies*, 29, 403–7.

Sandercock, L. 1977, *Cities for sale: property, politics and urban planning in Australia* (Melbourne University Press, Melbourne).

Sandercock, L. 1997, 'From main street to fortress: the future of malls as public spaces – or "shut up and shop"', *Just Policy*, 9, 27–34.

Sanders, N. and Bell, C. 1980, *A time to care: Tasmania's endangered wilderness* (Chris Bell, Blackmans Bay, Tas.).

Sandrey, R. 1997, 'Pacific Rim trade and APEC', pp. 46–61 in Watters, McGee and Sullivan (eds) (1997).

Sandrey, R. and Reynolds, R. (eds) 1990, *Farming without subsidies* (GP Books and Ministry of Agriculture, Wellington).

Saunders, B.G.R. (ed.) 1987, *Manawatu and its neighbours* (Massey University, Palmerston North).

Saunders, P. 1994, *Welfare and inequality: national and international perspectives on the Australian welfare state* (Cambridge University Press, Melbourne).

Savage, J. and Bollard, A. (eds) 1990, *Turning it around: closure and revitalisation in New Zealand industry* (Oxford University Press, Auckland).

Schedvin, C.B. 1976, 'The formation of post-war motor vehicles policy in Australia', *Monash Papers in Economic History*, 2.

Schedvin, C.B. 1990, 'Staples and regions of Pax Britannica', *Economic History Review*, 2nd series, 43, 533–59.

Schultz, J. 1985, *Steel city blues* (Penguin Books Australia, Ringwood, Victoria).

Schumm, S.A. 1977, *The fluvial system* (John Wiley and Sons, New York).

Schwebel, D.A. 1983, 'Quaternary dune systems', pp. 15–24 in Tyler, Twidale, Ling and Holmes (eds) (1983).

Schwimmer, E. (ed.) 1968, *The Maori people in the 1960s* (Blackwood/Janet Paul, Auckland).

Scott, R. (ed.) 1980, *Interest groups and public policy: cases from the Australian states* (Macmillan, Melbourne).

Searle, G.H. 1996, *Sydney as a global city* (New South Wales Department of Urban Affairs and Planning, Sydney).

Searle, G.H. 1998, 'Changes in producer services location, Sydney: globalisation, technology and labour', *Asia Pacific Viewpoint*, 39, 237–55.

Seebohm, K. 1994, 'The nature and meaning of the Sydney Mardi Gras in a landscape of inscribed social relations', pp. 193–222 in Aldrich (ed.) (1994).

Selby, M.J. 1985, *Earth's changing surface* (Clarendon Press, Oxford).

Selby, M.J. and Lowe, D.J. 1992, 'The middle Waikato Basin and hills', pp. 233–55 in Soons and Selby (eds) (1992).

Sheard, M.J. 1983, 'Volcanoes', pp. 7–14 in Tyler, Twidale, Ling and Holmes (eds) (1983).

Shirley, I.E. 1994, 'Social policy', pp. 130–45 in Spoonley, Pearson and Shirley (eds) (1994).

Shirley, I.E. and St John, S. 1997, 'Family policy and the decline of the welfare state in New Zealand', *British Review of New Zealand Studies*, 10, 39–62.

Simpson, A. 1997, *The immigrants: the great migration from Britain to New Zealand, 1830–1890* (Godwit Publishing, Auckland).

Sinclair, K. (ed.), 1990, *The Oxford illustrated history of New Zealand* (Oxford University Press, Auckland).

Singleton, G. (ed.) 1997, *The second Keating government* (Centre for Research in Public Sector Management, University of Canberra, Canberra).

Smith, D.I. 1998, *Water in Australia: resources and management* (Oxford University Press, Melbourne).

Smith, G.H. 1992, 'Education: biculturalism or separatism', pp. 157–65 in Novitz and Willmott (eds) (1992).

Smith, J. 1997, 'Livestock improvement', pp. 177–81 in Stokes and Begg (eds) (1997).

Smith, L. 1980, 'New black town or black new town?: the urbanisation of Aborigines', in Burnley, Pryor and Rowland (eds) (1980).

Smith, R. 1996, *The nuclear free and independent Pacific movement* (Cassell/Mansell, London).

Smolicz, J.J. 1995, 'The emergence of Australia as a multicultural nation: an international perspective', *Journal of Intercultural Studies*, 16(1–2), 3–23.

Solomon, R.J. 1959, 'Broken Hill – the growth of settlement, 1883–1956', *Australian Geographer*, 7, 181–92.

Solomon, R.J. and Dell, A.R. 1967, 'The Hobart bushfires of February, 1967', *Australian Geographer*, 10, 306–8.

Soons, J.M. 1992, 'The west coast of the South Island', pp. 439–55 in Soons and Selby (eds) (1992).

Soons, J.M. and Selby, M.J. (eds) 1982, *Landforms of New Zealand* (Longman Paul, Auckland), 1st edition.

Soons, J.M. and Selby, M.J. (eds) 1992, *Landforms of New Zealand* (Longman Paul, Auckland), 2nd edition.

Sorrenson, M.P.K. 1992, 'Maori and Pakeha', pp. 141–66 in Rice, G.W. (ed.) (1992).

Sorrenson, M.P.K. 1995, 'The Waitangi Tribunal and the resolution of Maori grievances', *British Review of New Zealand Studies*, 8, 21–36.

Sorrenson, M.P.K. 1998, *Waitangi: New Zealand's enduring struggle* (Australia–New Zealand Studies Center, Pennsylvania State University, University Park, PA).

Sparke, E. 1988, *Canberra: 1954–1980* (Australian Government Publishing Service, Canberra).

Spearritt, P. 1995, 'Suburban cathedrals: the rise of the drive-in shopping centre', in Davison, G., Dingle, T. and O'Hanlon, S. (eds) (1995).

Specht, R.L. 1970, 'Vegetation', in Leeper, G.W. (ed.) (1970).

Spoonley, P. 1994, 'Racism and ethnicity', pp. 81–97 in Spoonley, Pearson and Shirley (eds) (1994).

Spoonley, P. and Berg, L.D. 1997, 'Refashioning racism: immigration, multiculturalism and an election year', *New Zealand Geographer*, 53(2), 46–50.

Spoonley, P., Pearson, D. and Shirley, I.E. (eds) 1994, *New Zealand society: a sociological introduction* (Dunmore Press, Palmerston North).

Stace, H.C.T., Hubble, G.D., Brewer, R., Northcote, K.H., Sleeman, J.R., Mulcahy, M.J. and Hallsworth, E.G. 1968, *A handbook of Australian soils* (Rellim Technical Publications, Glenside, South Australia).

Starzecka, D.C. (ed.) 1996, *Maori art and culture* (British Museum Press, London).

Statistics New Zealand (SNZ) 1994, *New Zealand now: Maori* (SNZ, Wellington).

SNZ 1996, *Demographic trends 1996* (SNZ, Wellington).

SNZ 1997a, *New Zealand official yearbook* (GP Publications, Wellington).

SNZ 1997b, *People and places* (SNZ, Wellington).

SNZ 1998, *Business Demography Statistics* (SNZ, Wellington).

Stilwell, F. 1997, *Globalisation and cities: an Australian political–economic perspective* (Urban Research Program, Australian National University, Canberra).

Stimson, R.J. 1991, *Brisbane: magnet city. A report on strategies for the development, growth and management of Brisbane City into the 21st century* (Brisbane City Council, Brisbane).

Stimson, R.J. 1992, 'A place in the sun? policies, planning and leadership for the Brisbane region', *Urban Futures: Issues for Australian Cities*, 2(2), 51–67.

Stimson, R.J. 1997, 'Labour markets in Australia: trends and spatial patterns in the context of global and local processes', pp. 151–77 in Rimmer (ed.) (1997b).

Stimson, R.J., Jenkins, O.H., Roberts, B.H. and Daly, M.T. 1998, 'The impact of Daikyo as a foreign investor on the Cairns–Far North Queensland regional economy', *Environment and Planning A*, 30, 161–79.

Stimson, R.J. and Taylor, S.P. 1997, '"Hot spots" in the Australian space economy', *Australian Journal of Regional Studies*, 3, 19–33.

Stoker, G. (ed.) 1997, *The politics of identity in Australia* (Cambridge University Press, Melbourne).

Stokes, E. 1993, 'The Treaty of Waitangi and the Waitangi Tribunal: Maori claims in New Zealand', *Applied Geography*, 12, 176–91.

Stokes, E. and Begg, M. (eds) 1997, *Belonging to the land: people and places in the Waikato region* (Waikato Branch, NZ Geographical Society, Hamilton).

Stokes, G. 1997, 'Citizenship and aboriginality: two conceptions of identity in aboriginal political thought', pp. 158–71 in Stoker (ed.) (1997).

Stone, R.C.J. 1987, *The father and his gift: John Logan Campbell's later years* (Auckland University Press, Auckland).

Storey, K. 1998, *Commuter mining: a bibliography* (Department of Geography, Memorial University of Newfoundland, St Johns).

Storper, M. 1992, 'The limits to globalization: technology districts and international trade', *Economic Geography*, 68, 60–93.

Stratford, E. (ed.) (1999), *Australian cultural geographies* (Oxford University Press, Melbourne).

Stretton, H. 1989, *Ideas for Australian cities* (Transit Australia, Sydney).

Stretton, H. 1994, 'Transport and the structure of Australian cities', *Australian Planner*, 31(3), 131–6.

Stubbs, J.G. and Barnett, J.R. 1991, 'The geography of private contracting: the case of New Zealand public hospital ancillary services', *Area*, 23, 330–8.

Sturman, A.P. and Tapper, N.J. 1996, *The weather and climate of Australia and New Zealand* (Oxford University Press, Melbourne).

Sutherland, W. 1982, 'SPARTECA and continued problems of dependence', *Review*, 9, 4–11.

Taiuepa, T., Lyrer, P., Horsley, P., Davis, J., Bragg, M. and Moller, H. 1997, 'Co-management of New Zealand's conservation estate by Maori and Pakeha: a review', *Environmental Conservation*, 24, 236–50.

Taskforce on Regional Development, 1993, *Developing Australia: a regional perspective* (AGPS, Canberra), 3 volumes.

Tatz, C. 1995, 'The sport of racism', *Australian Quarterly*, 67(1), 38–48.

Taylor, J. 1998, 'Measuring short-term population mobility among indigenous Australians: options and implications', *Australian Geographer*, 29, 125–38.

Taylor, J. and Arthur, W.S. 1993, 'Spatial redistribution of the Torres Strait islander population: a preliminary analysis', *Australian Geographer*, 24, 26–38.

Taylor, M. (ed.) 1984, *The geography of Australian corporate power* (Croom Helm, Sydney).

Taylor, M. 1991, 'Hauling Australia into the 21st century: geographical perspectives on transport reform', *Australian Geographer*, 22, 120–3.

Taylor, M. 1996, 'Agriculture in recession: the regional financial performance of Australian broadacre livestock farming', *New Zealand Geographer*, 52, 46–55.

Taylor, M. and Garlick, S. 1989, 'Commonwealth Government involvment in regional development in the 1980s', pp. 79–103 in Higgins, B. and Zagorski, K. (eds) *Australian regional developments: readings in regional experiences, policies and prospects* (Office of Local Government, Department of Immigration, Local Government and Ethnic Affairs, AGPS, Canberra).

Thakur, R. 1995, 'In defence of multiculturalism', pp. 255–81 in Greif (ed.) (1995).

Theophanous, A.C. 1995, *Understanding multiculturalism and Australian identity* (Elikion Books, Melbourne).

Thorns, D.C. 1992, *Fragmenting societies? A comparative analysis of regional and urban development* (Routledge, London and New York).

Thorns, D.C. 1994, 'Urban', pp. 39–54 in Spoonley, Pearson and Shirley (eds) (1994).

Timms, B.V. 1992, *Lake geomorphology* (Gleneagles Publishing, Adelaide).

Tooth, S. and Nanson, G.C. 1995, 'The geomorphology of Australia's fluvial systems: retrospect, perspect and prospect', *Progress in Physical Geography*, 19, 35–60.

Toyne, P. and Vachon, D. 1984, *Growing up in the country: the Pitjantjatjara struggle for their land* (McPhee Gribble, Fitzroy).

Tranter, P.J. 1995, 'Behind closed doors: women, girls and mobility', pp. 25–30 in TransAdelaide (ed.) *On the move: debating the issues that affect women and Public Transport.* (TransAdelaide, Adelaide).

Tranter, P.J. 1996, 'Children's independent mobility and urban form in Australasian, English and German cities', pp. 31–44 in Hensher, D., King, J. and Oum, T. (eds) (1996).

Tranter, P.J. and Whitelegg, J. 1994, 'Children's travel behaviours in Canberra: car dependent lifestyles in a low density city', *Journal of Transport Geography*, 2(4), 265–73.

Trapeznik, A. 1995, 'Recent European migration to New Zealand', pp. 77–98 in Greif (ed) (1995).

Trengrove, A. 1975, *What's good for Australia . . .! The story of BHP* (Cassell Australia, Stanmore, NSW).

Troy, P.N. (ed.) 1995, *Australian cities: issues, strategies and policies for urban Australia in the 1990s* (Cambridge University Press, Cambridge).

Troy, P.N. 1996, *The perils of urban consolidation: a discussion of Australian housing and urban development policies* (The Federation Press, Sydney).

Tsokhas, K. 1986, *Beyond dependence: companies, labour processes and Australian mining* (Oxford University Press, Melbourne).

Tutua-Nathan, T. 1992, 'Maori tribal rights to ownership and control: the geothermal resource in New Zealand', *Applied Geography*, 12, 192–8.

Twidale, C.R. 1968, *Geomorphology, with special reference to Australia* (Nelson Australia, Sydney).

Twidale, C.R. and Campbell, F.M. 1993, *Australian Landforms: Structure, Process and Time* (Gleneagles Publishing, Adelaide).

Tyler, L. and Lattimore, R. 1990, 'Assistance to agriculture', pp. 60–79 in Sandrey and Reynolds (eds). (1990)

Tyler, M.J., Twidale, C.R., Ling, J.K. and Holmes, J.W. (eds,) 1983, *Natural history of the south east* (Royal Society of South Australia Inc., Adelaide).

Ufkes, F. 1993, 'Trade liberalisation, agro-food politics and the globalisation of agriculture', *Political Geography*, 12, 215–31.

Valentine, G. 1993, 'Heterosexing space: lesbian perceptions and experience of everyday spaces', *Environment and Planning D: Society and Space*, 11, 395–413.

Valentine, G. 1997, 'Oh yes I can. Oh no you can't: children and parents' understandings of kids' competence to negotiate public space safely', *Antipode* 29(1), 65–89.

Vance, J.E. 1970, *The merchant's world: the geography of wholesaling* (Prentice-Hall, Englewood Cliffs, NJ).

Veitch, D. 1997, *Hansonism: trick or treat?* (David Syme College of National Economics, Flemington, Vic.).

Viviani, N. 1996, *The Indochinese in Australia 1975–1995* (Oxford University Press, Melbourne).

Waitt, G.R. 1994, 'The Republic of Korea's foreign direct investments in Australia: the Chaebols Down Under', *Australian Geographical Studies*, 32, 191–213.

Waitt, G.R. 1996, 'Korean students' assessment of Australia as a holiday destination', *Australian Geographer*, 27, 249–70.

Waitt, G.R. and Hartig, K. 1997, 'Grandiose plans, but insignificant outcomes: the development of colonial ports at Twofold Bay, New South Wales', *Australian Geographer*, 28, 201–18.

Walcott, R.I. 1978, 'Plate tectonics and Late Cenozoic evolution of New Zealand', *Geophysical Journal of the Royal Astronomical Society*, 52, 137–64.

Waldegrave, C. and Stuart, S. 1997, 'Out of the rat race: the migration of low income urban families to small town Wairarapa', *New Zealand Geographer*, 53(1), 22–8.

Waldren, M. and Carruthers, F. 1998, 'Myths of the ghetto', *The Australian*, 21 March, p. 18.

Walker, R. 1992, 'Maori people since 1950', pp. 498–519 in Rice (ed.) (1992).

Walker, R. 1995, 'Immigration policy and the political economy of New Zealand', pp. 282–302 in Greif (ed.) (1995).

Wall, M. 1997, 'Stereotypical constructions of the Maori "race" in the media', *New Zealand Geographer*, 53(2), 41–5.

Walmsley, D.J. 1988, 'Space and government', pp. 57–67 in Heathcote (ed.) (1988).

Walmsley, D.J., Epps, W.R. and Duncan, C.J. 1998, 'Migration to the New South Wales north coast 1986–1991: lifestyle motivated counterurbanisation', *Geoforum*, 29, 105–18.

Walmsley, D.J. and Sorenson, A.D. 1993, *Contemporary Australia: explorations in economy, society and geography* (Longman Cheshire, Melbourne), Second edition.

Walmsley, D.J. and Weinand, H.C. 1997a, 'Fiscal federalism and social well-being in Australia', *Australian Geographical Studies*, 35(3), 260–70.

Walmsley, D.J. and Weinand, H.C. 1997b, 'Is Australia becoming more unequal?' *Australian Geographer*, 28(1), 69–88.

Wahlquist, H. 1997, "The Murray–Darling river system is collapsing under human abuse. Can science save it?' *The Australian Magazine*, 13–14 September, 17–19.

Wardle, P. 1991, *Vegetation of New Zealand* (Cambridge University Press, Cambridge).

Warner, R.F. (ed.) 1988 *Fluvial geomorphology of Australia* (Academic Press Australia, Sydney).

Warner, R.F. 1986, 'Hydrology', pp. 49–79 in Jeans, D.N. (ed.) (1986).

Wasson, R.J., Olive, L.J. and Rosewell, C.J. 1996, *Rates of erosion and sediment transport in Australia* (International Association of Hydrological Sciences Publ. no. 236, Proceedings of the Exeter Symposium, Erosion and Sediment Yield: Global and Regional Perspectives) pp. 139–48.

Watkin, T. 1997, 'A dying breed', *New Zealand Listener*, 157, No. 2968, 22–28 March , 20–3.

Watson, S. and Gibson, K. (eds) 1993, *Postmodern cities conference proceedings*, (Department of Urban and Regional Planning, University of Sydney, Sydney).

Watters, R., McGee, T. and Sullivan, G. (eds) 1997, *Asia–Pacific: new geographies of the Pacific Rim* (UBC Press, Vancouver).

Wheelwright, E. 1984, 'The political economy of foreign domination', pp. 208–25 in Lloyd, P.J. (ed.) (1984).

White, S. and Williams, L. 1996, 'Initial location decisions of immigrants', pp. 70–82 in Newton and Bell (eds) (1996b).

Whiteford, P. 1997, 'Measuring poverty and inequality in Australia', *Agenda*, 4(1), 39–50.

Whitehead, D. and Kelliher, F.M. 1991, 'Modeling the water balance of a small *Pinus radiata* catchment', *Tree Physiology*, 9, 17–33.

Whitehouse, I.E. and Pearce, A.J. 1992, 'Shaping the mountains of New Zealand', pp. 144–60 in Soons and Selby (eds) (1992).

Whittow, J. 1980, *Disasters: the anatomy of environmental hazards* (Penguin Books, Harmondsworth).

Williams, D.B. (ed.) 1990, *Agriculture in the Australian economy* (Sydney University

Press/Oxford University Press Australia, Sydney), 3rd edition.

Williams, G.E. 1970, 'The central Australian stream floods of February–March 1967', *Journal of Hydrology*, 11, 185–200.

Williams, M. 1974, *The making of the South Australian landscape* (Academic Press, London).

Williams, M. 1979, 'The perception of the hazard of soil degradation in South Australia: a review', pp. 275–89 in Heathcote and Thom (eds) (1979).

Williams, P.W. 1976, 'Impact of urbanisation on the hydrology of Wairau Creek, North Shore, Auckland', *Journal of Hydrology (NZ)*, 15, 81–99.

Williams, R.J. and Costin, A.B. 1994, 'Alpine and subalpine vegetation', pp. 467–500 in Groves (ed.) (1994).

Willis, R.P. 1984, 'Farming in New Zealand and the EEC – the case of the dairy industry', *New Zealand Geographer*, 40, 3–11.

Wilson, G.A. 1994, 'Wood chipping of indigenous forest on private land in New Zealand, 1969–1993', *Australian Geographical Studies*, 32, 256–73.

Wilson, H.D. 1994, *Field guide: Stewart Island plants* (Manuka Press, Christchurch).

Wilson, O. 1994, '"They changed the rules": farm family responses to agricultural deregulation in Southland, New Zealand', *New Zealand Geographer*, 50(1), 3–13.

Wilson, O. 1995, 'Farm household responses to agricultural deregulation: preliminary findings from a South Island case study', *Sociologia Ruralis*, 35, 756–66.

Winchester, H.P.M. and Costello, L.N. 1995, 'Living on the street: social organisation and gender relations of Australian street kids', *Environment and Planning D: Society and Space*, 13, 329–48.

Wiseman, J. 1998, *Global nation? Australian and the politics at globalisation.* (Cambridge University Press, Cambridge).

Woodford, J. 1997, 'The meaning of Wik: a user-friendly guide', *Sydney Morning Herald*, May 3, pp. 40–1.

Workman, M. and Moran, W. 1993,

'Enterprises in the geographic organisation of the wine industry', *Proceedings of the New Zealand Geographical Society Conference, 1993*, pp. 336–42.

Yates, J. 1997, 'Changing directions in Australian housing policies: the end of muddling through?', *Housing Studies*, 12(2), 265–77.

Yeates, N.R. 1997, 'Creating a global city: recent changes to Sydney's economic structure', pp. 178–96 in Rimmer (ed.) (1997b).

Yerex, D. 1989, *Empire of the dairy farmers* (Ampersand and Dairy Exporter Books, Petone).

Yerex, D. 1992, *The farming fiasco: why New Zealand, our farmers and the world's poor all lose out* (GP Publications, Wellington).

Yeung, Y. (ed.) 1993, *Pacific Asia in the 21st century: geographical and developmental perspectives* (Chinese University Press, Hong Kong).

Yeung, Y. (ed.) 1996, *Global change and the Commonwealth* (Hong Kong Institute of Asia–Pacific Studies, Chinese University of Hong Kong, Hong Kong).

York, B. 1996, 'From assimilation to multiculturalism: the Australian experience 1945–1989', *Studies in Australian Ethnic History, Centre for Immigration and Multicultural Studies, ANU*, No. 14.

Young, A.R.M. 1996, *Environmental change in Australia since 1788* (Oxford University Press, Melbourne).

Young, E.A. 1988, 'Aborigines and land in Northern Australian development', *Australian Geographer*, 19, 105–16.

Young, E.A. 1992, 'Aboriginal land rights in Australia: expectations, achievements and implications', *Applied Geography*, 12, 146–62.

Young, E.A. 1993, 'Managing the land: land and Aboriginal community development', pp. 218–29 in Cant, Overton and Pawson (eds) (1993).

Young, E.A. 1995, *Third World in the First: development and indigenous peoples* (Routledge, London and New York).

Yu, X., Tasplin, R. and Gilmour, A. 1997, 'Overseas market development: a strategy for Australian renewable energy industries', *Australian Geographer*, 28, 159–72.

Zerger, A. 1997, 'Cyclone hazard in the Australian tropical zone', pp. 465–70 in Bliss (ed) (1997).

Zhang, B.Q. and Webber, M. 1992, 'The Australian clothing industry: competition, productivity and scale', *Australian Geographer*, 23, 50–65.

Zodgekar, A. 1994, 'Population', pp. 308–23 in Spoonley, Pearson and Shirley (eds) (1994).

SUBJECT INDEX

INDEX OF PLACES

Places in this index are listed separately under Australia, Aotearoa/New Zealand and Other/Countries/Locations.

Aotearoa/New Zealand

Other Countries/Locations

AUSTRALIA AND NEW ZEALAND